Computational Granular Dynamics

Thorsten Pöschel
Thomas Schwager

Computational Granular Dynamics

Models and Algorithms

With 133 Figures

 Springer

Thorsten Pöschel
Thomas Schwager
Humboldt-Universität zu Berlin
Charité
Monbijoustraße 2
10117 Berlin, Germany

Library of Congress Number: 2004111169

ISBN-10   3-540-21485-2 Springer Berlin Heidelberg New York

ISBN-13   978-3-540-21485-4 Springer Berlin Heidelberg New York

Springer is a part of Springer Science+Business Media

springeronline.com

© Springer-Verlag Berlin Heidelberg 2005
Printed in The Netherlands

Typesetting: by the authors and TechBooks using a Springer LATEX macro package
Cover design: *design & production* GmbH, Heidelberg

Printed on acid-free paper   SPIN 10932445   54/3141/jl   5 4 3 2 1 0

Preface

Computer simulations belong to the most important methods of theoretical investigation of granular materials. Such simulations of technological relevance became possible only in the past few years due to the enormous advances in computer technology. We believe that the progress in computer technology contributed much to the renaissance of the scientific interest in granular matter as a subject of research in physics and engineering.

The present book is based on a lecture which was held at the Humboldt-University Berlin. It is intended to serve as an introduction to the application of numerical methods to systems of granular particles. The book emphasizes a general understanding of the subject rather than the presentation of latest advances in numerical algorithms. In favor of a clear and intuitive presentation and to keep the algorithms short, at some places we have intentionally abstained from presenting highly optimized algorithms which contain all numerical tricks that belong to the state of the art. Frequently we refer to the literature for information on further optimization of the presented methods and algorithms.

The book is accompanied by the web site www.springeronline.com/3-540-21485-2 where additional material can be found, such as the code of the programs, animated sequences of simulations, further programs which are not discussed in the book mainly due to their length, and other information.

For the understanding of the numerical methods and algorithms, basic knowledge of C++ is needed. We tried to avoid usage of elegant but complicated programming techniques to keep the text accessible to readers who are not experts in C++ or who prefer a different programming language.

The authors are indebted to a number of friends and colleagues. First of all we thank Hans Herrmann, one of the pioneers of numerical simulations of granular materials, for introducing one of the authors (TP) to this important and interesting field of research. Further we thank N. Brilliantov, V. Buchholtz, S. Esipov, A. Formella, J. Freund, I. Goldhirsch, C. Salueña, A. Santos, and L. Schimansky-Geier for many discussions and continued interest in our work.

July 2004   *T. Pöschel*
                                                                                   T. Schwager

Contents

1

Introduction

Granular materials are of utmost importance in many industrial processes, in the natural sciences and even in our everyday life. Different from other materials, it is often extremely complicated to handle granular materials since granular materials reveal very different behavior under different circumstances: excited (fluidized) granular material often resembles a liquid. It flows through pipes, reveals surface waves, flows inside an hour glass, etc. In other situations it behaves more like a solid: a heap of sand does not dissolve by itself but may be plastically deformed.

There are interesting intermediate stages: standing on the beach, we sink through the sand as in a viscous liquid, but only for a few centimeters. Then we come to a halt, almost independent of our weight. Even an elephant would not sink much deeper. Vibrated granular material tends to show convective flows, i.e., it flows like a liquid. On the other hand it may spontaneously form heaps, i.e., surface structures which do *not* dissolve. Under certain conditions a granular material resembles a gas rather than a liquid or a solid body. The energy of a raised dust cloud (a granular gas) is dissipated quickly, therefore, such a cloud transitions quickly into either a fluid or a solid state. In the absence of gravity, however, granular material can persist for a long time in the gaseous state. Granular gases are found, e.g., in the rings which encircle the large planets of our solar system.

Yet more exciting than the states of aggregation of granular matter are the transitions between these states which sometimes proceed very differently from those of common materials. So far there is not much known about the kinetics of these transitions.

Granular material as a subject of physical and engineering research has a long standing history. Eminent scientists such as Coulomb, Faraday, Hagen, Hertz, Huygens, Reynolds, Terzaghi, and others contributed to the body of research. From the physicist's point of view, there is presently no generally acknowledged theory of granular matter which is based on first principles, but only fragments of a theory, in contrast to solids, gases, and liquids where such theories exist. Engineers have at their command an enormous font of technical, experimental, and theoretical knowledge on granular materials. However, it can also be asserted that the technical understanding of granular materials is poor as compared with that of other materials: silos and hoppers continue to collapse, although they have been constructed and built according to the present state of the art.

It will be a task for the coming decades to develop a theory of granular matter up to a level which allows for reliable prediction of its behavior, similar to those for other materials. This problem will not be solved by physicists or engineers alone but will be the subject of intensive interdisciplinary cooperation.

In this introduction our goal is not to motivate research on granular materials nor to present the current body of knowledge which certainly goes beyond the scope of the book. Instead we focus on two important questions: Why do we need numerical simulations of granular materials? Why did we write this book the way we did?

1.1 Why Do We Need Numerical Simulations of Granular Materials?

There is a need for numerical simulations of granular materials for several reasons. Since there is no comprehensive theory on granular materials which reliably predicts the behavior of such materials in technical devices, numerical simulations can be used to predict and to optimize the function of machinery in powder technology before the device has been actually built. Experiments with engineering devices are frequently expensive, time consuming, and sometimes even dangerous. In such cases, numerical simulations can supplement and partially replace experiments.

Still more important is the improvement of the present theory by means of computer simulations since such improvements may also be universally applied to related problems. We illustrate this idea by means of an example, comminution of granular materials in ball mills, where a long standing problem has been solved using computer simulations. This solution in turn may also be generalized to related problems. Here, we neither discuss the details of the simulations nor any quantitative results, the example is only intended to motivate numerical simulations and to illustrate the way of thinking when performing simulations. For details on the discussed system and quantitative results see [49].

Ball mills are common devices in powder technology for the comminution of granular materials, e.g., ores. They consist essentially of a long revolving cylinder with a horizontal axis that is filled with granular material. Such devices may be as large as several meters in diameter and several tens of meters in length. In the case of autogenous comminution, the cylinder is filled with grist only; for the case of heterogeneous comminution, large and heavy bodies, such as steel spheres, are added. When the cylinder revolves, the granular material is ground by intensive interaction between the grist particles or the grist particles and the steel spheres. Ball mills are rather complicated devices, but this simplified description is sufficient for our introductory example. From the large size of a ball mill which may be filled with many tons of grist, it becomes clear that such machines consume large amounts of electrical power. Moreover, ball mills are widely used in many fields of powder technology and mining, hence, their optimization with respect to diameter and length of the mill as well as filling ratio, rotation speed, and others is an important technological problem. On the other hand, it is obvious from the geometric extensions of such machines that a pure experimental solution of this optimization is a costly problem by itself.

In 1992 Rolf and Rothkegel [246] performed experiments on the spatial distribution of pressure in a two-dimensional model ball mill. They used instrumented balls which emitted a flash of light whenever the strain exceeded a certain threshold. The positions of recorded flashes were digitized and from the statistics of the data they inferred the stationary strain distribution inside the ball mill. For various threshold values, they always found the peak strain deep inside the material (see Fig. 1.1). This was surprising because inside the material, the relative grain velocities were rather low in comparison with the relative velocities at the surface. One would have expected the opposite: high pressure at the surface where the particles collide rapidly.

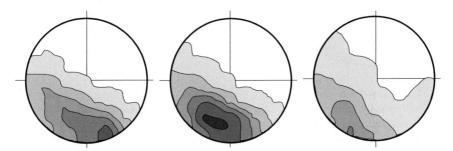

Fig. 1.1. Spatial distribution of the strain for different rotation velocities. Gray levels encode strain values. The regions of maximal strain are located inside the material. This figure is redrawn from [246]

This experimental finding called for an explanation. Based on experimental investigations alone the unexpected pressure distribution could not be clarified sufficiently. As demonstrated below this challenging open question could be answered using the powerful tools of Molecular Dynamics, namely the option to monitor physical quantities which are hardly accessible through experiments.

Figure 1.2 shows snapshots of typical configurations for two different rotation frequencies [49]. One can see that the location of the largest forces which can initiate fragmentation is close to the bottom of the rotating cylinder.

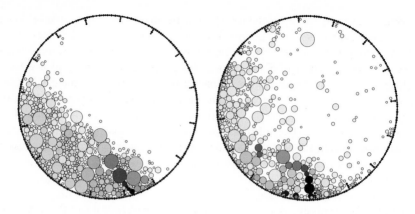

Fig. 1.2. Typical dynamical configurations for low (*left*) and high (*right*) rotational speed. The gray scale encodes the instantaneous maximum compression (normal force) experienced by each particle through contact with its neighbors (light: small normal force, dark: large force)

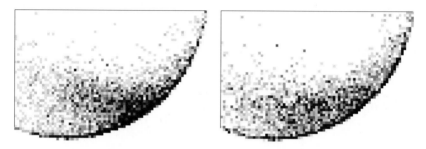

Fig. 1.3. Spatial distribution of fragmentation (i.e., high pressure) locations for two different rotation velocities (*left*: slow; *right*: fast). The pictures show that fragmentation occurs preferably close to the container walls but rarely in the bulk of the material

This property can be better visualized in a density plot. Figure 1.3 shows the spatial distribution of high pressure locations which lead to fragmentation for two different rotational velocities, analogous to the experimental result displayed in Fig. 1.1. The presentation has been restricted to the most important lower right segment of the ball mill. The gray values encode the frequency of fragmentation in a certain region, with dark meaning high frequency and light meaning low frequency, respectively. Preferred fragmentation regions are found near the bottom of the mill, i.e., deep inside the granular material.

This numerical simulation is in good agreement with the experimental results. Hence, further analysis can be applied to find an explanation of the experimentally measured pressure distribution.

Both the experiment as well as Fig. 1.3 display time-averaged data. Let us now visualize the pressure which acts on the particles at a certain time instant. This is a major difference in comparison with the experiment: in the experiment only a few particles were equipped with a pressure sensor and only those particles emitted a flash if the local pressure exceeded a predefined threshold. Moreover, only parts of the surface (not the entire surface) of these particles were sensitive. To produce the density plots in Fig. 1.1 the positions of the flashes were recorded over a long period of time. To measure experimentally the distribution of the *momentary* pressure (Fig. 1.4) it would be necessary to equip *all* particles with pressure sensors which are sensitive at the *entire* surface of the particles. Certainly, such an experiment would require much larger effort.

Most importantly, the pictures from these simulations reveal that grains experiencing peak stress often form linear chain-like structures. Using numerical analysis we will prove that the presence of those force chains is the crucial physical reason for the observed pressure distribution.

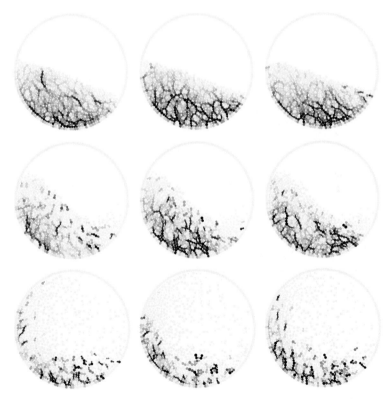

Fig. 1.4. Snapshots of the simulation for small (*top*) intermediate (*middle*) and large (*bottom*) rotation velocity. The gray scale encodes the local pressure

We define *force chains* by a set of three plausible conditions: Grains i, j, and k are considered to be members of the same force chain if

1. Particles i, j and j, k are next neighbors.
2. The pressure acting on each of the grains exceeds a certain threshold.
3. The connecting lines between i, j and j, k form an angle larger than 150°, i.e., the centers of three grains almost fall on a line.

These three conditions were evaluated by a computer algorithm and it was found that most of the harder stressed grains were part of force chains [49], i.e., the main part of the static and dynamic pressure propagated along force chains. The pressure acting on a highly stressed grain inside a force chain was up to 100 times larger as compared with the average pressure of the neighboring grains not belonging to a force chain, i.e., the force distribution was strongly inhomogeneous.

The existence of such force chains is not restricted to grains in a ball mill but has been observed in several granular systems experimentally, e.g., [25, 73, 161], and numerically, e.g., [75, 257, 304]. It could be shown that (at least) one of the granular phases is characterized by the appearance of force chains [75] (see Sect. 3.10.3). Thus, one might conclude that the occurrence of force chains is an inherent feature of granular matter, which may substantially contribute to its specific characteristics.

Based on the results drawn in Fig. 1.4, it might be suspected that the presence of force chains is responsible for the spatial pressure distribution. To prove this hypothesis we draw an analogous plot to Fig. 1.3, but with an important difference: Figure 1.5 shows the spatial distribution of the number of particles which experience a pressure above a certain threshold (left: slow rotation; right: faster rotation). In the bottom row, only particles belonging to force chains are considered, whereas in the top row only particles *not* belonging to a force chain are drawn. This figure can be compared directly with the figures by Rothkegel [246] (see Fig. 1.1); both are in good agreement.

Obviously most of the heavier loaded particles belong to force chains and are found near the wall of the mill inside the material, as observed in the experiment. This is an exclusive consequence of the presence of force chains. Grains not belonging to a force chain contribute only weakly to the comminution.

While the results presented in Fig. 1.4 could have been produced, in principle, experimentally with large effort, the data shown in Fig. 1.5 cannot be generated directly by an experiment since there is no direct way to decide in an experiment whether a particle belongs to a force chain.

Hence, by means of computer simulations we have identified force chains to be responsible for the spatial pressure distribution in ball mills. Consequently, it may be expected that the efficiency of ball mills is affected largely by the presence of force chains. The sketched simulation results explain the experiment by Rothkegel [246] which was the starting point of our investigation. For a more careful quantitative discussion of the sketched results, see [49, 53].

Fig. 1.5. Spatial distribution of particles exposed to high pressure. Dark pixels stand for high frequency. The *top* row only accounts for particles which do not belong to a force chain while, in contrast, the *bottom* row includes only particles that are members of a force chain. The *left* column shows simulation results for small rotational speed and the *right* one for larger speed

The effort to obtain these results is small: The described results require altogether a few hours of computer time on a common desktop computer. Once the simulation program is developed it may be used for the simulation of many different systems with relatively small changes. Frequently, real-world experiments do not have that favorable property.

Let us present again the way of thinking that led us to the final result:

1. The starting point was an unexplained experimental result. We did not understand the spatial distribution of pressure in a model ball mill.
2. The experimental result was reproduced using Molecular Dynamics.
3. By inspecting several physical properties of the particles, where some of them are hardly accessible in an experiment, we formulated the hypothesis that force chains are related to the pressure distribution.
4. The term "force chain" was defined mathematically and a program was developed to identify such force chains in simulations.
5. Finally, the pressure distribution was plotted again, but with regard to whether a particle belongs to a force chain or not. The result confirmed the hypothesis and explains the experimental result from which the investigation started.

We have to pose the question whether this result is useful. There are two different answers: Natural scientists are mainly interested in *understanding* phenomena. We understood an experiment and in this sense the result was useful. On the other hand, engineers try to *apply* scientific results.[1] So the question can be reformulated: If force chains are so essential to comminution processes, how can their formation be enhanced? The problem of optimizing the efficiency of milling can be reduced to the simpler problem of the generation of force chains. Therefore, we believe that engineers may also benefit from numerical simulations.

1.2 Organization of the Book

Molecular Dynamics plays the most important role among the simulation methods for granular systems. Molecular Dynamics is the time-dependent numerical solution of Newton's equation of motion for all particles of which the granular material consists. In the present literature, this simulation technique is by far the most prominent method for the numerical investigation of granular materials. It is the subject of Chap. 2.

Under certain preconditions, Molecular Dynamics can be simplified considerably if at each time instant none of the particles is in mechanical contact with more than one other particle (hard sphere approximation), thus reducing the many-particle system to a set of two-particle systems. This assumption is justified if the time lag between two successive collisions of a particle is, on average, much larger than the duration of a collision. In this case, *event-driven Molecular Dynamics*, a variation of Molecular Dynamics, is a highly efficient technique for simulating such systems. It allows for the simulation of much larger granular systems than conventional Molecular Dynamics. This powerful method is described in Chap. 3.

Direct Simulation Monte Carlo, as discussed in Chap. 4, is yet more efficient than event-driven Molecular Dynamics, although, further simplifying assumptions are required. The necessary precondition of uncorrelated motion of particles (molecular chaos assumption) is approximatively valid for dilute granular systems, also called granular gases. As granular gases have been studied intensively by means of kinetic theory, Direct Simulation Monte Carlo is suited for direct comparison with the results of kinetic theory.

In Chap. 5 we describe the Rigid-Body Dynamics (also called Contact Dynamics) and its application to sharp edged granular particles. With much higher numerical effort this method allows for highly realistic simulation of granular systems of such particles.

Besides Molecular Dynamics, event-driven Molecular Dynamics, Direct Simulation Monte Carlo, and Rigid-Body Dynamics there exist a number of

---

[1] Here we understate the contribution of engineers to natural sciences. Many phenomena which nowadays seem to belong evidently to natural sciences were first detected by engineers. Some of their results became cornerstones of physical and chemical sciences, although at the time of their discovery they were mainly interested in purely technical applications.

further methods and algorithms for the simulation of granular systems. Some of these methods turn out to be significantly more efficient than the above mentioned ones, although they are restricted to certain types of problems. The most important further methods are Cellular Automata, bottom-to-top reconstruction and Langevin dynamics. These simulation techniques and the according algorithms are discussed in Chaps. 6–8.

1.3 Why Did We Write This Book As Is?

Each chapter in this book starts with the description of the theory which is necessary for the understanding of the presented numerical methods. Then the according numerical algorithms are discussed and in most cases an implementation is presented. The presented code is discussed in detail in the text. The numerical models are demonstrated by examples and, finally, the properties of the methods are discussed, i.e., their efficiency, limitations, and so on. All presented algorithms have been tested carefully.

We owe the reader answers to some important questions:

1. **Why do we give the program implementations in C++ ?**

   Many scientists (still) prefer FORTRAN rather than C or C++ for a simple reason: For many problems FORTRAN is sufficient. Why should we torture ourselves with the complexity of C++ if the program can be written in a much simpler language? C++ grants the programmer much more freedom than FORTRAN—for sophisticated and elegant programming as well as for programming errors. FORTRAN is much more restrictive—many bugs in the code which cannot be detected by a C++ compiler become apparent in a FORTRAN program as early as during compilation.

   However, FORTRAN also has serious drawbacks: It is an old fashioned language which is not well adapted to modern programming techniques: there are no dynamical variables; there are no pointer variables; structured programming is sometimes awkward; etc.

   We have chosen C++ since this language is very powerful. In particular the *Standard Template Library* (STL) provides structures that allow for the implementation of complex algorithms at relatively low programming effort. Many of the algorithms which are contained in this book could hardly be presented as FORTRAN code since their notation in FORTRAN would require many more lines than the equivalent C++ programs. Eventually, however, the choice of the programming language is mainly a matter of personal taste.

   We have tried to keep the C++ code as simple as possible and have tried to avoid sophisticated techniques which make the code more elegant but which contribute little to its functionality. Wherever possible, the code is written in a way to allow any programmer to translate it into his/her favored programming language.

2. **Do we need code listings?**

Undoubtedly it would be possible to leave out all listings from the book and to provide the code on the internet. In fact, all code which is contained in this book is downloadable (see next question). However, it did not appear reasonable to us to leave out the program code for an important argument: The listed programs (or program fragments) serve for further explanation of the presented theory and algorithms. In this respect we consider them to be analogous to formulae. Let us explain this statement by an example. The following three descriptions are equivalent:

(a) To obtain an approximation for π, add the reciprocal values of the fourth powers of the first fifty odd numbers, multiply the sum by ninety six and compute the fourth root of it.

(b) $\pi \approx \sqrt[4]{96 \sum_{i=0}^{49} \frac{1}{(2i+1)^4}}$

(c)
```
pi.cc
 1  #include <iostream>
 2  #include <math.h>
 3
 4  int main(){
 5     double sum=0;
 6     for(int i=0; i<50; i++){
 7        sum+=1/pow(2*i+1,4);
 8     }
 9     sum=pow(sum*96,0.25);
10     cout << sum << endl;
11  }
```

However, these representations of π are equivalent only *in principle*. Each mathematically educated reader will agree that it is much simpler to read the formula than the text, although the text carries precisely the same information. We believe that for a book on numerical methods, at some places it is necessary to give the notation in form of the program code in addition to the formula notation. Whereas the first two notations are indeed fully equivalent, the third notation carries some extra information which is needed for an actual implementation:

(a) For the output of the result the library `iostream` has to be included. The power function needs the header file `math.h`.

(b) The variable `sum` has to be declared as `double` since the mantissa of a `float` number is insufficient to add the numbers 3.1415926536 and $1/99^4 \approx 10^{-8}$.

(c) The argument of the function `pow()` is cast automatically from `int` to `double`.

If we intend to transmit this information without providing the program code, we would have to write down the above information, as we have done here. Insofar the notation of the formula and of the three items above is equivalent to the notation of the code. While mathematicians may prefer the second notation to the first one, many programmers prefer the listing of the code to the verbal description of the program.

We believe that this simple example gives a convincing argument that
a book on numerical methods should contain program code for the
description of algorithms. Of course, besides serving us as an example,
the described method to determine π is very inefficient—it is just an example.

3. **Do I have to retype the programs?**

   No. All described programs and further ones which have not been
   listed in the book due to their length can be found in the internet at
   `www.springeronline.com/3-540-21485-2/`. The @ sign on the page
   margin announces the availability of downloadable material. On the web
   site, moreover, some optimized algorithms are provided which take advantage of more sophisticated numerical tricks and go beyond the scope
   of this book.

4. **Are the programs correct?**

   Hopefully yes! The programs have been tested carefully, however, of
   course we cannot exclude the presence of remaining bugs. The authors
   are grateful for any bug report and for improvements of the presented
   programs. Any errata will be published at `www.springeronline.com/`
   `3-540-21485-2/` as well.

2

Molecular Dynamics

2.1 Idea of Molecular Dynamics

2.1.1 Equations of Motion

Granular materials consist of a large number of particles whose typical size ranges from micrometers to centimeters. These particles interact via short-range forces, i.e., only via mechanical contact. Long-range forces, such as electrostatic forces, are not considered here. The dynamics of a granular material is governed by Newton's equation of motion for the center-of-mass coordinates and the Euler angles of its particles i $(i = 1, \ldots, N)$:

$$
\begin{aligned}
\frac{\partial^2 \vec{r}_i}{\partial t^2} &= \frac{1}{m_i} \vec{F}_i \left(\vec{r}_j, \vec{v}_j, \vec{\varphi}_j, \vec{\omega}_j \right) \\
\frac{\partial^2 \vec{\varphi}_i}{\partial t^2} &= \frac{1}{\hat{J}_i} \vec{M}_i \left(\vec{r}_j, \vec{v}_j, \vec{\varphi}_j, \vec{\omega}_j \right) , \qquad (j = 1, \ldots, N) .
\end{aligned} \tag{2.1}
$$

The force $\vec{F}_i$ and the torque $\vec{M}_i$, which act on particle i of mass m_i and the tensorial moment of inertia $\hat{J}_i$, are (sometimes complicated) functions of the particle positions $\vec{r}_j$, their angular orientations $\vec{\varphi}_j$, and the corresponding velocities $\vec{v}_j$ and $\vec{\omega}_j$. First, two-dimensional systems are considered, hence, the angular orientation of a particle is described by a single (scalar) angle φ_i and the moment of inertia reduces to a scalar value J_i. Newton's equation of motion then read

$$
\begin{aligned}
\frac{\partial^2 \vec{r}_i}{\partial t^2} &= \frac{1}{m_i} \vec{F}_i \left(\vec{r}_j, \vec{v}_j, \varphi_j, \omega_j \right) \\
\frac{\partial^2 \varphi_i}{\partial t^2} &= \frac{1}{J_i} \vec{M}_i \left(\vec{r}_j, \vec{v}_j, \varphi_j, \omega_j \right) , \qquad (j = 1, \ldots, N) .
\end{aligned} \tag{2.2}
$$

The computation of the forces and torques is the central part of each Molecular Dynamics simulation. It is discussed in detail in Sects. 2.3–2.7.

In general the system of coupled nonlinear differential equations (2.1) cannot be solved analytically. The approximative numerical solution of these equations, i.e., the computation of the trajectories of all particles of the system, is called *Molecular Dynamics*. The idea of Molecular Dynamics goes back to Alder and Wainwright who in 1957 investigated the physics of hard sphere gases [1–3]. Since then the Molecular Dynamics technique has been steadily developed and today it belongs to the most powerful and well established numerical tools of physicists and engineers. It goes far beyond the scope of this book to give a complete overview on Molecular Dynamics here and on the results which have been obtained applying this method. Instead we refer to the literature, e.g., the books by Hoover [124], Rapaport [231], and by Frenkel and Smit [79]. The volume by Allen and Tildesley [5] belongs to the standard works of modern Molecular Dynamics; we refer to this book in several places.

At first Molecular Dynamics was developed for the numerical simulation of molecular gases and simple liquids [1]. Today systems of many millions of particles can be simulated, and even some billions of particles may be simulated for very short periods of time [245]. The simulation of granular materials, is computationally expensive due to the peculiar interaction of granular particles: particles exert forces on each other only when they are in mechanical contact. However, since granular particles are rather rigid, their repulsive interaction force grows steeply with the compression once the particles are in contact. To obtain reliable results, the hardness of granular particles requires a very small integration time step for the computation of the trajectories. Hence, the Molecular Dynamics simulation of granular materials is rather computer-time consuming. At present using personal computers, systems of typical size of 20,000 particles can be simulated over real time of some seconds to some minutes. Pioneering work in the field of Molecular Dynamics of granular materials, has been done by Cundall (e.g., [68]), Haff (e.g., [107]), Herrmann (e.g., [83]), Walton (e.g., [289]), and others. In the present chapter this numerical method is described in detail.

For granular particles in the absence of long-range fields, the force $\vec{F}_i$ and the torque $\vec{M}_i$ acting upon particle i are given by sums of the pairwise interaction of particle i with all other particles of the system:

$$\vec{F}_i = \sum_{j=1,j\neq i}^{N} \vec{F}_{ij}\ , \qquad \vec{M}_i = \sum_{j=1,j\neq i}^{N} \vec{M}_{ij}\ . \qquad (2.3)$$

The limitation to pairwise interaction is an abstraction which is justified if the particles deform each other only slightly. For stronger interactions one has to take multi-particle interaction into account. If the forces $\vec{F}_{ij}$ and the torques $\vec{M}_{ij}$ of the colliding particles are given as functions of the particle coordinates $(\vec{r}_i, \varphi_i)$ and $(\vec{r}_j, \varphi_j)$ and their time derivatives, Newton's equation of motion (2.1) can be integrated numerically. The interaction laws are model specific, and the most simple case of colliding spheres is discussed in Sect. 2.1.4.

Besides the computation of the pairwise forces $\vec{F}_{ij}$ and torques $\vec{M}_{ij}$, one has to deal essentially with three problems for a Molecular Dynamics simulation

1. Summation of the forces and torques according to (2.3)
2. Integration of the equations of motion (2.1)
3. Data extraction from the computed particle trajectories

Just as a problem in continuum mechanics needs initial and boundary conditions, the description of a particle system is complete only if the behavior of the particles at the boundary of the simulation area is described and if the initial particle coordinates and velocities are given.

2.1.2 Boundary Conditions

In many cases the dynamic and static properties of a granular system are substantially affected by interaction of the granular material with the system boundaries, i.e., by the properties of the container or the surface on which the material is located. Examples are the convective motion of granular material in vertically or horizontally vibrating containers, the formation of density waves in pipes, the motion of granular materials on conveyors, and the clogging of hoppers. In these cases and many others, one has to be careful with the definition of the interaction of the granular material with the container.

Of particular importance is the realistic modeling of the wall roughness. Unfortunately the mechanical interaction of a deformable body with a rough wall is only poorly understood. A very efficient method to define the wall properties is to build up the walls from particles which obey the same rules of interaction as the particles of the granular material themselves. By choosing appropriate sizes and positions of the wall particles, system borders of adjustable roughness can be described. Such walls can be easily incorporated into a Molecular Dynamics simulation since no extra forces need to be defined—the particle–particle force law can be applied to the wall too. Figure 2.1 shows an example.

In many cases the motion of the container does not depend on the motion of the granular material inside, but is controlled externally. Hence, the dynamics of the wall particles is not determined by Newton's equation (2.1). The forces between the wall particles and the mobile particles affect only the motion of the latter. The container often moves in time along a predetermined path, e.g., for the case of a vertically vibrating container the y-coordinates of the wall particles follow a sine function. The corresponding time-dependent positions, velocities, angular orientations, and angular velocities of the wall particles are determined independent of the integration scheme.

The described method is not the only way for the definition of boundary conditions. Other methods can be found, e.g., in [83, 273].

Of particular importance are periodic boundary conditions, i.e., the periodic extension of the simulation area in one or more dimensions. Any particle which leaves the system at one side of the system is reinserted at the

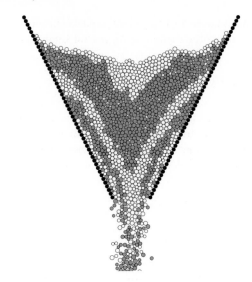

Fig. 2.1. Simulation of an outflowing hopper. The immobile particles which form the container are drawn in black

opposite side. Correspondingly, the interaction forces between particles at opposite sides of the simulation area have to be taken into account. For a two-dimensional system with one periodic dimension, the motion can be thought to take place on the surface of a cylinder; and for two periodic dimensions on the surface of a torus. Such boundary conditions are used to mimic infinitely extended systems. Since in granular systems there are only short-range particle interactions, the algorithmic implementation of periodic boundary conditions is rather simple (for the simulation of periodic systems with long-range forces see, e.g., [5]).

When using periodic boundary conditions the system has to be large enough to avoid artifacts due to non-physical correlations between the periodically extended borders of the simulation area. For a detailed discussion see Sect. 3.8.2.

2.1.3 Initial Conditions

The initial conditions define the values of the coordinates $\vec{r}_i(t = 0)$, the velocities $\vec{v}_i(t = 0)$, the Euler angles $\vec{\varphi}_i(t = 0)$, and the angular velocities $\vec{\omega}_i(t = 0)$ for $i = 1, \ldots, N$. For the important case of spherical particles the interaction forces are independent of the spatial orientation of the particles. Therefore, initial conditions are needed for $\vec{\omega}_i$, but not for $\vec{\varphi}_i$.

For the numerical integration of Newton's equation, a predictor–corrector scheme is applied (see Sect. 2.1.5) which also requires initial conditions for higher-order time derivatives, $\vec{r}_i^{(n)}$ and $\vec{\varphi}_i^{(n)}$, where $n = 2, 3 \ldots$. For $n > 2$, these quantities have no direct physical meaning.

Initial conditions are used only once for each simulation, hence, there is no need to develop sophisticated algorithms for their generation. For many problems the long-term behavior of the system is independent of the concrete initial conditions. If one is interested in stationary characteristics of a system, e.g., the flow velocity profile of granular material in a pipe, initial conditions can be generated by placing the particles piece by piece randomly in the system where one has to take care that they do not (or only slightly) deform each other initially. A simple random strategy is suitable for this purpose a particle is placed at a randomly chosen position until a position is found where it does not overlap with any of the previously placed particles. This step is performed until all particles have found their initial positions.

After positioning all particles a number of simulation steps is performed, i.e., the system is numerically integrated until it finds its steady state. To accelerate this procedure it may be appropriate to temporarily tune certain material parameters, e.g., elastic and dissipative material constants, to unrealistic values. If the steady state is found, the integration has to be performed further for a short time with the realistic material parameters which will be used for the Molecular Dynamics simulation. After this relaxation the complete phase space, i.e., all coordinates and their time derivatives (including the higher-order derivatives which are needed for the predictor–corrector integration scheme), $\vec{r}_i^{(n)}$ and $\vec{\varphi}_i^{(n)}$, and $n = 0 \ldots$, are stored to a file. This initialization file can later be read at the beginning of each simulation.

In general, after modifying parameters of the simulation, a number of relaxation steps have to be performed to find the new stationary state of the system, before extracting data. The particle trajectories which are generated during this relaxation procedure must not be used as simulation results.

This simple method is not suited for the generation of initial conditions of systems with a high filling factor (total volume of particles by volume of the simulation area). For such systems more sophisticated non-universal methods are applied.

2.1.4 Models for Spherical Particles

Contact of Spheres

The most simple model for a granular particle is a sphere. Here we always talk about spheres, although for the case of simulations in two dimensions, spheres are reduced to circular disks. Simulations using spherical particles are numerically very efficient since particle collisions may be identified in a very simple way: two particles are in mechanical contact if

$$\xi_{ij} \equiv R_i + R_j - |\vec{r}_i - \vec{r}_j| > 0 \,, \tag{2.4}$$

i.e., if the sum of their radii exceeds the distance of their centers. We call the quantity ξ_{ij} the mutual *compression* of particles i and j. For any other particle shape the detection of contacts is more complicated than for spheres.

The force between contacting particles is described by

$$\vec{F}_{ij} = \begin{cases} \vec{F}_{ij}^{\mathrm{n}} + \vec{F}_{ij}^{\mathrm{t}} & \text{if } \xi_{ij} > 0 \\ 0 & \text{otherwise}. \end{cases} \tag{2.5}$$

For two-dimensional systems, the normal and tangential components can be written in the form

$$\vec{F}_{ij}^{\mathrm{n}} = F_{ij}^{\mathrm{n}} \, \vec{e}_{ij}^{\mathrm{n}} , \qquad \vec{F}_{ij}^{\mathrm{t}} = F_{ij}^{\mathrm{t}} \, \vec{e}_{ij}^{\mathrm{t}} , \tag{2.6}$$

with the unit vectors

$$\vec{e}_{ij}^{\mathrm{n}} = \frac{\vec{r}_j - \vec{r}_i}{|\vec{r}_j - \vec{r}_i|} , \qquad \vec{e}_{ij}^{\mathrm{t}} = \begin{pmatrix} 0 & -1 \\ 1 & 0 \end{pmatrix} \cdot \vec{e}_{ij}^{\mathrm{n}} . \tag{2.7}$$

For simplicity of the notation in the following, the indices ij of the forces are dropped. The normal force F^{n} causes changes of the translational motion of the particles; the tangential force F^{t} causes changes of the rotational motion. Both components of the force are functions of the relative position of the particles $\vec{r}_i - \vec{r}_j$ and of the relative velocity $\vec{v}_i - \vec{v}_j$.

The next two subsections describe some approaches for modeling normal and tangential forces between colliding spheres. A more comprehensive review on interaction forces can be found in [252, 253].

Normal Forces

When granular particles collide, part of the kinetic energy of their relative motion is dissipated, i.e., it is transformed into heat. The deformations of the particles are assumed to be small and their spherical shape is assumed to be conserved on average after many collisions. The latter assumption seems to be plausible, although there are cases when it is not justified [56]. Moreover, the change of temperature of the particles is neglected. Temperature variations could cause a change of the material properties during the simulation. The assumption of constant temperature is justified if the generated heat is discharged from the material or if the heat generated by viscous particle deformations is much smaller than the total thermal energy of the system. In several technologically important processes, this assumption is not justified à priori. In such cases the values of the material parameters have to be adjusted due to the increase of temperature, i.e., according to the amount of dissipated mechanical energy.

Linear Dashpot Force

The normal force F^{n} consists of a dissipative and a conservative part. The simple linear ansatz,

$$F^{\mathrm{n}} = Y\xi + \gamma^{\mathrm{n}}\frac{\mathrm{d}\xi}{\mathrm{d}t} , \tag{2.8}$$

with the elastic and dissipative constants Y and γ^n, is used as a model in many Molecular Dynamics simulations. For pairwise collisions, the force represented by (2.8) causes a decrease of the relative normal velocity of the particles by a factor ε. This factor, defined as $\varepsilon \equiv g'/g$, is called the *coefficient of restitution*, where g is the absolute normal relative velocity before the collision and g' is the corresponding post-collision value. By integrating Newton's equation of motion it is found that the linear force (2.8) corresponds to the coefficient of restitution

$$\varepsilon = \exp\left(-\frac{\pi\gamma^n}{2m^{\mathrm{eff}}}\Bigg/\sqrt{\frac{Y}{m^{\mathrm{eff}}}-\left(\frac{\gamma^n}{2m^{\mathrm{eff}}}\right)^2}\right) , \qquad (2.9)$$

with $m^{\mathrm{eff}} \equiv m_i m_j/(m_i + m_j)$ as the effective mass of the colliding particles. The coefficient of restitution is an important characteristic of the material properties (see Sect. 3.2 for a detailed discussion). As seen in (2.9), the linear force law leads to a coefficient of restitution that is independent of the impact velocity. In experiments (e.g., [36]), however, a pronounced impact velocity dependence of ε was found. Moreover, $\varepsilon = $ const violates basic laws of the theory of viscous bodies [229] (this claim is justified in Sect. 3.5.1). A detailed discussion of the mechanics of colliding spheres and the according coefficient of restitution can be found in [42].

Many analytical results on continuum mechanics and kinetic theory of granular systems have been derived with the assumption $\varepsilon = $ const, hence, the results of Molecular Dynamics simulations using the force law (2.8) may be used as a direct comparison with these analytical results.

Viscoelastic Spheres

The interaction force of *elastic* spheres was derived by Heinrich Hertz [119] as a function of the deformation ξ and the material parameters Y (Young modulus) and ν (Poisson ratio):

$$F_{\mathrm{el}}^n = \frac{2Y\sqrt{R^{\mathrm{eff}}}}{3\left(1-\nu^2\right)} \xi^{3/2} , \qquad (2.10)$$

with R^{eff} as the effective radius of the colliding spheres

$$\frac{1}{R^{\mathrm{eff}}} = \frac{1}{R_i} + \frac{1}{R_j} . \qquad (2.11)$$

This result was later generalized to the contact of viscoelastic (damped) particles [44]:

$$F^n = \frac{2Y\sqrt{R^{\mathrm{eff}}}}{3\left(1-\nu^2\right)} \left(\xi^{3/2} + A\sqrt{\xi}\,\frac{d\xi}{dt}\right) , \qquad (2.12)$$

with the dissipative constant A being a function of the material viscosity (see [44] for details). The dissipative term in (2.12) follows from the solution of the viscoelastic equations for deformed spheres. The functional form of the dissipative term, $\sim \sqrt{\xi}\dot{\xi}$, has also been derived using other techniques [148, 190] which, however, are not able to specify A as a function of basic material parameters. The force (2.12) is in significantly better agreement with experimental results than (2.8), see [229]. The dissipative constant A is related to the rolling friction coefficient [37] which allows for an interesting technique to experimentally determine A (see Sect. 3.5.1, (3.28)).

Equations (2.10) and (2.12) apply only if both spheres are made from the same material. In case of different material properties the situation is more complicated. The elastic part reads

$$F_{el}^{n} = \frac{4\sqrt{R^{eff}}}{3} \left(\frac{1 - \nu_i^2}{Y_i} + \frac{1 - \nu_j^2}{Y_j} \right)^{-1} \xi^{3/2} , \tag{2.13}$$

i.e., the combination $(1 - \nu^2)/Y$ is added for both particles to obtain the respective prefactor for the elastic force law. For $\nu_i = \nu_j$ and $Y_i = Y_j$, (2.10) is recovered. The dissipative part of the interaction cannot be resolved as simply as the elastic part though. In general there is no easy way to add combinations of the dissipative properties to obtain the damping parameter A in the force law, as we did for the elastic law. One idea would be to treat the combination $YA/(1 - \nu^2)$ as the dissipative analogue of $Y/(1 - \nu^2)$ and perform the same addition of the reciprocals. However, even if only one of the particles deforms conservatively ($A = 0$) the collision is conservative too, i.e., there is no energy loss due to the dissipative deformation of the other collision partner. To avoid this problem we propose using the arithmetic mean of A as the damping constant, yielding

$$F^{n} = \frac{4\sqrt{R_{ij}^{eff}}}{3} \left(\frac{1 - \nu_i^2}{Y_i} + \frac{1 - \nu_j^2}{Y_j} \right)^{-1} \left(\xi^{3/2} + \frac{A_i + A_j}{2} \dot{\xi}\sqrt{\xi} \right) . \tag{2.14}$$

Again, for particles of the same material where $\nu_i = \nu_j$, $Y_i = Y_j$, and $A_i = A_j$ (2.14) reduces to (2.12).

Model by Walton and Braun

A further phenomenologically motivated force law has been proposed by Walton and Braun [287, 289]:

$$F^{n} = \begin{cases} Y_l \, \xi & \text{if} \quad d\xi/dt \geq 0 \\ Y_u \, (\xi - \xi_0) & \text{if} \quad d\xi/dt < 0 . \end{cases} \tag{2.15}$$

Here the constant of the restoring force Y adopts different values, depending on whether the particles approach or depart from each other.

Certainly the viscoelastic Hertz law (2.12) requires a larger numerical effort than the other described force laws (see [253] for further models), but the evaluation of the function $F\left(\vec{r}_i, \vec{v}_i, \vec{\varphi}_i, \vec{\omega}_i, \vec{r}_j, \vec{v}_j, \vec{\varphi}_j, \vec{\omega}_j\right)$ accounts only for a small fraction of the total computational costs.[1] Therefore, the computer time required for a Molecular Dynamics simulation depends only very slightly on the chosen force model. We prefer the force model (2.12) since it agrees very well with experimental results.

Duration of Collisions

The described force laws lead to an artifact which has to be eliminated in simulations. Consider a collision of particles as sketched in Fig. 2.2 (top row) and assume any of the force laws, e.g., (2.12). As long as the particles approach each other, i.e., $\dot{\xi} > 0$ (snapshots 1–3) and most of the decompression phase (snapshots 4–5), the term $(\xi^{3/2} + A\sqrt{\xi}\dot{\xi})$ in (2.12) is positive, leading to a repulsive (positive) normal force. At a certain stage of the collision (sixth snapshot in the sketch) when the deformation ξ is positive and the deformation rate $\dot{\xi}$ is negative, this term may change its sign, leading to a negative (attractive) force. When particles of a granular material collide, there are, however, no attractive forces, hence, this effect is an artifact of the considered force law. As a consequence, the particles separate from each other at a velocity that is too low. Note that the velocity arrows in the sixth snapshot are longer than in the final one. Analogously the force law (2.8) and many others lead to attractive forces too.

Considering the collision more realistically, the problem may be resolved (see bottom row in Fig. 2.2). Remember that particles do not *overlap*, they *deform*. In the expansion phase (snapshots 3–6) the particles gradually recover their spherical shape. At the time instant when $(\xi^{3/2} + A\sqrt{\xi}\dot{\xi})$ becomes negative, the centers of the particles separate too fast from each other to allow the surfaces to keep in touch while recovering their shape. Hence, although $\xi > 0$, the particles do not touch each other anymore, as sketched in the sixth snapshot. Consequently, the particles stop interacting and their velocities stay constant. Therefore, the arrows in the last two snapshots are of the same length.

From this discussion follows that the force law (2.12) and all other similar force laws have to be modified:

$$F^{\mathrm{n}} = \max\left\{0, \left[\frac{2}{3}\frac{Y\sqrt{R^{\mathrm{eff}}}}{(1-\nu^2)}\left(\xi^{3/2} + A\sqrt{\xi}\frac{d\xi}{dt}\right)\right]\right\}. \tag{2.16}$$

Taking this correction into account for the case of the linear dashpot model (2.8), the coefficient of restitution is slightly larger as compared with (2.9).

---

[1] The total force computation according to (2.5) includes the identification of contacting pairs with $\xi_{ij} > 0$, which consumes by far more computer time than the evaluation of F.

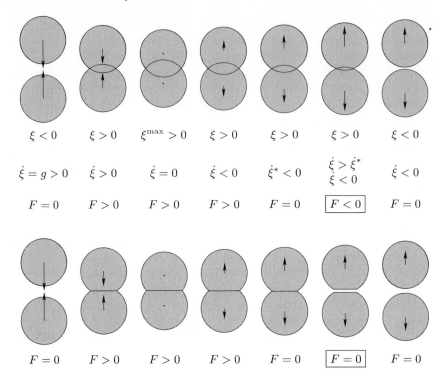

Fig. 2.2. Sketch of a particle collision. Due to an artifact of the considered force models, attractive forces may appear (boxed notation) at a certain stage of the collision. See text for explanation

In general, when deriving the coefficient of restitution from force laws, the reduced duration of the collision has to be considered. For a detailed discussion, see [264].

Tangential Forces

Granular particles are never perfect spheres but reveal a complicated surface texture. Therefore at oblique collisions, besides normal forces there are tangential forces F^t too[2] (also called shear force, see (2.5)): one cannot build a heap of perfect spheres on a flat surface. For that the particles and the surface must be rough enough. This force is mainly determined by the surface properties of the granular particles and is of essential importance for the realistic simulation of a granular system.

The modeling of forces between spherical particles is not consistent: for the description of the normal forces we assumed *exact* spherical shape of the

---

[2] Even perfectly smooth spheres exert tangential forces due to their bulk viscosity, see Sect. 3.5.3.

particles; now we justify the tangential forces by a surface texture, i.e., by deviations from the spherical shape. Sections 2.6–2.7 describe methods to partially eliminate this inconsistency at the price of larger numerical complexity.

Tangential forces are modeled in an intuitive way: Obviously the relevant velocity for the tangential force is the tangential relative velocity of the particle surfaces at the point of contact, v_{rel}^t. The point of contact is an approximation since for the description of the normal forces, a certain compression ξ of the contacting spheres is assumed, which implies a contact surface or a contact line in two dimensions. For realistic material parameters, in particular Y, however, the contact area is always much smaller than the radii of the colliding particles. Therefore, in a collision of two spheres i and j, the point of contact for i (or j) is defined as the intersection of the (undeformed) surface of the sphere i (or j) with the vector $\vec{r}_i - \vec{r}_j$ connecting the centers of the spheres. For the case of hard spheres this definition describes the point of contact exactly; for the case of deformable spheres it is an approximation.

The relative velocity of the spheres at the point of contact results from the relative velocity of the centers of the spheres and from their rotation:

$$v_{rel}^t = (\vec{v}_j - \vec{v}_i) \cdot \vec{e}_{ij}^t + R_i \omega_i + R_j \omega_j \,. \tag{2.17}$$

Model by Haff and Werner [107]

The force law

$$F^t = -\text{sign}\left(v_{rel}^t\right) \cdot \min\left(\gamma^t \left|v_{rel}^t\right|, \mu \left|F^n\right|\right) \tag{2.18}$$

was successfully used in many simulations. For small relative velocity v_{rel}^t or large normal force F^n, the tangential force according to (2.18) is a linear shear damping which grows linearly with the relative velocity. The shear force is limited by Coulomb's friction law:

$$\left|F^t\right| \le \mu \left|F^n\right| \,, \tag{2.19}$$

with the friction parameter μ. For large relative velocity or small normal force, when $\gamma^t \left|v_{rel}^t\right|$ exceeds $\mu \left|F^n\right|$, the tangential force $\left|F^t\right| = \mu \left|F^n\right|$ is selected by (2.18). This force law is hence in agreement with Coulomb's friction law.

The force law (2.18) gives reliable results in Molecular Dynamics simulations, in particular in systems where the particles collide mainly at finite velocities but do not rest statically on each other as for the case of a sand heap. Problems appear in simulations of static systems when the relative velocities at the contact point v_{rel}^t and, hence, F^t vanish, since the model (2.18) does not incorporate static friction. Therefore, this model is not suited for the simulation of static granular systems—a heap of such particles dissolves slowly. Only heaps of identical particles on a commensurable surface are stable. The straightforward solution of this problem is to choose more complicatedly shaped particles. Non-spherical particles which can be used for the simulation of dynamic and static granular systems are discussed in Sects. 2.6–2.7.

A further problem concerns the damping constant γ^t. The tangential force is mainly determined by surface properties, i.e., by very small asperities at the particle surface. Therefore, there is no experimentally measurable material constant from which γ^t could be derived. Instead this coefficient can only be determined à posteriori from the comparison of simulation results with experiments.

Model by Cundall and Strack [68]

This model compromises relative small computational effort and good agreement with experiments for the simulation of static behavior. Static friction is described by a spring acting in a direction tangential to the contact plane (imagine the particles as gear wheels with flexible gears). This spring is initialized at the time t_K of first contact of the particles and it exists until the particle surfaces separate from each other. Its elongation,

$$\zeta(t) = \int_{t_K}^{t} v_{rel}^t (t')\, dt' ,$$ (2.20)

determines the restoring tangential force, again limited by Coulomb's friction law:

$$F^t = -\text{sign}\left(v_{rel}^t\right) \cdot \min\left(\left|\kappa^t \zeta\right|, \mu\left|F^n\right|\right) ,$$ (2.21)

i.e., the term $\gamma^t \left|v_{rel}^t\right|$ in (2.18) is replaced by a constant κ^t times the elongation ζ. As for the previous model there is no microscopic mechanism which justifies the assumption of the model spring. The constant κ^t has to be determined from the comparison of simulations with experimental results.

Model by Walton and Braun [289]

This tangential force (and its normal complement (2.15)) is somewhat more sophisticated. It is not only determined by the particle positions and velocities at the present time, it also depends on the history of the interaction. Suppose at the previous time step of the simulation there acts the tangential force F^t. If both particles slip with respect to each other the tangential force changes to $F^t + \Delta F^t$ according to

$$\Delta F^t = k^t \Delta \zeta ,$$ (2.22)

where $\Delta \zeta$ is the change of the relative tangential displacement of the surfaces of both particles, i.e., how far the particles have slid since the last time step. The factor k^t is not a constant but depends on how the present state was prepared. Suppose two particles collide at a certain angle. After initial contact they have a nonzero tangential velocity, hence, they slide against each other. This sliding motion increases the frictional force (one could imagine an emerging spring as in the model by Cundall and Strack). The force is limited by the Coulomb law. If the contact lasts long enough the motion stops eventually. If (e.g., by external influence) the particle is now moved in opposite direction, the frictional force decreases, i.e., it is not immediately reversed. However, the factor k^t has now a different value than for the first part of the motion. The functional form of k^t is chosen in such a way that it enforces the Coulomb law, i.e., $k^t = 0$ when the frictional force assumes its maximal value. It furthermore depends on the tangential force F^* when the last reversal of slipping occurred. Initially this value is set to zero. Thus,

$$
k^t = \begin{cases} k_0^t \left(\dfrac{\mu F^n - F^t}{\mu F^n - F^*} \right)^\gamma & \text{if slipping increases } F^t \\[3mm] -k_0^t \left(\dfrac{\mu F^n + F^t}{\mu F^n + F^*} \right)^\gamma & \text{if slipping decreases } F^t \, , \end{cases} \tag{2.23}
$$

with μ as the Coulomb friction constant, k_0 as the initial tangential stiffness, and γ as a numerical parameter of typical value $1/3$. The parameters have to be determined experimentally. The difference of the tangential stiffnesses for the two phases is the source of energy dissipation.

The models described in this section and further models are discussed and compared with each other in much detail in [252, 253].

Tangential Force for Different Materials

Let us very briefly describe the tangential force between particles of different material. For the normal force the computation of effective parameters Y and A has been discussed in Sect. 2.1.4 (see (2.14)). For the effective friction coefficients μ and γ, used for the computation of the tangential force, we apply a different heuristics: if the parameters of the colliding particles differ, we take their minima. Let us discuss some arguments for its justification: in a strict sense, friction is a property of *two* contacting surfaces. Actually there is a friction coefficient for each combination of materials. If a particle is assigned a certain friction coefficient, it is meant that this coefficient is valid for a collision between two particles with identical mechanical properties. In general, the coefficients are not known for all possible combinations. So a method is needed to compute the coefficients for particles of different materials. The proposed minimum-rule comes from a common experience: if a polished steel surface slides on a moderately rough steel surface, approximately the same friction is observed as when the same smooth surface slides on a very rough steel surface, i.e., the frictional force is mainly determined by the smooth surface. We have to admit that this heuristics is not universal, hence, it may fail in certain situations. Other rules may be justified as well. Whether any of those rules is suitable can only be assessed by comparison of the simulation results with experiments.

2.1.5 Gear's Integration Scheme

There is a variety of methods for the numerical integration of systems of coupled differential equations (e.g., [224]). In this section we present only one efficient method which has proven to be powerful for Molecular Dynamics.

The integration of Newton's equation of motion (2.1) for granular particle systems is numerically difficult due to the short range interaction with extremely steep gradients. The Gear algorithm [5, 87, 88] is particularly suited for this problem, mainly because of its numerical stability. It is somewhat more complicated when compared with other integration schemes. However, a computer-time profiling reveals that the actual numerical integration of the equations of motion consumes only small part of the total computational costs. The main part of the computer time is spent on the evaluation of the forces which act on the particles in each time step. Consequently, using a less complicated integration algorithm cannot save much computer time. On the contrary, computer time is wasted by using a less stable numerical integrator since to achieve a comparable accuracy, a smaller time step is needed which increases the number of force evaluations.

The Gear algorithm has another important advantage over many other integration schemes: in each time step only one evaluation of the interaction forces is required. Hence, there is a huge gain in efficiency since the computationally expensive force evaluation is performed less frequently.

The Gear algorithm consists of two steps. First the *predictor* computes the particle positions, velocities, and higher-order time derivatives at time $t + \Delta t$ by extrapolating the present values using a Taylor expansion.[3] In most cases we use the Gear algorithm of fifth order, i.e., the numerical error of one step grows as $(\Delta t)^5$. For the extrapolation all time derivatives up to $\mathrm{d}^4/\mathrm{d}t^4$ are required:

$$\vec{r}_i^{\,\mathrm{pr}}(t + \Delta t) = \vec{r}_i(t) + \Delta t\, \vec{v}_i(t) + \frac{1}{2}\,\Delta t^2\, \ddot{\vec{r}}_i(t) + \frac{1}{6}\,\Delta t^3\, \vec{r}_i^{\,(3)}(t) + \cdots$$

$$\vec{v}_i^{\,\mathrm{pr}}(t + \Delta t) = \qquad\quad \vec{v}_i(t) + \quad \Delta t\, \ddot{\vec{r}}_i(t) + \frac{1}{2}\,\Delta t^2\, \vec{r}_i^{\,(3)}(t) + \cdots$$

$$\ddot{\vec{r}}_i^{\,\mathrm{pr}}(t + \Delta t) = \qquad\qquad\qquad\quad \ddot{\vec{r}}_i(t) + \quad \Delta t\, \vec{r}_i^{\,(3)}(t) + \cdots \qquad (2.24)$$

$$\vdots$$

The predicted coordinates and time derivatives are now used for the computation of the forces $\vec{F}_i\left(\vec{r}_j^{\,\mathrm{pr}},\, \vec{v}_j^{\,\mathrm{pr}},\, \omega_j^{\,\mathrm{pr}}\right)$ and torques $M_i\left(\vec{r}_j^{\,\mathrm{pr}},\, \vec{v}_j^{\,\mathrm{pr}},\, \omega_j^{\,\mathrm{pr}}\right)$. From the forces and torques, the linear and angular accelerations $\ddot{\vec{r}}_i^{\,\mathrm{corr}}(t + \Delta t)$ and $\ddot{\varphi}_i^{\,\mathrm{corr}}(t + \Delta t)$ are obtained. (Again, only the description for spheres is given. For more complex particle models, this procedure has to be performed for all degrees of freedom which are subject to Newton's equation of motion.)

---

[3] For simplicity only the integration of the translational degrees of freedom is described. The algorithm is perfectly analogous for the particle rotation and for other degrees of freedom which are used for more complex particle models.

In general these accelerations differ from the predicted quantities $\ddot{\vec{r}}^{\text{pr}}$. If the prediction was accurate, the predicted coordinates, velocities, and accelerations would satisfy the equation of motion, hence, $\ddot{\vec{r}}_i^{\text{corr}} = \ddot{\vec{r}}_i^{\text{pr}}$. Therefore, the difference $\Delta \ddot{\vec{r}} \equiv \ddot{\vec{r}}_i^{\text{corr}} - \ddot{\vec{r}}_i^{\text{pr}}$ is a measure for the deviation of the predicted coordinates, velocities, and higher-order derivatives from their true values. The second step of the integration algorithm, the *corrector*, improves the predicted values by adding a number which is proportional to the deviation $\Delta \ddot{\vec{r}}$. This correction is weighted differently for the different time derivatives:

$$
\begin{pmatrix} \vec{r}_i^{\text{corr}}(t + \Delta t) \\ \vec{v}_i^{\text{corr}}(t + \Delta t) \\ \ddot{\vec{r}}_i^{\text{corr}}(t + \Delta t) \\ \dddot{\vec{r}}_i^{\text{corr}}(t + \Delta t) \\ \vdots \end{pmatrix} = \begin{pmatrix} \vec{r}_i^{\text{pr}}(t + \Delta t) \\ \vec{v}_i^{\text{pr}}(t + \Delta t) \\ \ddot{\vec{r}}_i^{\text{pr}}(t + \Delta t) \\ \dddot{\vec{r}}_i^{\text{pr}}(t + \Delta t) \\ \vdots \end{pmatrix} + \begin{pmatrix} c_0 \\ c_1 \dfrac{1}{(\Delta t)^1} \\ c_2 \dfrac{2}{(\Delta t)^2} \\ c_3 \dfrac{6}{(\Delta t)^3} \\ \vdots \end{pmatrix} \frac{\Delta t^2}{2} \Delta \ddot{\vec{r}}. \quad (2.25)
$$

All other variables are corrected in precisely the same way.

The coefficients c_i depend on the order of the algorithm and on the type of the differential equation. For an algorithm of sixth order to integrate Newton's equation (a second order differential equation), these coefficients are [5]

$$
c_0 = \frac{3}{16}, \quad c_1 = \frac{251}{360}, \quad c_2 = 1, \quad c_3 = \frac{11}{18}, \quad c_4 = \frac{1}{6}, \quad c_5 = \frac{1}{60}. \quad (2.26)
$$

For an algorithm of fifth order they are

$$
c_0 = \frac{19}{90}, \quad c_1 = \frac{3}{4}, \quad c_2 = 1, \quad c_3 = \frac{1}{2}, \quad c_4 = \frac{1}{12}; \quad (2.27)
$$

and for a fourth order algorithm we have

$$
c_0 = \frac{1}{6}, \quad c_1 = \frac{5}{6}, \quad c_2 = 1, \quad c_3 = \frac{1}{3}. \quad (2.28)
$$

Finally the system time is updated, $t := t + \Delta t$, and the Molecular Dynamics algorithm proceeds again with the predictor step.

The Gear algorithm can be, in principle, implemented at any desired order. Naturally its accuracy is limited by the compiler-dependent floating point number representation.

Since particle models are described by different dynamical variables, the integration algorithm is specific to the chosen particle model. The implementation is explained for different particle models in Sects. 2.3, 2.6, and 2.7.

2.1.6 Sketch of the Molecular Dynamics Algorithm

The Molecular Dynamics simulation is performed according to the following algorithm (see Fig. 2.3):

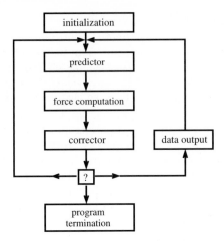

Fig. 2.3. Sketch of the Molecular Dynamics algorithm

1. **Initialization:** The coordinates of the particles $(\vec{r}_i, \varphi_i)$ and their time derivatives are read from the initialization file. The initialization file further specifies the type of each particle, i.e., whether it is a particle of the granular material or a wall particle (Sect. 2.1.3).
2. **Predictor:** Computation of the coordinates and time derivatives of the particles at time $t + \Delta t$ as a Taylor expansion of the values at the present time t. The predictor is the first part of the predictor–corrector integration scheme (see Sect. 2.1.5).
3. **Forces:**
   (a) Selection of the interaction pairs. This step affects the overall efficiency of the simulation significantly (see Sect. 2.4).
   (b) Computation of the pairwise interaction forces between the particles and between the particles and the walls, based on the predicted coordinates and velocities. The computation of the forces is unique to the particle model.
4. **Corrector:** The second step of the integration scheme corrects the predicted coordinates and time derivatives using the forces which have been computed in step 3.
5. **Data extraction:** The desired data are recorded either in predefined time intervals or upon occurrence of certain events. There are two fundamentally different procedures: Either the desired values are computed by the Molecular Dynamics program itself or the coordinates and velocities (and possibly further values) of all particles are recorded in a file in certain time intervals and the data are processed later.
6. **Program termination:** The program is terminated at a predefined time t^{end} or upon occurrence of a certain event. Otherwise the simulation is continued at step 2 and so on.

2.1.7 Vector Processing

The characteristics of a particle, e.g., its position and velocity, can be described by vectors. In the case of two-dimensional particles, the position vector of particle i contains the components x_i, y_i and the orientation φ_i. Since all mathematical operations may be written as vector operations it is convenient to adopt this notation for the program code as well. To this end we have defined the C++ class **Vector**, suitable for two-dimensional simulations. Its extension to three dimensions is straightforward.

For its application it is not necessary to understand its implementation in detail. Here the vector class is introduced by means of some self-explanatory examples. The full description can be found in Sect. 2.11. The first two components of a vector are the x- and y-component of a two-dimensional position or direction, and the third component is an orientation angle. Hence, the scalar and vector product only use the first two components. Furthermore, since the vector product of two vectors from the xy-plane always point in z-direction, the vector product is only a scalar, interpreted as the z-component of the resulting vector. The program below performs the following computation:

$$\vec{a} = \begin{pmatrix} x \\ y \\ \varphi \end{pmatrix} := \begin{pmatrix} 1 \\ 2 \\ 3 \end{pmatrix} , \qquad \vec{b} := \begin{pmatrix} 4 \\ 5 \\ 6 \end{pmatrix} , \qquad d := 7 ,$$

$$\vec{c} := \vec{a} + \vec{b} ,$$

$$\vec{c} := \vec{a} - \vec{b} ,$$

$$\vec{c} := \vec{c} + \vec{a} ,$$

$$\vec{c} := \vec{c} - \vec{a} ,$$

$$\vec{c} := d\vec{c} , \qquad\qquad\qquad\qquad\qquad\qquad (2.29)$$

$$d := |\vec{a}| = \sqrt{a_x^2 + a_y^2} ,$$

$$d := \vec{a} \cdot \vec{b} = a_x b_x + a_y b_y ,$$

$$d := \left[\vec{a} \times \vec{b} \right]_z = a_x b_y - a_y b_x ,$$

$$\vec{c} := \begin{pmatrix} 0 \\ 0 \\ 0 \end{pmatrix} .$$

```
────────────────── vectordemo.cc ──────────────────
1  #include "Vector.h"
2
3  int main()
4  {
5      Vector a(1,2,3), b(4,5,6), c;
6      double d=7;
7      c=a+b;
8      c=a-b;
9      c+=a;   c=c+a;
10     c-=a;   c=c-a;
11     c*=d;   c=c*d;
```

```
12    d=norm2d(a);
13    d=scalprod2d(a,b);
14    d=vecprod2d(a,b);
15    c=Vector(0,0,0); c=null;
16  }
```

In line 5 the vectors $\vec{a}$ and $\vec{b}$ are initialized and the vector $\vec{c}$ is declared. Line 10 shows two equivalent notations of the operation $\vec{c} := \vec{c} + \vec{a}$. Similar equivalent notations for $\vec{c} := \vec{c} - \vec{a}$ and for the multiplication with a constant, $\vec{c} := d\vec{c}$ are given in the next lines. Vector operations can, hence, be written in the same way as operations on numbers. In lines 12–14 the absolute value of a vector as well as the inner and outer products are computed. Finally in line 15, two equivalent notations are shown for assigning a vector the vectorial zero.

2.2 Overview of the Presented Simulation Programs

In Sects. 2.3–2.7 a number of Molecular Dynamics simulation programs will be presented. We will deal with three different particle models:

- Spherical particles
- Particles composed of spheres
- Sharp edged particles composed of triangles

The performance bottleneck of Molecular Dynamics simulations is the force computation. We will present four different approaches:

- A very simple brute-force approach, which is only intended to provide a complete program at minimal programming effort
- The Verlet list method
- The link cell algorithm
- A lattice algorithm

The programs have a modular structure. The first implementation (simple force computation, spherical particles) will serve as a reference. The code for more efficient methods (Verlet lists, link cell, and lattice methods) is based on the former program where parts of the simple program have to be replaced by more efficient functions. In a similar way the basic particle type `Sphere` may be replaced by more complicated particle types.

This modular design principle allows us to assemble all combinations of particle types and force computation methods to set up, e.g., a program for simulating spheres using the Verlet method or a program for simulating particles composed of triangles using the link cell method. The modular structure can also be preserved when implementing further particle types or force computation algorithms. The table below explains the necessary modifications to the basic program for all 12 combinations presented here.

Except for `SingleTriangle.cc` and `CompositeTriangle.cc` which can only be found for download on the book's web site, the implementations are discussed in detail in the following sections.

	Simple	Verlet list Sect. 2.4.2	Link cell Sect. 2.4.3	Lattice Sect. 2.4.4
Spheres Sect. 2.3	gcc common.cc Sphere.cc simple.cc	gcc common.cc Sphere.cc verlet.cc	gcc common.cc Sphere.cc linkcell.cc	gcc common.cc Sphere.cc lattice.cc
	Replace all instances of "Sphere" by "CompositeParticle" in common.cc and in			
Composite (spheres) Sect. 2.6	gcc common.cc Sphere.cc **CompositeParticle.cc** simple.cc	gcc common.cc Sphere.cc **CompositeParticle.cc** verlet.cc	gcc common.cc Sphere.cc **CompositeParticle.cc** linkcell.cc	gcc common.cc Sphere.cc **CompositeParticle.cc** lattice.cc
	Replace all instances of "Sphere" by "CompositeTriangle" in common.cc and in			
Composite (triangles) Sect. 2.7	gcc common.cc **SingleTriangle.cc** **CompositeTriangle.cc** simple.cc	gcc common.cc **SingleTriangle.cc** **CompositeTriangle.cc** verlet.cc	gcc common.cc **SingleTriangle.cc** **CompositeTriangle.cc** linkcell.cc	gcc common.cc **SingleTriangle.cc** **CompositeTriangle.cc** lattice.cc

Remark 1: gcc stands only as an example of an compiler call.

Remark 2: Instead of replacing Sphere by the used particle model we could have written the program more elegantly by using templates. Here we have avoided templates in order to keep the code simple and also accessible for non-specialists of C++.

Remark 3: Using a UNIX like operation system such as LINUX, the replacements can be done conveniently using the stream editor sed, e.g.,

```
mv common.cc commonA.cc
cat commonA.cc | sed s/Sphere/CompositeParticle/ > common.cc
rm commonA.cc
```

2.3 Molecular Dynamics Using Spherical Particles

2.3.1 Basic Structure and Main Program

We want to specify the general description of the Molecular Dynamics algorithm, given in Sect. 2.1.6, for the simulation of a granular system of spherical particles. In this section, a simple but complete implementation of the program is presented, without emphasizing numerical efficiency. This program serves later as a basis for

- Improvements with respect to more efficient algorithms (Sect. 2.4) and
- Molecular Dynamics simulations using more complex particle models (Sects. 2.6 and 2.7).

These modifications require small modifications of the presented programs which are explained in the corresponding later sections.

Let us begin our discussion on the implementation of the most basic Molecular Dynamics algorithm. The code of this program is organized in five files:

`common.h` and `common.cc`	Contain code and declarations of all global variables and functions which are used in more than one source file. These files are common to all combinations of force summation algorithms and particle models which are presented in this chapter.
`simple.cc`	Contains the code to perform a simple (but inefficient) force computation that will be improved afterwards.
`Sphere.h` and `Sphere.cc`	Code and declarations of the particle class `Sphere`. This class contains all information which characterize a spherical particle and the particle-type-specific functions to compute its dynamics. These files also contain some particle-specific functions which are friends of the class `Sphere`.

The organization of the code in these five files reflects the modular structure of the program. As explained in the previous section, programs for Molecular Dynamics simulation using different force computation methods and different particle models can be constructed easily by combining the appropriate source files.

This organization has, however, a drawback: the files cannot be described one after the other for the systematic explanation of the code. Instead, for the sake of a systematic discussion, the code is described disregarding its organization in the files. In the text the file name is given for all described functions in square brackets. The line numbering of the listings runs independently for all described files.

common.h

The header file `common.h` contains all declarations which are common to all simulation programs, i.e., for all particle types and all force computation

methods. Declarations which are only used in a single source file are not
contained in common.h.

```
————————————————— common.hh —————————————————
 1  #ifndef _common_h
 2  #define _common_h
 3
 4  #include <math.h>
 5  #include <fstream>
 6  #include <vector>
 7  #include <set>
 8  #include <string>
 9
10  using namespace std;
11
12  extern double lx, ly;
13  extern double x_0, y_0;
14
15  extern unsigned int no_of_particles;
16  extern double Time;
17  extern ofstream fphase;
18  extern ofstream fenergy;
19
20  void step();
21  void make_forces();
22  void integrate();
23  void init_algorithm();
24  void phase_plot(ostream & os);
25  #endif
```

After including the necessary standard headers, the system size lx, ly and the
position of the lower-left corner x_0 and y_0 are defined. Then the global variables are declared: no_of_particles (number of particles), Time[4] (elapsed
real time) and the output files fphase and fenergy for dumping the phase
space and logging the energy of the system.[5] Finally the header contains the
prototypes of functions which are used in more than one source file. These
functions are explained later in this section. The following table describes
their function in key words.

---

[4] This variable is spelled with capital T to avoid confusion with the standard function time().

[5] C++ distinguishes between *declaration* and *definition* of a variable. Declaring a
variable means indicating its syntax, while defining means actually allocating
memory and, if specified, initializing it (see §4.9. of [270] for further discussion
of this distinction). Since at this point the variables are only declared (strictly
speaking they *have to* be declared since variables must not be defined in a header
file), the keyword extern has to be prepended before the variable declarations.

Function [file]	Description	Page
`init_algorithm()` [`simple.cc`]	Initializes all data specific to the force summation algorithm (see Sect. 2.4)	36
`step()` [`simple.cc`]	Performs one time step of the simulation	36
`integrate()` [`common.cc`]	Integrates Newton's equation of motion using the Gear algorithm (predictor–corrector scheme)	37
`make_forces()` [`simple.cc`]	Computes the total forces acting on each particle	38
`phase_plot()` [`common.cc`]	Dumps the current positions, velocities, and further values into the output file	47

Head of common.cc

The code of the parts common to all programs is contained in the file `common.cc`. First it includes the header file **Sphere.h** of the particle class **Sphere**. This class describes the properties of spherical particles and the particle-model-specific functions for their manipulation (see Sect. 2.3.3).

```
                              ———————— common.cc ————————
1  #include "common.h"
2  #include "Sphere.h"                    /* <===== replace this line if necessary */
3  using namespace std;
4
5  double Time=0, timestep;
6  int nstep, nprint, nenergy;
7  Vector G;
8  double lx, ly, x_0, y_0;
9  vector<Sphere> particle;               /* <===== replace this line if necessary */
10 unsigned int no_of_particles;
11
12 ofstream fphase("phase.dat"), flast("lastframe.dat"), fenergy("energy.dat");
13
14 void init_system(char * fname);
15 double total_kinetic_energy();
                              ———————— to be continued ————————
```

The constants `nstep`, `timestep`, `nprint`, and `nenergy` denote the number of time steps to be executed, the size of the time step itself, and the phase space dump interval, i.e., the number of time steps between two successive dumps of the particle coordinates, velocities, and further values. The interval between two computations of the total energy in the system is controlled by the same constant. The variable `G` represents gravity.

The vector `particle` contains all information about the particles (of type **Sphere**, see Sect. 2.3.3). When modeling particles of different type this line has to be edited as specified in Sect. 2.2. Moreover, the variables `no_of_particles`, `Time`, `lx`, `ly`, `x_0`, and `y_0` are defined, which have been declared in `common.h`. Note that all these parameters are not specified here, but are determined from an input file (see p. 46). The output files `fphase` and `fenergy` are assigned the names `"phase.dat"` and `"energy.dat"`, respectively.

Finally, the head of the file contains the function prototypes for init_system() for initializing the simulation from the aforementioned input file and total_kinetic_energy() for computing the kinetic energy of the system. These functions are used only in this file, therefore, there is no need to include them in common.h.

main() [common.cc]

The main program first checks the correctness of the program call (lines 18–21), which must contain the name of the initialization file as its single parameter, e.g.,

    mdsimu inputfile

The file inputfile contains the initial positions and velocities of all particles. If the program is called erroneously with another number of arguments, a brief description of the syntax is displayed:

    usage: mdsimu particle_initfile

and the program terminates.

```
                              common.cc
16  int main(int argc, char ** argv)
17  {
18    if(argc!=2){
19      cerr << "usage: " << argv[0] << " particle_initfile\n";
20      exit(0);
21    }
22    fenergy.precision(10);
23    init_system(argv[1]);
24    init_algorithm();
25    phase_plot(fphase);
26    for(int i=0;i<nstep;i++){
27      step();
28      if((i+1)%nprint==0){
29        cout << "phase_plot: " << i+1 << "  " << particle.size() << " particles\n";
30        phase_plot(fphase);
31      }
32      if((i+1)%nenergy==0){
33        fenergy << Time << "\t" << total_kinetic_energy() << endl;
34      }
35    }
36    phase_plot(flast);
37  }
                           to be continued
```

Next init_system() is called to read the initial particle positions, velocities, and higher-order time derivatives from the input file (which is the argument of the program call). Moreover it initializes some system-specific variables. The description of this function is postponed to Sect. 2.3.3.

For the case of the simple algorithm, the function init_algorithm() which is called next is empty. For more sophisticated algorithms, as described in Sect. 2.4, this function serves for the initialization of some algorithm-specific variables.

In the main loop of the program, starting at line 27, the function step() is called for all nstep time steps. In intervals of nprint time steps, `phase_plot()` is called to dump the current phase space, i.e., the particle positions, velocities, and some further values to the output file. As an example for on-the-fly data processing, the total kinetic energy of the system is computed by `total_kinetic_energy()` and written to the file `fenergy`.

There is a separate section, Sect. 2.3.4, devoted to data extraction where the functions `phase_plot()` and `total_kinetic_energy()` are described.

2.3.2 Simple Algorithm for the Computation of Forces

Essential parts of the simple force algorithm are implemented in the file `simple.cc`. This file starts with the declaration of the variables and function prototypes:

```
                          ─── simple.cc ───
1  #include "common.h"
2  #include "Sphere.h"              /* <===== replace this line if necessary */
3
4  using namespace std;
5
6  extern vector<Sphere> particle;    /* <===== replace this line if necessary */
                          ─── to be continued ───
```

Except for `particle`[6] there are no algorithm- or particle-specific declarations.

init_algorithm() and step() [simple.cc]

For the simple implementation of a Molecular Dynamics algorithm presented in this section, the functions `init_algorithm()` and `step()` look redundant:

```
                          ─── simple.cc ───
7   void init_algorithm()
8   {
9   }
10
11  void step()
12  {
13     integrate();
14  }
                          ─── to be continued ───
```

The idea of these functions becomes clear when efficient force computation algorithms are discussed in Sect. 2.4.

integrate() [common.cc]

The function `integrate()`, called by `step()`, solves Newton's equation of motion for mobile particles according to the Gear algorithm (see Sect. 2.1.5). As specified in Sect. 2.1.2 *mobile* particles are those whose motion is determined by forces. The type of a particle can be determined by `ptype()`. Mobile

---

[6] According to our general rule the vector `particle` should have been declared in `common.h` since it is used here and in `common.cc`. As an exception, here it is declared twice to keep the the number of files small, which have to to be edited for modifying the particle type.

particles are of type 0, i.e., the expressions `particle[i].ptype()==0` in lines 41 and 50 are `true`.

The Molecular Dynamics simulations presented in this chapter are always performed with periodic boundary conditions. By defining (stationary or moving) walls as described later (p. 42), other boundary conditions may be applied.

```
                              ——— common.cc ———
38  void integrate()
39  {
40    for(unsigned int i=0;i<particle.size();i++){
41      if(particle[i].ptype()==0) {
42        particle[i].set_force_to_zero();
43        particle[i].predict(timestep);
44      } else {
45        particle[i].boundary_conditions(i,timestep, Time);
46      }
47    }
48    make_forces();
49    for(unsigned int i=0;i<particle.size();i++){
50      if(particle[i].ptype()==0) particle[i].correct(timestep);
51    }
52    for(unsigned int i=0;i<particle.size();i++){
53      particle[i].periodic_bc(x_0, y_0, lx, ly);
54    }
55    Time+=timestep;
56  }
                              ——— to be continued ———
```

For each particle the force is first reset to zero (line 42). For particles of type 0 (normal mobile particle) the function `predict()` is called to perform the first part of the Gear algorithm.

Then using the predicted particle positions and velocities, `make_forces()` evaluates the forces which act on the particles.

The second step of the predictor–corrector algorithm, `correct()`, is executed for mobile particles (type 0). This function corrects the predicted particle positions, velocities, and higher-order time derivatives by means of the computed forces. Finally the system time `Time` is updated.

Particles that rest or move according to a predefined schedule are used to model the system boundaries; we call them *wall particles*. Their type is different from 0. The motion of wall particles is not governed by Newton's equation of motion and, hence, their positions are not determined by the integration scheme. To determine the time-dependent positions and velocities of such particles, the function `boundary_conditions()` in line 45 is called for all particles of type different from 0.

The subdivision of the wall particles into different types (1,2,...) allows propagation of different wall particles along different trajectories. These trajectories are specified by the simulation problem. The usage of the particle types is described together with the implementation of `boundary_conditions()` on p. 42. Different types of particles (e.g., spheres or more complicated particles) require different types of boundary conditions, therefore, `boundary_conditions()` is a member of the particle class.

make_forces() [simple.cc]

Based on the predicted particle positions and velocities, the function `make_forces()` computes the forces that act on all particles. The most simple implementation of this function is presented here.

```
                                          simple.cc
15 | void make_forces()
16 | {
17 |   for(unsigned int i=0;i<particle.size()-1;i++){
18 |     for(unsigned int k=i+1;k<particle.size();k++){
19 |       if((particle[i].ptype()==0) || (particle[k].ptype()==0)){
20 |         force(particle[i],particle[k],lx,ly);
21 |       }
22 |     }
23 |   }
24 | }
                                      to be continued
```

For each pair of particles (i,k) where either i or k is mobile (type 0), `force()` is called, which computes its interaction force and adds it to the total forces of both particles. The function `force()` is particle-model-specific, hence, it is discussed in Sects. 2.3.3 and 2.6 where the particle classes are described.

This simple method of computing the interaction forces is presented here only to obtain a complete Molecular Dynamics program that represents, at minimum effort, the basis for more efficient algorithms. It is suitable only for very small systems of a few dozen particles. For larger numbers of particles, it becomes very inefficient since most of the $N(N-1)/2$ pairs of particles do not interact and can be excluded from the force computation by simple means. Such methods, which lead to a significant improvement of the simulation performance, are the subject of Sect. 2.4.

2.3.3 Particle Class Sphere

Spherical particles are represented as objects of the class **Sphere**. This class contains all information which characterizes a particle and all methods which compute its dynamics. The implementation of the particle class consists of two files, **Sphere.h** and **Sphere.cc**. Besides the particle class, **Sphere.h** defines the function **normalize()** (see below).

Sphere.h

The header file **Sphere.h** contains all declarations of the particle class **Sphere**. The operators >>, << and the functions **Distance()**, and **force()** are declared as **friends**, which allows them to access the private data of the particle class. These latter functions are not members of the particle class since they do not operate on one particle but on pairs of particles. The function **Distance()** computes the spatial distance of two particles. As an exception, we spell this function with capital D to avoid conflict with the function **distance()** of the Standard Template Library. Since periodic boundary conditions are simulated, the function `normalize()` (lines 8–13) is defined to select the suitable periodic image of a coordinate.

```
                          ———— Sphere.h ————
 1  #ifndef _Sphere_h
 2  #define _Sphere_h
 3
 4  #include <iostream>
 5  #include "Vector.h"
 6  using namespace std;
 7  inline double normalize(double dx, double L)
 8  {
 9    while(dx<-L/2) dx+=L;
10    while(dx>=L/2) dx-=L;
11    return dx;
12  }
13
14  class Sphere {
15    friend istream & operator >> (istream & is, Sphere & p);
16    friend ostream & operator << (ostream & os, const Sphere & p);
17    friend double Distance(const Sphere & p1, const Sphere & p2,
18                           double lx, double ly){
19      double dx=normalize(p1.rtd0.x()-p2.rtd0.x(),lx);
20      double dy=normalize(p1.rtd0.y()-p2.rtd0.y(),ly);
21      return sqrt(dx*dx+dy*dy);
22    }
23
24    friend void force(Sphere & p1, Sphere & p2, double lx, double ly);
25  public:
26    Sphere(): rtd0(null), rtd1(null), rtd2(null), rtd3(null), rtd4(null)
27    {}
28    Vector & pos() {return rtd0;}
29    Vector pos() const {return rtd0;}
30    double & x() {return rtd0.x();}
31    double x() const {return rtd0.x();}
32    double & y() {return rtd0.y();}
33    double y() const {return rtd0.y();}
34    double & phi() {return rtd0.phi();}
35    double phi() const {return rtd0.phi();}
36    double & vx() {return rtd1.x();}
37    double vx() const {return rtd1.x();}
38    double & vy() {return rtd1.y();}
39    double vy() const {return rtd1.y();}
40    double & omega() {return rtd1.phi();}
41    double omega() const {return rtd1.phi();}
42    const Vector & velocity() const {return rtd1;}
43    double & r() {return _r;}
44    double r() const {return _r;}
45    double m() const {return _m;}
46    int ptype() const {return _ptype;}
47    void predict(double dt);
48    void add_force(const Vector & f){_force+=f;}
49    void correct(double dt);
50    double kinetic_energy() const;
51    void set_force_to_zero(){_force=null;}
52    void boundary_conditions(int n, double timestep, double Time);
53    void periodic_bc(double x_0, double y_0, double lx, double ly);
54  private:
55    double _r, _m, _J;
56    int _ptype;
57    double Y,A,mu,gamma;
58    Vector rtd0,rtd1,rtd2,rtd3,rtd4;
59    Vector _force;
60  };
61  #endif
```

The private variables of the class describe the particle itself, i.e., its radius _r; the moment of inertia _J; its mass _m; the particle type _ptype; the material constants Y, A, mu, and gamma, which are needed for the computation of the particle interaction force (see (2.14) and (2.18)); as well as all information which describes its motion, i.e., the position vector rtd0, the velocity vector rtd1, and the vectors rtd2[7], rtd3, and rtd4 of the higher-order time derivatives used by the Gear integration scheme. The vector _force stands for the total force which acts on the particle due to its interaction with other mobile and wall particles. This force does not contain gravity, as accounted for in correct().

The position and velocity of the particle can be accessed as well as modified by functions of the form double & x() and double x() const. The operation of these functions is explained in Sect. 2.11. Finally, the declaration of the particle class contains the definition of some simple functions and the prototypes of more complex functions which are explained in the following. The table below describes them briefly:

Function	Description	Page
add_force()	Adds a value to the total force _force of the particle	39
ptype()	Returns the type of the particle. Particles of type 0 are subject to Newton's equation of motion, particles of other types belong to walls.	39
predict()	First step of the Gear algorithm	40
correct()	Second step of the Gear algorithm	40
force()	Computes the interaction force of two particles and adds the result to the total force accounts of both	41
boundary_conditions()	Computes the positions and velocities of particles of type other than 0	42
periodic_bc()	Enforces the periodic boundary conditions	44
kinetic_energy()	Computes the kinetic energy of the particle	45
<<	Output operator	45
>>	Input operator	45

predict() and correct() [Sphere.cc]

The functions predict() and correct() implement the Gear integration scheme; predict() computes the positions, velocities, and higher-order time derivatives as Taylor expansions of the present values (see (2.24)). In correct() these values are corrected according to (2.25) by means of the force which acts on the particle at its predicted position and velocity.

---

[7] The name stands for $\vec{r}$-time-derivative 2, i.e., the second time derivative of the position vector $\vec{r}$.

```
                        ───── Sphere.cc ─────
 1  #include "Sphere.h"
 2  #include <assert.h>
 3
 4  extern Vector G;
 5
 6  void Sphere::predict(double dt)
 7  {
 8    double a1=dt;
 9    double a2=a1*dt/2;
10    double a3=a2*dt/3;
11    double a4=a3*dt/4;
12
13    rtd0 += a1*rtd1 + a2*rtd2 + a3*rtd3 + a4*rtd4;
14    rtd1 += a1*rtd2 + a2*rtd3 + a3*rtd4;
15    rtd2 += a1*rtd3 + a2*rtd4;
16    rtd3 += a1*rtd4;
17  }
18
19  void Sphere::correct(double dt)
20  {
21    static Vector accel,corr;
22    double dtrez = 1/dt;
23    const double coeff0=double(19)/double(90)*(dt*dt/double(2));
24    const double coeff1=double(3)/double(4)*(dt/double(2));
25    const double coeff3=double(1)/double(2)*(double(3)*dtrez);
26    const double coeff4=double(1)/double(12)*(double(12)*(dtrez*dtrez));
27
28    accel=Vector((1/_m)*_force.x()+G.x(),
29                 (1/_m)*_force.y()+G.y(),
30                 (1/_J)*_force.phi()+G.phi());
31    corr=accel-rtd2;
32    rtd0 += coeff0*corr;
33    rtd1 += coeff1*corr;
34    rtd2 = accel;
35    rtd3 += coeff3*corr;
36    rtd4 += coeff4*corr;
37  }
                        ───── to be continued ─────
```

Using the vector operators + (or +=) and * is convenient for writing numerical expressions, although certain performance reduction has to be accepted. To keep the listings short here we present this version, a faster implementation (requiring longer notation) is provided on the book's website.

force() [Sphere.cc]

The function force() computes the force which is exerted by two spheres, p1 and p2, on each other. Since this function needs information on *two* particles, force() is not a member of the class Sphere. Note that this function has to be defined for all particle models used in the simulation.

```
                        ───── Sphere.cc ─────
38  void force(Sphere & p1, Sphere & p2, double lx, double ly)
39  {
40    double dx=normalize(p1.x()-p2.x(),lx);
41    double dy=normalize(p1.y()-p2.y(),ly);
42    double rr=sqrt(dx*dx+dy*dy);
43    double r1=p1.r();
44    double r2=p2.r();
45    double xi=r1+r2-rr;
46
```

```
47   if(xi>0){
48       double Y=p1.Y*p2.Y/(p1.Y+p2.Y);
49       double A=0.5*(p1.A+p2.A);
50       double mu = (p1.mu<p2.mu ? p1.mu : p2.mu);
51       double gamma = (p1.gamma<p2.gamma ? p1.gamma : p2.gamma);
52       double reff = (r1*r2)/(r1+r2);
53       double dvx=p1.vx()-p2.vx();
54       double dvy=p1.vy()-p2.vy();
55       double rr_rez=1/rr;
56       double ex=dx*rr_rez;
57       double ey=dy*rr_rez;
58       double xidot=-(ex*dvx+ey*dvy);
59       double vtrel=-dvx*ey + dvy*ex + p1.omega()*p1.r()-p2.omega()*p2.r();
60       double fn=sqrt(xi)*Y*sqrt(reff)*(xi+A*xidot);
61       double ft=-gamma*vtrel;
62
63       if(fn<0) fn=0;
64       if(ft<-mu*fn) ft=-mu*fn;
65       if(ft>mu*fn) ft=mu*fn;
66       if(p1.ptype()==0) {
67           p1.add_force(Vector(fn*ex-ft*ey, fn*ey+ft*ex, r1*ft));
68       }
69       if(p2.ptype()==0) {
70           p2.add_force(Vector(-fn*ex+ft*ey, -fn*ey-ft*ex, -r2*ft));
71       }
72   }
73 }
```
—————————————— to be continued ——————————————

In lines 48–51 the effective material parameters for the collision are computed from the material parameters of the colliding particles, as described on p. 20 for the normal force and on p. 25 for the tangential component.

The function force() proceeds with implementing the force law (2.16, 2.18) which has been described in Sect. 2.3. Finally, for particles of type 0 the interaction forces are added to their total forces by add_force().

boundary_conditions() [Sphere.cc]

In our simulations the container walls are built of particles to model surface roughness. The motion of such particles is not governed by their interaction with other particles but follows a predescribed trajectory. This concerns not only the outer walls of the simulation area but also other objects which do not move according to Newton's law, such as stirrers, obstacles, fixtures, and other objects, which are all considered as walls in the context of Molecular Dynamics simulations.

For fixed walls, these particles cannot move at all, no matter what forces act on them. For oscillating walls (e.g., in vibrating containers), the wall particles follow a sinusoidal trajectory. To distinguish normal particles whose motion is determined by interaction forces from particles obeying boundary conditions, the variable _ptype is used (accessed via the function ptype()). By convention (see discussion on p. 37 and the table on p. 40), the mobile particles of the granular material are of type 0; the other types are used to distinguish the particles that are subject to boundary conditions.

The function boundary_conditions() expects three parameters: the particle index n, the integration time step, and the current simulation time.

```
                         ─── Sphere.cc ───
74  void Sphere::boundary_conditions(int n, double timestep, double Time)
75  {
76    switch(ptype()){
77    case(0): break;
78    case(1): break;
79    case(2): {
80        x()=0.5-0.4*cos(10*Time);
81        y()=0.1;
82        vx()=10*0.4*sin(Time);
83        vy()=0;
84      } break;
85    case(3): {
86        double xx=x()-0.5;
87        double yy=y()-0.5;
88        double xp=xx*cos(timestep)-yy*sin(timestep);
89        double yp=xx*sin(timestep)+yy*cos(timestep);
90
91        x()=0.5+xp;
92        y()=0.5+yp;
93        vx()=-yp;
94        vy()= xp;
95        omega()=1;
96      } break;
97    case(4): {
98        x()=0.5+0.1*cos(Time) + 0.4*cos(Time+2*n*M_PI/128);
99        y()=0.5+0.1*sin(Time) + 0.4*sin(Time+2*n*M_PI/128);
100       vx()=-0.1*sin(Time) - 0.4*sin(Time+2*n*M_PI/128);
101       vy()= 0.1*cos(Time) - 0.4*cos(Time+2*n*M_PI/128);
102       omega()=1;
103     } break;
104   case(5): {
105       y()=0.1+0.02*sin(30*Time);
106       vx()=0;
107       vy()=0.02*30*cos(30*Time);
108     } break;
109   case(6): {
110       int i=n/2;
111       y()=i*0.02+0.1+0.02*sin(30*Time);
112       vx()=0;
113       vy()=0.02*30*cos(30*Time);
114     } break;
115   default: {
116       cerr << "ptype: " << ptype() << " not implemented\n";
117       abort();
118     }
119   }
120 }
                      ─── to be continued ───
```

The seven implemented types of particles serve only as examples. In general, each problem requires its own set of particle types according to the boundary condition of the problem under investigation. These seven example types are explained in detail below:

type 0: The particle is a normal granular particle whose motion is determined by its interaction with other particles, according to Newton's equation of motion. The function `boundary_conditions()` returns immediately without doing anything. The motion of such particles is computed by `predict()` and `correct()`, thus solving the equation of motion for the particle. This case is, strictly speaking, redundant (but left in for completeness) since this function should not be called for particles of type 0.

type 1: The function `boundary_conditions()` exits immediately, the particle position and velocity remain unchanged. These particles can, thus, represent static walls (the functions `predict()` and `correct()` are not called for these particles, in contrast to case 0!).

type 2: The particle oscillates in the horizontal direction, following the trajectory $x(t) = [0.5 - 0.4\cos(10\,t/\text{sec})]\,$m.

type 3: The particles revolve with frequency 1/sec around the center $x = 0.5\,$m, $y = 0.5\,$m, thus, realizing a rotating container of arbitrary shape. To ensure the correct rotation of the particles synchronous to the cylinder rotation, we set `omega()=1`.[8]

type 4: These particles are located on a circle (radius $= 0.4\,$m) whose center performs an eccentric rotation of radius $0.1\,$m with frequency 1/sec. This kind of motion is used to model a circular cylinder that performs a swirling motion.

type 5: The particle oscillates vertically with frequency $\omega = 30/\text{sec}$ and amplitude 2 cm around $y = 10\,$cm. The bottom wall in Fig. 2.6 is modeled by this type of particles (all particles have the same y-coordinate).

type 6: The particles perform the same motion as particles of type 5, but around the vertical coordinate $y = (2\,i + 10)\,$cm with i being the particle index divided by 2 (fractions are rounded down). The particle indices at the left wall are even numbers, at the right wall they are odd. This particle type was used to simulate the vertical walls in Fig. 2.6. Boundary conditions which contain the particle index require enumeration of the particles in a certain sequence. Here the vertical container walls have the first index numbers in the system.

One can easily define more particle types by adding more cases to the `switch` statement in `boundary_conditions()`. The examples shown are only meant to demonstrate how to implement the desired boundary conditions.

periodic_bc() [Sphere.cc]

The function `periodic_bc()` enforces the periodic boundary conditions, i.e., it places the particle at the position of its periodic image inside the simulation area if it crossed the boundary of the system.

```
                                    Sphere.cc
121 | void Sphere::periodic_bc(double x_0, double y_0, double lx, double ly)
122 | {
123 |   while(rtd0.x()<x_0) rtd0.x()+=lx;
124 |   while(rtd0.x()>x_0+lx) rtd0.x()-=lx;
125 |   while(rtd0.y()<y_0) rtd0.y()+=ly;
126 |   while(rtd0.y()>y_0+ly) rtd0.y()-=ly;
127 | }
                                    to be continued
```

---

[8] Since the value of `omega()` is not changed, it could be set to 1 only once at the start of the program. However, this relies on proper initialization. The (redundant) method is safer and serves to demonstrate the proper boundary condition.

kinetic_energy() and I/O operators [Sphere.cc]

As an example for on-the-fly data processing, the member function
`kinetic_energy()` returns the kinetic energy of the particle.

```
                            ———— Sphere.cc ————
128  double Sphere::kinetic_energy() const
129  {
130    return _m*(rtd1.x()*rtd1.x()/2 + rtd1.y()*rtd1.y()/2)
131      + _J*rtd1.phi()*rtd1.phi()/2;
132  }
                        ———— to be continued ————
```

The operators `<<` and `>>` serve for convenient input and output of particle
data, i.e., the particle coordinates, their time derivatives, and further particle
specific values.

In the program the moment of inertia for three-dimensional spheres, $J = 2/5mR^2$, is applied even though the motion of the particles is restricted to
two dimensions. This is consistent with the chosen interaction force (the Hertz
law (2.12)) which also applies to three-dimensional spheres.

```
                            ———— Sphere.cc ————
133  istream & operator >> (istream & is, Sphere & p)
134  {
135    is >> p.rtd0 >> p.rtd1
136       >> p._r >> p._m >> p._ptype
137       >> p.Y >> p.A >> p.mu >> p.gamma
138       >> p._force
139       >> p.rtd2 >> p.rtd3 >> p.rtd4;
140    p._J=p._m*p._r*p._r/2;
141    return is;
142  }
                        ———— to be continued ————
```

The operator `<<` sends all available information about the sphere into an output stream (a file or the screen). This function is in turn used in `phase_plot()`
(see p. 47) to dump the current system state to an output file which can be
used later to extract simulation results or to restart the simulation using the
current particle coordinates, velocities, etc., as initial conditions.

```
                            ———— Sphere.cc ————
143  ostream & operator << (ostream & os, const Sphere & p)
144  {
145    os << p.rtd0 << " " << p.rtd1 << " ";
146    os << p._r << " " << p._m << " " << p._ptype << " ";
147    os << p.Y << " " << p.A << " " << p.mu << " " << p.gamma << " ";
148    os << p._force << " ";
149    os << p.rtd2 << " " << p.rtd3 << " " << p.rtd4 << "\n" << flush;
150    return os;
151  }
                        ———— to be continued ————
```

To complete the description of the basic program for Molecular Dynamics of
spherical particles in two dimensions, some peripheral functions which serve
the system initialization and the data extraction are described in the following
sections.

init_system() [common.cc]

The function `init_system()` initializes the simulation system. It first reads the global system parameters `G` (gravity), `Time`, `nstep`, `timestep`, `nprint`, `nenergy`, `lx`, `ly`, `x_0`, and `y_0` from the initialization file. These parameters must be specified in the form

> `#gravity:` *value*
>
> `#Time:` *value*
>
> etc.

```
————————————————————— common.cc —————————————————————
57  void init_system(char * fname)
58  {
59    ifstream fparticle(fname);
60    while(fparticle.peek()=='#'){
61      string type;
62      fparticle >> type;
63      if(type=="#gravity:"){
64        fparticle >> G.x() >> G.y() >> G.phi();
65        fparticle.ignore(100,'\n');
66        cout << "gravity: " << G << endl;
67      } else if(type=="#Time:"){
68        fparticle >> Time;
69        fparticle.ignore(100,'\n');
70        cout << "Time: " << Time << endl;
71      } else if(type=="#nstep:"){
72        fparticle >> nstep;
73        fparticle.ignore(100,'\n');
74        cout << "nstep: " << nstep << endl;
75      } else if(type=="#timestep:"){
76        fparticle >> timestep;
77        fparticle.ignore(100,'\n');
78        cout << "timestep: " << timestep << endl;
79      } else if(type=="#nprint:"){
80        fparticle >> nprint;
81        fparticle.ignore(100,'\n');
82        cout << "nprint: " << nprint << endl;
83      } else if(type=="#nenergy:"){
84        fparticle >> nenergy;
85        fparticle.ignore(100,'\n');
86        cout << "nenergy: " << nenergy << endl;
87      } else if(type=="#lx:"){
88        fparticle >> lx;
89        fparticle.ignore(100,'\n');
90        cout << "lx: " << lx << endl;
91      } else if(type=="#ly:"){
92        fparticle >> ly;
93        fparticle.ignore(100,'\n');
94        cout << "ly: " << ly << endl;
95      } else if(type=="#x_0:"){
96        fparticle >> x_0;
97        fparticle.ignore(100,'\n');
98        cout << "x_0: " << x_0 << endl;
99      } else if(type=="#y_0:"){
100       fparticle >> y_0;
101       fparticle.ignore(100,'\n');
102       cout << "y_0: " << y_0 << endl;
103     } else {
104       cerr << "init: unknown global property: " << type << endl;
105       abort();
106     }
107   }
108   while(fparticle){
109     Sphere pp;
```

```
110        fparticle >> pp;
111        if(fparticle){
112          particle.push_back(pp);
113        }
114      }
115      no_of_particles=particle.size();
116      cout << no_of_particles << " particles read\n" << flush;
117    }
```
——————————— to be continued ———————————

After reading the last parameter, i.e., when finding the first line which does not start with "#" the particle **pp** is read from the input stream (e.g., a file) **fparticle** in the line

```
    fparticle >> pp;
```

The compiler replaces this line by the call of the operator (which is basically a special function with prescribed syntax)

```
    istream & operator >> (istream &, const Sphere &)
```

which performs the data input according to the desired structure of the initialization file (see p. 45).

2.3.4 Data Extraction

Depending on the actual problem, the program writes output data which are functions of the particle coordinates, velocities, accelerations, etc., and of the particle properties, such as mass, radius, and moment of inertia.

total_kinetic_energy() [common.cc]

As an example of an output function, `total_kinetic_energy()` returns the total kinetic energy of the system. For each particle, this program calls the element function `kinetic_energy()` of the particle class.

——————————— common.cc ———————————
```
118  double total_kinetic_energy()
119  {
120    double sum=0;
121    for(unsigned int i=0;i<particle.size();i++){
122      if(particle[i].ptype()==0){
123        sum+=particle[i].kinetic_energy();
124      }
125    }
126    return sum;
127  }
```
——————————— to be continued ———————————

phase_plot() [common.cc]

Frequently it is desired to analyze the result of a simulation not at run time, but independent of the Molecular Dynamics run. The offline data analysis is particularly advantageous if the result of a simulation is not foreseeable. Furthermore, online data visualization is frequently not possible—due to the high computational effort the individual frames are produced at too slow a rate to generate a fluent animation (there may be minutes and more between two frames). Therefore, the function `phase_plot()` is needed to dump the particle coordinates, velocities, radii, masses, and the higher-order derivatives (i.e., all information which is read from the init file) by means of the operator `<<`. The function `phase_plot()` is called by the main program in regular time intervals.

```
                             ──────── common.cc ────────
128  void phase_plot(ostream & os)
129  {
130    os << "#NewFrame\n";
131    os << "#no_of_particles: " << no_of_particles << endl;
132    os << "#compressed: no\n";
133    os << "#type: SphereXYPhiVxVyOmegaRMFixed25\n";
134    os << "#gravity: " << G.x() << " " << G.y() << " " << G.phi() << endl;
135    os << "#Time: " << Time << endl;
136    os << "#timestep: " << timestep << endl;
137    os << "#EndOfHeader\n";
138    for(unsigned int i=0;i<particle.size();i++){
139      os << particle[i];
140    }
141    os << flush;
142  }
```

At the beginning of `phase_plot()`, global information is written in humanly readable format. This serves two purposes. First, it is often beneficial if the simulation data and the corresponding parameters are inseparable since it helps with record-keeping of simulation results. If the parameters and the simulation results are stored in different files, the results become worthless if the parameter file gets lost. Second, the header can be used to convey information to data processing tools. As an example, an identification string (`#type`) can be written to be used by visualization tools to determine the method of data display.

2.3.5 Examples

The described program is applied to two sample problems. First, the velocity profile in an outflowing hopper is computed. We use the opportunity to demonstrate how to generate an initialization file for the Molecular Dynamics simulation.

This file should contain the initial state of all particles. At the beginning of the simulation the particles are expected to stay at rest inside the hopper. Since it is practically impossible to generate stable particle positions in one step, the initialization file is generated as follows:

1. The coordinates of the wall particles (of type 1) are computed according to the desired shape of the hopper. Additional wall particles close the hopper from below. The mobile particles are arranged above the hopper on a regular lattice, simulating an incoming stream of particles.

2. The Molecular Dynamics program simulates the filling of the hopper, i.e., the particles fall down from their lattice positions. Thus the hopper is filled by the grains. At the end of this step the program dumps the phase space to a file.

3. The additional wall particles which close the hopper from below are removed.

Let us describe this procedure in more detail: the first step is performed by the following program.

```
                              init_hopper.cc
1  #include <fstream.h>
2  #include <stdlib.h>
3
4  void dump_particle(ostream & os,
5                     double x, double y, double vx, double vy,
6                     double radius, double mass, int type)
7  {  os << x << "\t" << y << "\t0\t" << vx << "\t" << vy << "\t0\t"
8         << radius << "\t" << mass << "\t"
9         << type << "\t0\t0\t0\t0\t0\t0\t0\t0\n";
10 }
11
12 int main()
13 {
14   ofstream fout("closed_hopper.random");
15   fout << "#gravity: 0 -9.81 0\n";
16   fout << "#timestep: 1e-6\n";
17   fout << "#nstep: 5000000\n";
18   fout << "#nprint: 1000\n";
19   fout << "#nenergy: 1000\n";
20   fout << "#Time: 0\n";
21   for(int i=0;i<11;i++)
22     dump_particle(fout,0.45+i*0.01,0.19,0,0,0.005,1,1);
23
24   for(int i=0;i<50;i++){
25     dump_particle(fout,0.55+(i+0.5)*0.005,i*0.01+0.2,0,0,0.005,1,1);
26     dump_particle(fout,0.45-(i+0.5)*0.005,i*0.01+0.2,0,0,0.005,1,1);
27   }
28   const double Rmax=0.006, Rmin=0.004;
29
30   for(int i=0;i<67;i++){
31     for(int k=0;k<30;k++){
32       double centerx=0.3115+0.013*i;
33       double centery=0.6+0.013*k;
34       double z=drand48();
35       double r=Rmin*Rmax/(Rmax-z*(Rmax-Rmin));
36       dump_particle(fout,centerx,centery,0,0,r,r*r/(Rmax*Rmax),0);
37     }
38   }
39 }
```

The function dump_particle() writes the particle information to the output stream os (e.g., a file) in the format as required by the simulation program. The file closed_hopper.random is opened and 11 particles are placed to form the bottom of the hopper (lines 21–22) and 50 particles for each of the inclined walls (lines 24–27). Then 2,010 regular particles are positioned on a lattice on top of the hopper (lines 30–38). Initially the particles do not touch each other.

The radii of the particles are chosen from the interval (R_{min}, R_{max}) in such a way that the total mass of all particles from a certain size interval is the same for all sizes, thus ensuring that neither large nor small particles dominate the system. This property is given if the radii are chosen according to the probability distribution

$$p(R) = \frac{R_{min}R_{max}}{R_{max} - R_{min}} \frac{1}{R^2} . \tag{2.30}$$

Random numbers according to the distribution (2.30) can be generated from equi-distributed random numbers $z \in [0, 1)$ via the transformation

$$R = \frac{R_{min}R_{max}}{R_{max} - z\,(R_{max} - R_{min})} . \tag{2.31}$$

In lines 34–35 the transformation (2.31) is applied to initialize the particle radii. The derivations of the distribution function (2.30) and the transformation (2.31) are explained in detail on pp. 274–276. The configuration of particles as obtained after this first step of the initialization procedure is drawn in Fig. 2.4 (a).

In the second step of the initialization the file `closed_hopper.random` is read by the Molecular Dynamics program. The particles fall down into the closed hopper and relax there. This initial simulation terminates after 5 million time steps (5 sec in real time), when the particles are at rest. From the final dump of the phase space, we remove the entries for the 11 wall particles which close the hopper from below, using a text editor to obtain the input file for the simulation of the outflowing hopper. Equivalently, a new particle type and corresponding boundary conditions could have been defined for these 11 particles, say as type 7, which assure that these particles are removed immediately before the final phase space dump. Figure 2.4 shows snapshots from the initialization and relaxation procedure.

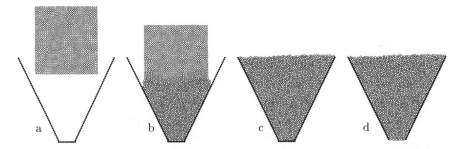

Fig. 2.4. Generation of the initialization file. Wall particles are drawn filled. (**a**) In the first step the mobile particles are arranged on a regular lattice on top of the hopper (only part of these particles is shown). The hopper is closed from below by additional immobile particles. (**b**) The particles are released and fall into the closed hopper. (**c**) After complete relaxation. (**d**) The bottom particles are removed and the positions of the particles are stored to a file

Using the just generated initialization file, the Molecular Dynamics simulation can be started to determine the flow profile. The parameters are:

Number of mobile particles	$N = 2{,}010$
Particle radii	$R = 0.4\,\text{cm--}0.6\,\text{cm}$
Young modulus	$Y = 10^9\,\text{Pa}$
Coulomb friction parameter	$\mu = 0.5$
Damping constant	$A = 0.01\,\text{sec}$
Tangential damping	$\gamma^t = 10\,\text{Nsec/m}$
Material density	$\rho = 8\,\text{g/cm}^3$
Integration time step	$\Delta t = 10^{-6}\,\text{sec}$

For this demonstration it is enough just to shade the particles as shown in Fig. 2.5. The flow profile can be deduced from the deformation of the stripes. For a quantitative analysis, of course more sophisticated methods of data processing are employed. An animated sequence of the simulation can be found on the book's website.

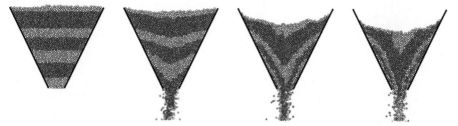

Fig. 2.5. Snapshots of the outflowing hopper. The deformation of the stripes of marked particles visualizes the flow profile

As a second example, a polydisperse granular material comprising $N = 855$ particles in a vertically vibrating container is simulated. The other parameters are the same as in the previous example, except for the radii, which lie in the interval $0.8\,\text{cm--}1.2\,\text{cm}$ according to the distribution (2.30). A large particle ($R = 2.4\,\text{cm}$) is added at the bottom of the container (see Fig. 2.6, left).

The initialization file was produced in a similar way as described above. The wall particles are of type 5 and 6 (see p. 44 for reference). The container oscillates vertically according to $B\cos(\Omega t)$ with $\Omega = 30/\text{sec}$ and $B = 2\,\text{cm}$. After a short simulation time the large particle moves upward (see Fig. 2.6). Moreover, in the animated sequence a pronounced convection flow can be observed. Both effects are related to each other and are well investigated, e.g., [139]. The first Molecular Dynamics simulations describing this effect can be found in [83, 272].

The presented simulation method can be generalized to three-dimensional systems with moderate effort. Analogous to Fig. 2.5, Fig. 2.7 shows snapshots of an outflowing hopper simulated in three dimensions.

Fig. 2.6. Particle size segregation in vertically vibrated granular material. *Left*: beginning of the simulation; *middle*: after about 6 sec real time; *right*: after about 8.6 sec

Fig. 2.7. An outflowing hopper in three dimensions. The number of mobile particles was 11,892, the wall was built from 2,176 particles. The snapshots were recorded after 0, 0.75, and 1.5 sec (real time)

2.3.6 Critical Discussion of the Particle Model

Many Molecular Dynamics simulations using spherical particle models with different interaction force laws can be found in literature. Most of these simulations describe in particular the *dynamical* behavior of various systems very reliably. Qualitative and in some situations also quantitative agreement with experiments has been reported.

The great advantage of the spherical particle model is the numerical efficiency due to a very simple collision detection: to check whether two particles i and j touch and, hence, exert forces on each other, it is sufficient to compare the distance of their centers $|\vec{r}_i - \vec{r}_j|$ with the sum of their radii. This comparison can be performed in two and also in three dimensions at very low computational costs. Therefore, simulations of systems of many thousand spherical particles are feasible.

The efficiency of such simulations allows for the detailed numerical investigation of granular systems, including variation of different material properties, container geometries, etc. Certain system properties, such as flow profiles, can be determined up to high accuracy by averaging over long periods in real time.

Typical examples of Molecular Dynamics simulations using spherical particles include granular pipe flow (e.g., [103, 212]), irregularly outflowing silos and hoppers (e.g., [141, 240]), the motion of granular material inside rotating cylinders (e.g., [49, 59, 239]), and on mechanical conveyors ([84]), the convective motion in vertically (e.g., [82, 83, 272]) or horizontally (e.g., [155, 249]) vibrating containers, and many others.

Although spherical particles are adequate for many problems, there are also limitations of this simple model:

1. The normal force between colliding spheres has been derived from basic material properties (at least for simple stress–strain relations), i.e., there is a well founded theoretical basis. For the tangential component, however, there are only phenomenological laws incorporating parameters which cannot be derived theoretically from material properties (see discussion in Sect. 2.1.4). These parameters cannot be determined by experimental investigations of materials either, but only indirectly by comparison of the Molecular Dynamics results with the corresponding experiment. Hence, to determine appropriate values of these parameters, e.g., the damping coefficient γ^t, the parameters are varied until the simulation and experimental results agree.

2. For small relative particle velocities the tangential force (2.18) vanishes. It is therefore difficult to apply the spherical particle model to *static* systems, such as resting sand heaps or completely and persistently clogged pipes and hoppers. There exist special force laws to overcome this drawback (see the paragraph on tangential forces in Sect. 2.1.4). However, this requires the introduction of further mechanisms and parameters which have no counterpart in the mechanics of materials.

In Sects. 2.6–2.7, particle models are proposed to partially resolve these problems at growing computational and algorithmic complexity.

2.4 Efficient Force Summation

2.4.1 Algorithmic Complexity of the Force Summation

The Molecular Dynamics program as described in Sect. 2.3 is complete; however, so far it is computationally inefficient. The efficiency of any Molecular Dynamics program is mainly determined by the efficiency of the computation of the forces which act on the particles. Let us explain this by means of an example: Assume a simulation of $N = 1,000$ particles. In each time step all possible pairs of particles have to be considered with respect to their interaction force, hence, $N(N-1)/2 \approx 500,000$ force computations are needed. For short-range particle interactions the majority of these force evaluations is unnecessary since the corresponding particles are located very distant from each other. For approximatively equal-sized particles each single particle can be in contact with not more than about 6 other particles, hence, only about $3N = 3,000$ force computations are necessary. For the case of the naive force computation scheme (which was applied in the previous section), at least 166 times more pair interactions are computed than necessary – what a waste of computer time! Therefore, methods are needed to reduce the number of pair interactions which are considered at each time step. Since the forces between granular particles are of short range, the force computation can be restricted to pairs of particles which are close neighbors.

Deciding which particles are close neighbors is, however, not trivial since *all* pairs of close particles have to be considered. Even if only one of them is missed, serious consequences for the validity of the simulation results may follow. The intuitive solution to simply check whether all pairs of particles are neighbors or not, leads again to about 500,000 computations which was to be avoided. This section describes three methods for the efficient computation of the forces, the *Verlet algorithm*, the *link cell* algorithm, and a *lattice algorithm*.

2.4.2 Verlet Lists

The idea of the Verlet algorithm is based on a simple property of particle dynamics: neighborhood relations between particles change only slowly, i.e., two particles which are close to each other at a given time will stay close neighbors, at least in the following few time steps.

At the time of initialization the neighborhood relations between the particles are determined, i.e., the distances of all close pairs of particles are computed. We call two particles close neighbors if the distance of their surfaces is smaller than a predefined constant `verlet_distance`. For spheres this criterion can be written as $(|\vec{r}_i - \vec{r}_j| - R_i - R_j) <$`verlet_distance`. For non-spherical particles this criterion has to be generalized (see below). For each particle there is a list (the *Verlet list*) in which all close neighbors of this particle are recorded. To initialize the Verlet lists efficiently, a grid that covers the simulation area is defined. Its mesh size is larger than the largest particle. For the construction of the lists only pairs whose particles reside in the same or adjacent grid cells are considered. This procedure guarantees detection of all close pairs.

Double entries in the Verlet lists (if A is a neighbor of B then B is a neighbor of A) are avoided by the restriction that the list of particle i contains only neighbors with index $j < i$. The individual Verlet list of a certain particle is, hence, a set of particle indices. As explained below, the required data structure must allow rapid decisions on whether a certain particle j is contained in the Verlet list of particle i. This requirement is fulfilled by the data structure `set` of the Standard Template Library (STL). Since the Verlet lists are sets of integers they are implemented as `set<int>`.[9] The Ver-

---

[9] As an alternative, one can implement the list as `vector<int>`. A `vector` allows faster access to its elements while a `set` is more appropriate for finding elements. Sets are faster in typical simulations.

let lists of all particles are organized as a `vector` of such individual lists, i.e., as `vector<set<int>> verlet`. The individual list of particle `i` is then `verlet[i]`.

For the computation of the particle interaction forces, only pairs which are recorded in one of the Verlet lists are considered. Hence the Verlet list of each particle i is scanned and the interaction force of i with each entry j in its list is computed.

The Verlet algorithm is implemented in the file `verlet.cc`.

```
                            ──── verlet.cc ────
1   #include "common.h"
2   #include "Sphere.h"              /* <===== replace this line if necessary */
3
4   extern vector<Sphere> particle;  /* <===== replace this line if necessary */
5
6   vector<Sphere> safe;             /* <===== replace this line if necessary */
7   double Timesafe;
8
9   double verlet_ratio = 0.6, verlet_distance = 0.00025, verlet_grid = 0.05;
10  double verlet_increase = 1.1;
11  int vnx,vny;
12  double dx,dy;
13  vector<vector<vector<int> > > celllist;
14  vector<set<int> > verlet;
15
16  bool verlet_needs_update();
17  bool make_verlet();
18  bool do_touch(int i, int k);
                        ──── to be continued ────
```

First the global header `common.h` and the particle class header `Sphere.h` are included and the array `particle` is declared, just as in the case of the simple algorithm discussed in the preceding section. The next declarations are specific to the Verlet algorithm. The array `safe` and the variable `Timesafe` serve as a backup (for `particle` and `Time`) in the rare case that the Verlet lists become incorrect (see below). The values of `verlet_ratio`, `verlet_increase`, and `verlet_distance` govern the performance of the algorithm. They are explained in detail below. Finally the functions `verlet_needs_update()`, `make_verlet()`, and `do_touch()` are declared.

The functions `init_algorithm()`, `step()`, and `make_forces()` which have been introduced in Sect. 2.3.2 have to be replaced by algorithm-specific functions. Since `make_forces()` contains the very essence of the Verlet algorithm we start with the implementation of this function. The previous section describes `make_forces()` (see p. 38) to compute the forces by considering all pairs of particles. For the computation of the forces by means of the Verlet lists this function has to be replaced by the following:

```
                            ──── verlet.cc ────
19  void make_forces()
20  {
21    for(unsigned int i=0;i<particle.size();i++){
22      set<int>::iterator iter;
23      for(iter=verlet[i].begin();iter!=verlet[i].end();iter++){
24        force(particle[i],particle[*iter], lx, ly);
25      }
26    }
27  }
                        ──── to be continued ────
```

Instead of considering all other particles as potential interaction partners for a given particle i, only the members in its Verlet list are considered now. For suitable choice of the parameter verlet_distance the Verlet list is very short, and the number of potential interaction partners is far smaller than for our simple algorithm. The Verlet lists are constructed by make_verlet().

```
————————————— verlet.cc —————————————
28  bool make_verlet()
29  {
30    bool ok=true;
31
32    verlet.resize(no_of_particles);
33    for(int ix=0;ix<vnx;ix++){
34      for(int iy=0;iy<vny;iy++){
35        celllist[ix][iy].clear();
36      }
37    }
38    for(unsigned int i=0;i<no_of_particles;i++){
39      int ix=int((particle[i].x()-x_0)/dx);
40      int iy=int((particle[i].y()-y_0)/dy);
41      celllist[ix][iy].push_back(i);
42    }
43    for(unsigned int i=0;i<no_of_particles;i++){
44      set<int> oldverlet=verlet[i];
45      verlet[i].clear();
46      int ix=int((particle[i].x()-x_0)/dx);
47      int iy=int((particle[i].y()-y_0)/dy);
48      for(int iix=ix-1;iix<=ix+1;iix++){
49        for(int iiy=iy-1;iiy<=iy+1;iiy++){
50          int wx = (iix+vnx)%vnx;
51          int wy = (iiy+vny)%vny;
52          for(unsigned int k=0;k<celllist[wx][wy].size();k++){
53            int pk=celllist[wx][wy][k];
54            if(pk<(int)i){
55              if(Distance(particle[i],particle[pk], lx, ly)<
56                  particle[i].r()+particle[pk].r()+verlet_distance){
57                if((particle[i].ptype()==0) ||
58                   (particle[pk].ptype()==0)){
59                  verlet[i].insert(pk);
60
61                  if(oldverlet.find(pk)==oldverlet.end()){
62                    if(do_touch(i,pk)) {
63                      ok=false;
64                    }
65                  }
66                }
67              }
68            }
69          }
70        }
71      }
72    }
73    return ok;
74  }
————————————— to be continued —————————————
```

First, the number of Verlet lists is specified in line 32. The particles are sorted into a grid of mesh size dx·dy. For each grid cell there is a list of particles residing in this cell. These lists are organized in the two-dimensional array celllist. In lines 33–36 the currently existing lists are discarded; these lists have been constructed in the preceding call of make_verlet(). Then, in lines 38–42 the particles are sorted into the grid.

The Verlet lists are constructed in the main loop (lines 43–72). Since new contacts have to be distinguished from already existing ones, the expired Verlet list of the currently considered particle i is copied to the variable oldverlet and is then emptied. The indices ix and iy, computed in lines 46–47, select the grid cell in which particle i resides. In the loops starting in lines 48 and 49 the cell (ix,iy) and its adjacent cells are visited and all particles pk from those cells are tested for close neighborhood. To avoid double entries, all particles with pk>i are disregarded. For all particles pk<i lines 55–56 check whether the distance of the surfaces of pk and i is smaller than verlet_distance, provided at least one of them is mobile, i.e., of type 0. If this is the case, the index pk is added to the list of particle i. If pk is not contained in the list oldverlet (i.e., pk is a new neighbor) *and* the particles touch each other (do_touch(i,pk)==true), an error has occurred since the particles have been recognized too late as neighbors. The reason of this error and methods to fix it are discussed below. As soon as this error occurs the Boolean variable ok is set to the value false. If this error does not take place for any of the particles, the construction of the Verlet lists is finished and the variable ok keeps its initial value true. The program make_verlet() returns the variable ok, i.e., true for successful construction of the Verlet list and false to report an error.

```
                          verlet.cc
75  bool do_touch(int i, int k)
76  {
77      return (Distance(particle[i],particle[k],lx,ly)
78             <particle[i].r()+particle[k].r());
79  }
                        to be continued
```

The function do_touch(int i, int k) yields true if the particles are in contact.

During the simulation, the neighborhood relations of the particles change, therefore, the neighborhood lists have to be rebuilt from time to time. Whether the Verlet lists need to be reconstructed is decided by verlet_needs_update(), depending on how far the particles have traveled since the time when the currently valid lists have been built. Let us explain the idea of this function: The Verlet list of a particle i must contain at any time all neighbors j with $j < i$. This assures that two particles i and j are never in touch if they are not known as neighbors, i.e., j is not in the list of i and i is not in the list of j. Hence,

$$|\vec{r}_i - \vec{r}_j| - R_i - R_j > 0 \tag{2.32}$$

is required for all pairs (i, j) of particles which are *not* known as neighbors. This condition provides a criterion for the necessity to reconstruct the Verlet lists. Figure 2.8 illustrates the idea: Assume at the instant when the Verlet lists are constructed, the surfaces of the particles have the distance $|\vec{r}_i - \vec{r}_j| - R_i - R_j >$verlet_distance, i.e., they are not classified as neighbors. If the Verlet lists are updated before one of these particles has traveled the distance

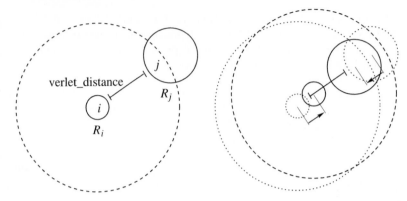

Fig. 2.8. Checking the validity of the Verlet lists. *Left*: the particles i and j are not recognized as neighbors since the distance of their surfaces is larger than `verlet_distance`. The radius of the dashed circle is $R_i + R_{max}+$ `verlet_distance`. *Right*: in the most unfortunate case the particles approach each other directly, traveling at the same velocity. As soon as one of the particles has traveled the distance `verlet_distance/2` (*arrows*), the Verlet lists have to be rebuilt. The particles i and j are now recognized as neighbors

`verlet_distance/2` since the lists were constructed, they can never collide without being recognized as neighbors first.

The positions of the particles at the time when the Verlet lists are built are stored in the variable `safe`. This variable is used to determine the distances which have been traveled by the particles since that time. Since the elements of `safe` are exact copies of the particles, the desired distance can be determined as any other distance by calling `Distance()`, i.e., `Distance(particle[i],safe[i])`.

The function `verlet_needs_update()` implements the described method with an additional optimization which is discussed below. In case the Verlet lists have to be rebuilt this function yields the value `true`.

```
――――――――――――― verlet.cc ―――――――――
1  bool verlet_needs_update()
2  {
3    for(unsigned int i=0;i<no_of_particles;i++){
4      if(Distance(particle[i],safe[i],lx,ly)>= verlet_ratio*verlet_distance){
5        return true;
6      }
7    }
8    return false;
9  }
―――――――――――― to be continued ―――――――――
```

The experience with Molecular Dynamics simulations shows that for practical purposes our criterion for list-updating is too strict. It assures that collisions are never missed, even in the worst case when

1. At the time of list construction two particles have a distance of only slightly more than `verlet_distance` (i.e., they are not recognized as neighbors), *and*

2. These particles approach each other directly, *and*
3. These two particles are the fastest particles of the system, *and*
4. Both particles move with the same absolute velocity (i.e., both of them travel the distance `verlet_distance/2`).

This case is, however, very unlikely. Therefore, the lists can be rebuilt using a weaker criterion: they are rebuilt when the fastest particle has traveled the distance `verlet_ratio * verlet_distance`, where `verlet_ratio > 0.5`. Finding the concrete value requires some experience; in many cases `verlet_ratio = 0.6` is appropriate. The impact of this optimization is, however, not very large since the construction of the Verlet lists is quite efficient already; usually only a few percent of the total computation time is spent on constructing the Verlet lists.

In spite of the weakened criterion for constructing the Verlet lists, the correct operation of the program must stay assured, even for the worst case. For `verlet_ratio > 0.5`, we cannot exclude that a collision occurs although the colliding particles were not recognized as neighbors before, i.e., the pair is not contained in either particle's Verlet list. This situation is fatal since such particles cannot experience any interaction force and will, hence, penetrate each other. Precautions have to be taken to avoid this situation. The problem, once detected in `make_verlet()`, is resolved by restoring the particle positions, velocities, and higher-order time derivatives using previously stored values and to restart the simulation from that point with a larger value of `verlet_distance` by multiplying it with the constant `verlet_increase`. Typically `verlet_increase = 1.1` is chosen.

The described method is implemented in `step()`, replacing the simple procedure `step()`, (see p. 36):

```
                         ──────── verlet.cc ────────
 1  void step()
 2  {
 3    bool ok=true, newverlet=false;
 4
 5    if(verlet_needs_update()){
 6      ok = make_verlet();
 7      newverlet=true;
 8    }
 9    if(!ok){
10      cerr << "fail: going back from " << Time << " to " << Timesafe << endl;
11      particle=safe;
12      Time=Timesafe;
13      verlet_distance*=verlet_increase;
14      make_verlet();
15    }
16    if(newverlet && ok){
17      safe=particle;
18      Timesafe=Time;
19    }
20    integrate();
21  }
                    ──────── to be continued ────────
```

In case the Verlet lists need updating, they are constructed in line 6. If `make_verlet()` returns the value `true` (no error, see p. 56), the system time is stored in `Timesafe` and the state of the particles (coordinates, velocities, and higher-order time derivatives) is stored in `safe`. If an error is reported (`make_verlet()` returns `false`), these data are used to restore an earlier correct state of the system. Then `verlet_distance` is increased by the factor `verlet_increase` and the simulation restarts from the restored data.

The Verlet algorithm can be further optimized by allowing a reduction of the `verlet_distance` if a speed gain is likely according to a properly defined criterion. The implementation of this idea is straightforward and, therefore, not shown here.

We conclude this section with the implementation of `init_algorithm()`.

```
                                    verlet.cc
22  void init_algorithm()
23  {
24    safe=particle;
25    Timesafe=Time;
26    vnx=int(lx/verlet_grid);
27    vny=int(ly/verlet_grid);
28    if(vnx==0) vnx=1;
29    if(vny==0) vny=1;
30    dx=lx/vnx;
31    dy=ly/vny;
32    celllist.resize(vnx);
33    for(int i=0;i<vnx;i++) celllist[i].resize(vny);
34    make_verlet();
35  }
```

Since the Verlet list is needed prior to the first time step, `make_verlet()` is called first. The return value is ignored since possible overlaps of particles cannot arise due to inconsistencies in the algorithm but result from the initial conditions. Finally, the status of the particles and the current time to the backup variables is stored.

By means of the Verlet lists, the number of force evaluations is reduced drastically as compared with the simple algorithm in Sect. 2.3. There are two crucial parameters which determine the performance of the algorithm, the number of cells N_c for the construction of the Verlet lists and the Verlet distance r_v.

For very large numbers of particles and, therefore, for very large numbers of grid cells there are two contributions to the total time spent in `make_verlet()`. First, the effort to store and access N_c cells; and second, the effort to test the neighborhood relations of the N particles and the $\sim N/N_c$ (on average) particles in their neighborhood. The total time spent in `make_verlet()` reads

$$T_{\text{verlet}} = aN_c + b\frac{N^2}{N_c} \,. \tag{2.33}$$

The consideration of the first contribution may seem a bit surprising but for very large numbers of particles (as simulated in Chapter 3) it is indeed of significance. Expression (2.33) is minimal for

$$\frac{N_c}{N} = \sqrt{\frac{b}{a}} \,. \tag{2.34}$$

The execution time of `make_verlet()` then grows as $\sim N$.

The main contributions to the *total* computing time T are the time T_{force} to compute the forces and the time T_{verlet} just computed to construct the Verlet lists. The time T_{force} is proportional to the number of interaction pairs. Each particle can only interact with other particles if they are in the small circular area of radius $2R$ (R is the typical radius of the particles) and of thickness r_{v} (Verlet distance). Therefore, the expected number of interaction partners of a particle is $\sim R r_{\text{v}}$ or, since R is a constant, $\sim r_{\text{v}}$. The time to compute the forces for all N particles is

$$T_{\text{force}} \sim N r_{\text{v}} \ .$$

The time to build the Verlet lists is, for optimal choice of N_{c}, proportional to N. The lists must be constructed when the fastest particle traveled the distance $\sim r_{\text{v}}$, i.e., every $\sim r_{\text{v}}/(\Delta v_{\text{max}} t)$ time steps. Thus, on average

$$T_{\text{totalverlet}} \sim \frac{N v_{\text{max}} \Delta t}{r_{\text{v}}} \sim \frac{N}{r_{\text{v}}}$$

per time step is needed for the construction of the Verlet lists. The total computing time per time step is

$$T = A N r_{\text{v}} + \frac{BN}{r_{\text{v}}} \ , \tag{2.35}$$

with unknown coefficients A and B. Differentiating with respect to r_{v} yields the optimal Verlet distance

$$\hat{r}_{\text{v}} = \sqrt{\frac{B}{A}} \ ,$$

which is independent of N. The total complexity of the algorithm is $O(N)$, i.e., for large numbers of particles the effort grows proportionally to N.

Analyses like the above are, however, to be taken with some healthy skepticism as it is not immediately clear how we can compare simulations with different numbers of particles. In order to make a fair comparison in the spirit of the above calculation, it would be necessary to reproduce the system simulated with N_1 particles for simulation with N_2 particles in such a way as to preserve the values of a, b, A, and B, which depend in a complicated way on the system properties, like size distribution of the particles, average number of contacts, distribution of the number of contacts, velocity distribution in the system, etc. Usually it is neither possible nor desirable to preserve these quantities when running simulations with different numbers of particles.

2.4.3 Link Cell Algorithm

An alternative to the Verlet algorithm for the efficient computation of the forces is the link cell algorithm. Here the simulation area of size `lx`$\times$`ly` is subdivided into `nx`$\times$`ny` identical rectangular boxes. The boxes are, hence, of equal size $(lx/nx) \times (ly/ny)$. Each of these boxes is assigned a list of particles which reside in the box. These lists are used to reduce the number of force evaluations. This concept has already been applied in the function `make_verlet()` in the previous section.

The link cell algorithm is implemented in `linkcell.cc`. The first few lines of the head of the file up to the declaration of **particle** agree with the simple and the Verlet algorithms. Then the constants `nx` and `ny` are defined which determine the number of boxes in both spatial dimensions. Moreover, the lists `lc` and **neighbors** and the prototypes of the functions `make_link_cell()`, `is_valid_neighbor()`, and `init_neighbors()` are declared.

```
                                  linkcell.cc
1  #include "common.h"
2  #include "Sphere.h"                /* <===== replace this line if necessary */
3
4  extern vector<Sphere> particle;    /* <===== replace this line if necessary */
5
6  const int nx=10, ny=10;
7  vector<vector<vector<int> > > lc;
8  vector<vector<vector<pair<int,int> > > > neighbors;
9
10 void make_link_cell();
11 bool is_valid_neighbor(int ix, int iy, int iix, int iiy);
12 void init_neighbors();
                                 to be continued
```

Let us now describe the computation of the lists of particles in each box. The indices of the box which contains the center of particle i are given by $(int(nx*x/lx), int(ny*y/ly))$ where (x,y) are the coordinates of the particle. The latter computation requires constant time, hence, the assignment of the particles to boxes needs time $\mathcal{O}(N)$. The lists are organized in the form `vector<vector<vector<int>>> lc`. The first two indices describe the box, and the last index enumerates the particles which are located in the box. The entry `lc[4][7][3]=9` means, for example, that particle 9 is located in box $(4,7)$ where it is stored in the third position. The lists are computed in `make_link_cell()`.

```
                                  linkcell.cc
13 void make_link_cell()
14 {
15   for(unsigned int ix=0;ix<lc.size();ix++){
16     for(unsigned int iy=0;iy<lc[ix].size();iy++){
17       lc[ix][iy].clear();
18     }
19   }
20   for(unsigned int i=0;i<no_of_particles;i++){
21     int ix=int(nx*(particle[i].x()-x_0)/lx);
22     int iy=int(ny*(particle[i].y()-y_0)/ly);
23     if((ix>=0) && (ix<nx) && (iy>=0) && (iy<ny)){
24       lc[ix][iy].push_back(i);
25     } else {
26       cerr << "Particle " << i << " outside simulation area\n";
27       exit(0);
28     }
29   }
30 }
                                 to be continued
```

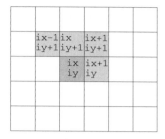

Fig. 2.9. The algorithm seeks for collision partners of particles from (ix,iy) in the same box and in the (gray-shaded) neighboring boxes

First the lists are emptied, then each particle is assigned to the box which contains its center. The indices are computed in lines 21 and 22.

The lists lc are used to identify collision partners. To this end the box extensions, lx/nx and ly/ny, must exceed the diameter of the largest particle. Any particle which belongs to box (ix,iy) can then interact only with other particles from the same box or with particles from the neighbor boxes (ix±1,iy±1), provided the box (ix,iy) is not located at the border of the simulation area. Using commutativity of the neighborhood relation (if i is a neighbor of j then j is a neighbor of i too), it is sufficient to restrict the search to the boxes (ix,iy), (ix-1,iy+1), (ix,iy+1), (ix+1,iy+1), and (ix+1,iy) (see Fig. 2.9).

To efficiently determine the possible collision partners of particles inside each box in make_forces(), each box (ix,iy) is assigned a list of neighboring boxes, taking periodic boundary conditions into account.

These lists of neighboring boxes of the system are implemented as vector< vector <vector<pair<int,int>>>> neighbors. The first two indices specify the box whose neighboring boxes are recorded as a vector of pairs. For each box (ix,iy) in the interior of the simulation area, this vector reads

```
neighbors[ix][iy][0]=(ix-1,iy+1)
neighbors[ix][iy][1]=(ix,iy+1)
neighbors[ix][iy][3]=(ix+1,iy+1)
neighbors[ix][iy][2]=(ix+1,iy)
```

as sketched in Fig. 2.9. For boxes at the borders of the simulation area, the appropriate periodic image must be used. These neighborhood lists are generated in init_neighbors() at initialization and stay invariant during the simulation.

```
                              linkcell.cc
31  void init_neighbors()
32  {
33    int iix,iiy;
34
35    neighbors.resize(nx);
36    for(int ix=0;ix<nx;ix++){
37      neighbors[ix].resize(ny);
```

```
38    }
39    for(int ix=0;ix<nx;ix++){
40      for(int iy=0;iy<ny;iy++){
41        for(int dx=-1;dx<=1;dx++){
42          for(int dy=-1;dy<=1;dy++){
43            iix=(ix+dx+nx)%nx;
44            iiy=(iy+dy+ny)%ny;
45            if(is_valid_neighbor(ix,iy,iix,iiy)) {
46              neighbors[ix][iy].push_back(pair<int,int>(iix,iiy));
47            }
48          }
49        }
50      }
51    }
52 }
```
──────────── to be continued ────────────

The function `is_valid_neighbor()` for each box (ix,iy) can be used to check
whether the boxes (ix±1,iy±1) have to be recorded in the list of neighboring
boxes of (ix,iy). In this case the function returns true.

──────────── linkcell.cc ────────────
```
53 bool is_valid_neighbor(int ix, int iy, int iix, int iiy)
54 {
55    if((iix==(ix-1+nx)%nx) && (iiy==(iy+1+ny)%ny)) return true;
56    if((iix==(ix+nx)%nx)   && (iiy==(iy+1+ny)%ny)) return true;
57    if((iix==(ix+1+nx)%nx) && (iiy==(iy+1+ny)%ny)) return true;
58    if((iix==(ix+1+nx)%nx) && (iiy==(iy+ny)%ny)) return true;
59    return false;
60 }
```
──────────── to be continued ────────────

The particle interaction forces can be computed in a simple way, using the list
`lc` of particles which are located in each box (ix,iy) and the list `neighbors`
of adjacent boxes. The function `make_forces()` below replaces the simple
`make_forces()` from p. 38 or the corresponding function `make_forces()` for
the summation of the forces using Verlet lists (p. 55), respectively.

──────────── linkcell.cc ────────────
```
61 void make_forces()
62 {
63    for(unsigned int ix=0;ix<lc.size();ix++){
64      for(unsigned int iy=0;iy<lc[ix].size();iy++){
65        for(unsigned int j=0;j<lc[ix][iy].size();j++){
66          int pj=lc[ix][iy][j];
67          for(unsigned int k=j+1;k<lc[ix][iy].size();k++){
68            int pk=lc[ix][iy][k];
69            force(particle[pj],particle[pk], lx, ly);
70          }
71          for(unsigned int n=0;n<neighbors[ix][iy].size();n++){
72            int iix=neighbors[ix][iy][n].first;
73            int iiy=neighbors[ix][iy][n].second;
74            for(unsigned int k=0;k<lc[iix][iiy].size();k++){
75              int pk=lc[iix][iiy][k];
76              force(particle[pj],particle[pk], lx, ly);
77            }
78          }
79        }
80      }
81    }
82 }
```
──────────── to be continued ────────────

For all particles `pj` which are contained in box (ix,iy), the interaction forces
with all other particles `pk` located in the same box are first computed by means

of force(). Then the program proceeds with the interaction with the particles located in the neighboring boxes whose indices are stored in neighbors.

Each time step of the simulation consists of the construction of the lists by make_link_cell() and of the integration integrate(), which in turn calls make_forces():

```
────────────────── linkcell.cc ──────────────────
83  void step()
84  {
85    make_link_cell();
86    integrate();
87  }
────────────────── to be continued ──────────────────
```

Although the lists are constructed in each simulation time step this operation is not too time consuming since it scales as $\mathcal{O}(N)$.

Apart from the particle initialization, the lists are initialized at the beginning of the simulation. First, the array lc is resized, the neighbor lists are initialized, and then make_link_cell() is called.

```
────────────────── linkcell.cc ──────────────────
88  void init_algorithm()
89  {
90    lc.resize(nx);
91    for(unsigned int ix=0;ix<lc.size();ix++) lc[ix].resize(ny);
92    init_neighbors();
93    make_link_cell();
94  }
```

2.4.4 Lattice Algorithm

For the link cell algorithm the size of the lattice cells must exceed the diameter of the largest particle. In this section, a complementary algorithm is described, where it is required that that no lattice site is occupied by the center of more than one particle. Consequently the size of the lattice sites gk must not be larger than $\sqrt{2}R_{\min}$. For this estimate the deformation of the particles is not taken into account, i.e., the lattice constant should be chosen slightly smaller. Figure 2.10 sketches the system. Each lattice site is assigned a number which is equal to the index of the particle whose center resides at this site or -1 if there is no such particle. The diameter of the largest particle of the system covers gm=int(2*rmax/gk)+1 lattice sites in each dimension. The centers of possible collision partners of a particle at site (i, j) are hence located at the sites $(i + k, j + l)$, with $k, l = \pm 1, \pm 2, \ldots, \pm gm$. To find the possible collision partners of a particle we have to loop through all of those sites $(i + k, j + l)$ and check whether any of the site values (in the program called pindex) is non-negative. If such sites are found, the corresponding particles are inserted into the list of particles interacting with the one at hand. This interaction list is constructed at the beginning of each time step. The algorithmic complexity of this construction process is $\mathcal{O}(N)$, i.e., for large particle numbers it is quite fast.

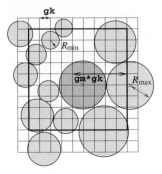

Fig. 2.10. Sketch of the system. Possible collision partners of the dark particle are contained in the lattice sites in the bold framed area

This algorithm is implemented in the file `lattice.cc`. At the beginning of the file some variables are defined: `gk`, `gm`, `nx`, `ny`, the vectors `pindex`, `partners`, and the prototypes of the functions `make_ilist()` and `clear_pindex()`. At this point the parameters `rmin` and `rmax` are also declared, representing the smallest and largest radii of the particles which are used to determine the lattice size.

```
———————————————————————— lattice.cc ————————————————————————
 1  #include "Sphere.h"                /* <===== replace this line if necessary */
 2  #include "common.h"
 3
 4  extern vector<Sphere> particle;    /* <===== replace this line if necessary */
 5
 6  int nx, ny, gm;
 7  double rmin, rmax, gk;
 8  vector<vector<int> > partners, pindex;
 9
10  void make_ilist();
11  void clear_pindex();
———————————————————————— to be continued ————————————————————————
```

When starting the simulation, the main program calls `init_system()` to initialize the particles. Then in `init_algorithm()` the lattice parameters, i.e., the lattice constant `gk`, the constant `gm`, and the size of the lattice in both spatial dimensions, `nx` and `ny`, are initialized.

```
———————————————————————— lattice.cc ————————————————————————
12  void init_algorithm()
13  {
14    rmin=particle[0].r();
15    rmax=particle[0].r();
16    for(unsigned int i=1;i<particle.size();i++){
17      if(particle[i].r()<rmin){
18        rmin=particle[i].r();
19      }
20      if(particle[i].r()>rmax){
21        rmax=particle[i].r();
22      }
23    }
24    gk=sqrt(2)*rmin;
25    gm=int(2*rmax/gk)+1;
26    nx=int(lx/gk)+1;
27    ny=int(ly/gk)+1;
```

```
28    partners.resize(no_of_particles);
29    pindex.resize(nx);
30    for(unsigned int ix=0;ix<pindex.size();ix++){
31      pindex[ix].resize(ny);
32    }
33    clear_pindex();
34    make_ilist();
35  }
```
———————— to be continued ————————

Then **partners** and **pindex** are initialized and **make_ilist()** is called to initially construct the interaction lists. For each particle, the vector **partners** contains the indices of possible collision partners (see below). For each lattice site, the vector **pindex** contains the index of the particle whose center resides in this cell or -1 if the cell is empty. Initially all cells are assigned the value -1:

———————— lattice.cc ————————
```
36  void clear_pindex()
37  {
38    for(unsigned int ix=0;ix<pindex.size();ix++){
39      for(unsigned int iy=0;iy<pindex[ix].size();iy++){
40        pindex[ix][iy]=-1;
41      }
42    }
43  }
```
———————— to be continued ————————

In **make_ilist()** the vector elements of **pindex** of all occupied sites are assigned the number of the particles involved. The lists of neighbor particles are emptied by **partners[i].clear()** for all particles i.

———————— lattice.cc ————————
```
44  void make_ilist()
45  {
46    for(unsigned int i=0;i<particle.size();i++){
47      double x=particle[i].x(), y=particle[i].y();
48      if((x>=x_0) && (x<x_0+lx) && (y>=y_0) && (y<y_0+ly)){
49        int ix=int((x-x_0)/gk);
50        int iy=int((y-y_0)/gk);
51        pindex[ix][iy]=i;
52        partners[i].clear();
53      }
54    }
55    for(unsigned int i=0;i<particle.size();i++){
56      double x=particle[i].x(), y=particle[i].y();
57      if((x>=x_0) && (x<x_0+lx) && (y>=y_0) && (y<y_0+ly)){
58        int ix=int((x-x_0)/gk);
59        int iy=int((y-y_0)/gk);
60        for(int dx=-gm;dx<=gm;dx++){
61          for(int dy=-gm;dy<=gm;dy++){
62            int k=pindex[(ix+dx+nx)%nx][(iy+dy+ny)%ny];
63            if(k>(int)i){
64              partners[i].push_back(k);
65            }
66          }
67        }
68      }
69    }
70    clear_pindex();
71  }
```
———————— to be continued ————————

In the second part of the function, starting from line 55, for each particle i located at (ix,iy), the adjacent sites up to the distance $\pm$gm are scanned for possible interaction partners and these particles are recorded in the vector `partners[i]`. Finally `clear_pindex()` initializes `pindex` for the next time step.

The interaction forces are computed in `make_forces()` by means of the vector `partners`.

```
──────────────────── lattice.cc ────────────────────
72 │ void make_forces()
73 │ {
74 │   for(unsigned int i=0;i<particle.size();i++){
75 │     for(unsigned int k=0;k<partners[i].size();k++){
76 │       int pk=partners[i][k];
77 │       force(particle[i],particle[pk], lx, ly);
78 │     }
79 │   }
80 │ }
81 │
──────────────────── to be continued ────────────────────
```

An integration time step is then performed by constructing the interaction lists and integrating Newton's equations of motion:

```
──────────────────── lattice.cc ────────────────────
82 │ void step()
83 │ {
84 │   make_ilist();
85 │   integrate();
86 │ }
```

This concludes the description of the lattice algorithm. The presented algorithm is completely vectorizable [50]. Therefore, it is suited in particular for Molecular Dynamics simulations using vector computers (see Sect. 2.10.1).

2.5 Quaternions for Three-Dimensional Simulations

The position of a particle in two dimensions is described by three scalar values: two coordinates of the center of mass and the angle φ of the angular orientation. In three dimensions, six coordinates are needed: three for the position of the center of mass and three angles Φ, Θ, and Ψ for the orientation of the particle.

For the description of the particle motion, several coordinate systems must be carefully distinguished. The *space-fixed system* S^s (the laboratory system) is particle-independent. It is an arbitrarily defined right-handed (as all other systems discussed here), time-independent, Cartesian inertial system. This is the native system of the simulation. Ultimately, all coordinates and other values describing the particles' state have to be expressed in this system.

To separate the conceptually simple translational motion from the more complicated rotation the *co-moving system* S^c is introduced. Its origin is located in the center of mass of the particle, while its axes are parallel to the axes of the space-fixed system. In this system the particle performs a pure rotation. The motion of S^c itself with respect to S^s is completely described by the motion of the center of mass of the particle. Newton's equation of motion for this translational motion can be solved in precisely the same way as for two-dimensional systems, therefore, it will not be discussed here.

The integration of the rotational degrees of freedom is more complicated: Newton's equation of motion is usually formulated in the space-fixed coordinate system S^s. In this coordinate system, however, the moments of inertia of the particles are time-dependent and contain off-diagonal elements. This complication is overcome by introducing the *body-fixed* coordinate system S^b whose origin coincides with the origin of the system S^c and whose axes are aligned along the particle's principal axes. In this system the moment of inertia tensor $\hat{J}$ assumes its simplest form, i.e., it is constant and diagonal:

$$\hat{J} = \begin{pmatrix} J_{xx} & 0 & 0 \\ 0 & J_{yy} & 0 \\ 0 & 0 & J_{zz} \end{pmatrix} . \tag{2.36}$$

The system S^b is, however, not inertial since it moves with yet unknown acceleration due to the interaction with other particles. Therefore, equations of motion cannot be formulated in this system. To resolve this problem we define a fourth coordinate system, the *frozen* system S^f, which is identical to the body-fixed coordinate system at time t.

Obviously each point in time corresponds to a different frozen system. The axes of the frozen system which belongs to time t are identical to the axes of the body-fixed system at the same time.[10] Thus in the frozen system the moment of inertia tensor adopts its simple form (2.36) as for the body-fixed coordinate system, while at the same time it is inertial, hence, allowing for a simple formulation of the equations of motion.

To compute the coordinates, orientation, and the corresponding velocities of a particle i at time $t + \Delta t$ we follow the procedure below:

- Compute the forces and torques which act on particle i due to interaction with other particles and the wall. These values are represented in the space-fixed system S^s.
- Transform all relevant variables describing particle i into the co-moving system S^c where the center of mass position is $(0, 0, 0)^T$.
- Construct the frozen system S^f whose base vectors point in the direction of the principal axes of particle i and transform the relevant particle properties into this system.
- Formulate the equation of motion for the orientation in S^f and compute the new orientation and the according velocities by integrating Newton's equation over the time interval Δt.
- Transform the variables back, first into the co-moving system S^c and then into the space-fixed coordinate system S^s.

---

[10] It became a bad habit in literature to omit the difference between the body-fixed and the frozen frame of reference. This usually leads to the question of how the angular velocity in the body-fixed coordinates may be different from zero. To avoid this confusion we introduce the frozen system.

- Compute the new position of the center of mass of i by integrating Newton's equation of motion for the translational degrees of freedom.

This procedure is repeated for all particles of the system. Note that the body-fixed coordinate system is not needed for the computation. Its practical use is indeed small since it is not inertial—it is only needed as an abstract concept to define the frozen system.

Let us add an important remark here: a vector, e.g., a particle velocity, is given by its absolute value and its direction. Its definition is independent of its representation in a certain coordinate system. In a coordinate system the vector can be characterized by its coordinates which, of course, depend on the orientation of the axes and the unit size. If a vector is represented in a specified coordinate system, it is implied that operations in this system have certain properties. For example, when manipulating the orientation vector $(\Phi, \Theta, \Psi)^T$ represented in S^f, it is implied that $\hat{J}$ is diagonal. This vector can also be manipulated in any other coordinate system, but then the diagonality of $\hat{J}$ must not be exploited.

The above listed steps are now described in detail. The frozen system S^f can be obtained from S^c via rotation by three angles Φ, Θ, and Ψ which are called *Euler angles*. The definition of the Euler angles is not unique in the literature. Here we follow [254] and define them as sketched in Fig. 2.11. To obtain the system S^f, S^c is first rotated by Φ around its z-axis. Then this intermediate system is rotated by Θ around its y-axis and finally this second intermediate system is rotated by Ψ around its z-axis again.

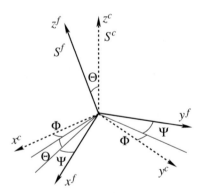

Fig. 2.11. Definition of the Euler angles Φ, Θ, and Ψ

The base vectors $\vec{e}_{x,y,z}^{\,f}$ of the frozen system can be obtained from their counterparts of the co-moving system using rotation matrices. Vectors in the frozen system carry the superscript f and vectors in the co-moving system the superscript c, just like the coordinate systems themselves. The rotation by the angle Φ around the z-axis is performed by the matrix $\hat{\Phi}$, the other rotations are done by the matrices $\hat{\Theta}$ and $\hat{\Psi}$:

$$\hat{\Phi} = \begin{pmatrix} \cos\Phi & -\sin\Phi & 0 \\ \sin\Phi & \cos\Phi & 0 \\ 0 & 0 & 1 \end{pmatrix} \quad \hat{\Theta} = \begin{pmatrix} \cos\Theta & 0 & \sin\Theta \\ 0 & 1 & 0 \\ -\sin\Theta & 0 & \cos\Theta \end{pmatrix} \quad \hat{\Psi} = \begin{pmatrix} \cos\Psi & -\sin\Psi & 0 \\ \sin\Psi & \cos\Psi & 0 \\ 0 & 0 & 1 \end{pmatrix}.$$

$$(2.37)$$

The new base vectors are

$$\vec{e}^{\,f}_{x,y,z} = \hat{\Psi}\,\hat{\Theta}\,\hat{\Phi}\ \vec{e}^{\,c}_{x,y,z}\,. \tag{2.38}$$

To transform vectors from S^c to S^f, one has to apply the *inverse* of the rotation operators $\hat{\Psi}$, $\hat{\Theta}$, and $\hat{\Phi}$ (if the xy-plane is rotated by Φ, the coordinates of an invariant vector are the same as if the vector itself was rotated by $-\Phi$ instead of the xy-plane). To simplify the notation we define the total rotation matrix $\hat{A}$ as

$$\hat{A} = \hat{\Psi}^{-1}\hat{\Theta}^{-1}\hat{\Phi}^{-1} \tag{2.39}$$

$$= \begin{pmatrix} \cos\Psi\cos\Theta\cos\Phi - \sin\Psi\sin\Phi & \cos\Psi\cos\Theta\sin\Phi + \sin\Psi\cos\Phi & -\cos\Psi\sin\Theta \\ -\sin\Psi\cos\Theta\cos\Phi - \cos\Psi\sin\Phi & -\sin\Psi\cos\Theta\sin\Phi + \cos\Psi\cos\Phi & \sin\Psi\sin\Theta \\ \sin\Theta\cos\Phi & \sin\Theta\sin\Phi & \cos\Theta \end{pmatrix}.$$

$$(2.40)$$

The representation of a vector u in S^f is obtained from its representation in S^c by

$$\vec{u}^f = \hat{A}\vec{u}^c\,. \tag{2.41}$$

The elements of the rotation matrix are the directional cosines of the body-fixed axis vectors in the co-moving frame: the angular velocity of the particle in the system S^c may be translated into time derivatives of the Euler angles. The three rotations by the Euler angles correspond to the angular velocities ω_Φ, ω_Θ, and ω_Ψ as sketched in Fig. 2.12. These angular velocities are best represented in the intermediate systems. In S^c and in the first intermediate

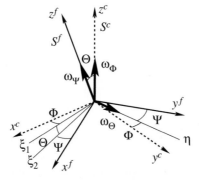

Fig. 2.12. Translation of the angular velocity in the co-moving system into time derivatives of the Euler angles

system $(\xi_1\eta z^c)$, $\vec{\omega}_\Phi$ points in the z-direction and is of length $\dot{\Phi}$. To find the representation of this vector in S^f one has to rotate the vector by $\hat{\Theta}^{-1}$ and then by $\hat{\Psi}^{-1}$, yielding

$$\vec{\omega}_\Phi^f = \hat{\Psi}^{-1}\hat{\Theta}^{-1}\begin{pmatrix} 0 \\ 0 \\ \dot{\Phi} \end{pmatrix} = \dot{\Phi}\begin{pmatrix} -\cos\Psi\sin\Theta \\ \sin\Psi\sin\Theta \\ \cos\Theta \end{pmatrix}. \tag{2.42}$$

In both the $\xi_1\eta z^c$-system and the $\xi_2\eta z^f$-system, $\vec{\omega}_\Theta$ has the length $\dot{\Theta}$ and points in the η-direction. To transform this vector into S^f, the operator $\hat{\Psi}^{-1}$ is applied.

$$\vec{\omega}_\Theta^f = \left(\hat{\Psi}^{-1}\right)\begin{pmatrix} 0 \\ \dot{\Theta} \\ 0 \end{pmatrix} = \dot{\Theta}\begin{pmatrix} \sin\Psi \\ \cos\Psi \\ 0 \end{pmatrix}. \tag{2.43}$$

Finally, the vector $\vec{\omega}_\Psi$ points in the z^f direction. Since it is already defined in the S^f system we immediately obtain

$$\vec{\omega}_\Psi^f = \dot{\Psi}\begin{pmatrix} 0 \\ 0 \\ 1 \end{pmatrix}. \tag{2.44}$$

The total angular velocity is thus

$$\vec{\omega}^f = \vec{\omega}_\Phi^f + \vec{\omega}_\Theta^f + \vec{\omega}_\Psi^f = \begin{pmatrix} \dot{\Theta}\sin\Psi - \dot{\Phi}\sin\Theta\cos\Psi \\ \dot{\Theta}\cos\Psi + \dot{\Phi}\sin\Theta\sin\Psi \\ \dot{\Phi}\cos\Theta + \dot{\Psi} \end{pmatrix}. \tag{2.45}$$

This system of equations may be solved for $\dot{\Phi}$, $\dot{\Theta}$, and $\dot{\Psi}$ to obtain the evolution of the Euler angles themselves [98]:

$$\dot{\Phi} = -\omega_x^f\frac{\cos\Psi}{\sin\Theta} + \omega_y^f\frac{\sin\Psi}{\sin\Theta}$$

$$\dot{\Theta} = \omega_x^f\sin\Psi + \omega_y^f\cos\Psi \tag{2.46}$$

$$\dot{\Psi} = \omega_x^f\frac{\cos\Psi\cos\Theta}{\sin\Theta} - \omega_y^f\frac{\sin\Psi\cos\Theta}{\sin\Theta} + \omega_z^f.$$

Until here the transformation of vectors from one coordinate system into another has been considered. Transforming time derivatives of vectors is more complicated. Consider a vector $\vec{u}$ which at time t has the coordinates $\left(u_x^f, u_y^f, u_z^f\right)^T$ in the frozen system. Seen from the same system S^f an infinitesimal time interval dt later, the vector has changed by $d\vec{u}^f$. It turns out that there are two contributions to this change. First, the vector may change in the body-fixed frame itself, e.g., if the physical object described by the vector changes. Second, there is a change due to the rotation of the coordinate axes.[11] In the frozen system the vector $\vec{u}$ is given in coordinates by

$$\vec{u}^{\mathrm{f}}(t) = u_x^{\mathrm{f}}\vec{e}_x^{\mathrm{f}} + u_y^{\mathrm{f}}\vec{e}_y^{\mathrm{f}} + u_z^{\mathrm{f}}\vec{e}_z^{\mathrm{f}}$$
$$\vec{u}^{\mathrm{f}}(t+\mathrm{d}t) = \left[u_x^{\mathrm{f}}\right]' \left[\vec{e}_x^{\mathrm{f}}\right]' + \left[u_y^{\mathrm{f}}\right]' \left[\vec{e}_y^{\mathrm{f}}\right]' + \left[u_z^{\mathrm{f}}\right]' \left[\vec{e}_z^{\mathrm{f}}\right]' \ . \tag{2.47}$$

The vectors $\vec{e}_{x,y,z}^{\mathrm{f}}$ are the base vectors of S^{f} and $\left[\vec{e}_{x,y,z}^{\mathrm{f}}\right]'$ are the base vectors of the frozen system at the time $t + \mathrm{d}t$, expressed in coordinates of S^{f}. The base vectors at time $t + \mathrm{d}t$ are related to the bases at time t via

$$\left[\vec{e}_{x,y,z}^{\mathrm{f}}\right]' = \vec{e}_{x,y,z}^{\mathrm{f}} + \mathrm{d}\vec{\varphi}^{\mathrm{f}} \times \vec{e}_{x,y,z}^{\mathrm{f}} \ , \tag{2.48}$$

i.e., the frozen system $\left[S^{\mathrm{f}}\right]'$ (at time $t + \mathrm{d}t$) is rotated by $\mathrm{d}\vec{\varphi}^{\mathrm{f}}$ with respect to the original frozen system S^{f}. With (2.47) and (2.48) we obtain

$$\begin{aligned}
\vec{u}^{\mathrm{f}}(t+\mathrm{d}t) &= \left(u_x^{\mathrm{f}} + \mathrm{d}u_x^{\mathrm{f}}\right)\left[\vec{e}_x^{\mathrm{f}}\right]' + \left(u_y^{\mathrm{f}} + \mathrm{d}u_y^{\mathrm{f}}\right)\left[\vec{e}_y^{\mathrm{f}}\right]' + \left(u_z^{\mathrm{f}} + \mathrm{d}u_z^{\mathrm{f}}\right)\left[\vec{e}_z^{\mathrm{f}}\right]' \\
&= u_x^{\mathrm{f}}\vec{e}_x^{\mathrm{f}} + u_y^{\mathrm{f}}\vec{e}_y^{\mathrm{f}} + u_z^{\mathrm{f}}\vec{e}_z^{\mathrm{f}} + \mathrm{d}u_x^{\mathrm{f}}\vec{e}_x^{\mathrm{f}} + \mathrm{d}u_y^{\mathrm{f}}\vec{e}_y^{\mathrm{f}} + \mathrm{d}u_z^{\mathrm{f}}\vec{e}_z^{\mathrm{f}} \\
&\quad + u_x^{\mathrm{f}}\left(\mathrm{d}\vec{\varphi}^{\mathrm{f}} \times \vec{e}_x^{\mathrm{f}}\right) + u_y^{\mathrm{f}}\left(\mathrm{d}\vec{\varphi}^{\mathrm{f}} \times \vec{e}_y^{\mathrm{f}}\right) + u_z^{\mathrm{f}}\left(\mathrm{d}\vec{\varphi}^{\mathrm{f}} \times \vec{e}_z^{\mathrm{f}}\right) \tag{2.49} \\
&= \vec{u}^{\mathrm{f}} + \mathrm{d}\vec{u}_{\mathrm{intern}}^{\mathrm{f}} + \mathrm{d}\vec{\varphi}^{\mathrm{f}} \times \vec{u} \tag{2.50}
\end{aligned}$$
$$\mathrm{d}\vec{u}^{\mathrm{f}} = \vec{u}^{\mathrm{f}}(t+\mathrm{d}t) - \vec{u}^{\mathrm{f}}(t) = \mathrm{d}\vec{u}_{\mathrm{intern}}^{\mathrm{f}} + \mathrm{d}\vec{\varphi}^{\mathrm{f}} \times \vec{u}^{\mathrm{f}}(t) \ . \tag{2.51}$$

In (2.49) all contributions proportional to products of $\mathrm{d}u_{x,y,z}$ and $\mathrm{d}\vec{\varphi}^{\mathrm{f}}$ are neglected, i.e., only first order contributions are considered. The vector $\mathrm{d}\vec{u}_{\mathrm{intern}}^{\mathrm{f}}$ describes the change of $\vec{u}$ at time $t + \mathrm{d}t$ as seen by a co-rotating observer (we nevertheless express this vector in the original system S^{f}). Replacing the differentials by derivatives, the representation of the derivative of the vector $\vec{u}$ in S^{f} reads

$$\dot{\vec{u}}^{\mathrm{f}} = \dot{\vec{u}}_{\mathrm{intern}}^{\mathrm{f}} + \vec{\omega}^{\mathrm{f}} \times \vec{u}^{\mathrm{f}} \ . \tag{2.52}$$

Equation (2.52) allows us to express the equation of motion of the particle in the frozen reference frame. In any inertial system (i.e., S^{s}, S^{c}, and S^{f}) the torque acting on a particle causes a change of its angular momentum $\vec{M} = \dot{\vec{L}}$. Expressing this relation in the system S^{f} yields

$$\vec{M}^{\mathrm{f}} = \dot{\vec{L}}^{\mathrm{f}} = \dot{\vec{L}}_{\mathrm{intern}}^{\mathrm{f}} + \vec{\omega}^{\mathrm{f}} \times \vec{L}^{\mathrm{f}} \ . \tag{2.53}$$

The relation between $\vec{L}^{\mathrm{f}}$ and $\vec{\omega}^{\mathrm{f}}$ is

$$\vec{L}^{\mathrm{f}} = \hat{J}\vec{\omega}^{\mathrm{f}} \ . \tag{2.54}$$

---

[11] The velocity of the axes rotation is finite only in the the body-fixed system. The axes of the frozen system do not rotate. The frozen system (which is only valid for the time for which it was constructed) is inertial. Technically, the frozen system at time t is different from the frozen system at time $t + \mathrm{d}t$.

Since in the body-fixed reference frame the moment of inertia is constant, the expression

$$\dot{\vec{L}}^{\mathrm{f}}_{\mathrm{intern}} = \hat{J}\dot{\vec{\omega}}^{\mathrm{f}} \tag{2.55}$$

can be inserted into (2.53), leading to

$$\vec{M}^{\mathrm{f}} = \hat{J}\dot{\vec{\omega}}^{\mathrm{f}} + \vec{\omega}^{\mathrm{f}} \times \left(\hat{J}\vec{\omega}^{\mathrm{f}}\right) . \tag{2.56}$$

Representing (2.56) in frozen coordinates (keeping in mind that the moment of inertia tensor is diagonal in this coordinate system) yields

$$
\begin{aligned}
J_{xx}\,\dot{\omega}^{\mathrm{f}}_x - \omega^{\mathrm{f}}_y\,\omega^{\mathrm{f}}_z\,(J_{yy} - J_{zz}) &= M^{\mathrm{f}}_x \\
J_{yy}\,\dot{\omega}^{\mathrm{f}}_y - \omega^{\mathrm{f}}_z\,\omega^{\mathrm{f}}_x\,(J_{zz} - J_{xx}) &= M^{\mathrm{f}}_y \\
J_{zz}\,\dot{\omega}^{\mathrm{f}}_z - \omega^{\mathrm{f}}_x\,\omega^{\mathrm{f}}_y\,(J_{xx} - J_{yy}) &= M^{\mathrm{f}}_z .
\end{aligned}
\tag{2.57}
$$

These equations describe the evolution of the angular velocity $\vec{\omega}^{\mathrm{f}}$ in the frozen reference frame. They can be solved in precisely the same way as Newton's equation for the linear motion. To this end we start with the torques in S^{c}, i.e., $\vec{M}^{\mathrm{c}}$, then compute the torques $\vec{M}^{\mathrm{f}}$ in the frozen system (remember, this step is necessary to use the simple form of the moment of inertia), then the angular velocities $\vec{\omega}^{\mathrm{f}}$ in the frozen system, and finally the evolution of the Euler angles, $\{\dot{\Phi}, \dot{\Theta}, \dot{\Psi}\}$. This sequence can be iterated to obtain a Molecular Dynamics algorithm for the rotational degrees of freedom of three-dimensional particles.

The numerical evaluation, however, turns out to be problematic: For $\sin\Theta = 0$, the equations (2.46) become singular. Hence, if Θ approaches 0, $\pm\pi$, $\pm 2\pi$, etc., the term $1/\sin\Theta$ diverges and the numerical errors become large.

There are several solutions to this problem (see, e.g., [5]); the quaternion method (e.g., [5, 98, 152]) is briefly discussed here. The three Euler angles are mapped to a quadruplet of coordinates, (q_0, q_1, q_2, q_3), which are called *quaternions*. The quaternions lie on a four-dimensional sphere

$$q_0^2 + q_1^2 + q_2^2 + q_3^2 = 1 . \tag{2.58}$$

There exist different equivalent definitions of the quaternions (e.g., [5, 288]). The most simple one of them [9] is

$$q_\alpha = \left(\cos\frac{\alpha}{2}, \vec{n}\sin\frac{\alpha}{2}\right)^{\mathrm{T}} = \left(\cos\frac{\alpha}{2}, n_x\sin\frac{\alpha}{2}, n_y\sin\frac{\alpha}{2}, n_z\sin\frac{\alpha}{2}\right)^{\mathrm{T}} \tag{2.59}$$

for an arbitrary rotation by the angle α around the axis $\vec{n}$. For example, in this definition it is trivial to construct the quaternion of the inverse rotation $q_\alpha^{-1} = q_{-\alpha}$. One can combine two rotations by multiplying the corresponding quaternions. With the notation $q = (s, \vec{v})$, where s is a scalar and $\vec{v}$ a three-dimensional vector, the product of two quaternions is defined as

$$q_1 q_2 = (s_1 s_2 - \vec{v}_1 \cdot \vec{v}_2 \ , \ s_1 \vec{v}_2 + s_2 \vec{v}_1 + \vec{v}_1 \times \vec{v}_2) \ . \tag{2.60}$$

The quaternion for the transition between S^c and S^f, represented by the Euler angles and the rotation matrix $\hat{A}$ now reads

$$
\begin{aligned}
q_0 &= \cos \frac{\Theta}{2} \cos \frac{\Phi + \Psi}{2} \\
q_1 &= \sin \frac{\Theta}{2} \sin \frac{\Phi - \Psi}{2} \\
q_2 &= - \sin \frac{\Theta}{2} \cos \frac{\Phi - \Psi}{2} \\
q_3 &= - \cos \frac{\Theta}{2} \sin \frac{\Phi + \Psi}{2} \ .
\end{aligned}
\tag{2.61}
$$

To calculate the evolution of the quaternion components, we compute the time derivatives of these equations $\dot{q}_n$ and replace the occurring time derivatives of the Euler angles by their evolution equations (2.46). Expressing all terms as products of quaternion components and components of the angular velocity, a singularity-free set of equations is obtained:

$$
\begin{pmatrix} \dot{q}_0 \\ \dot{q}_1 \\ \dot{q}_2 \\ \dot{q}_3 \end{pmatrix} = \frac{1}{2} \begin{pmatrix} q_1 & q_2 & q_3 \\ -q_0 & -q_3 & q_2 \\ q_3 & -q_0 & -q_1 \\ -q_2 & q_1 & -q_0 \end{pmatrix} \cdot \begin{pmatrix} \omega_x^f \\ \omega_y^f \\ \omega_z^f \end{pmatrix} \ . \tag{2.62}
$$

These equations can be solved in perfect analogy to Newton's equation for the center of mass motion, e.g., by Gear's algorithm. Instead of Euler angles the four quaternions are used to describe the orientation of the particle. Inverting (2.61) one can compute the Euler angles for any given time. One has to keep in mind, however, that the quaternions have to satisfy (2.58). Although the equation of motion of quaternions conserves this property, the limited precision of computers makes it necessary to renormalize the quaternions frequently to avoid accumulating numerical errors.

Examples of three-dimensional Molecular Dynamics simulation of granular systems using quaternions can be found, e.g., in [82, 288]. We do not provide the code of the simulation program here, since it is relatively long but conceptually simple. An implementation (in FORTRAN77) can be found in [5].

2.6 Composite Particles

2.6.1 Idea

For many applications the realistic description of the static behavior of a particle system is essential. Examples of technological relevance are clogging pipes and hoppers, the formation and development of avalanches and land slides, sintering, and solidification and sedimentation.

While the spherical particle model allows for efficient simulation of large systems, we gave arguments in Sect. 2.3.6 why this particle model is frequently insufficient when the static behavior of such systems is addressed. Problems arise mainly from the fact that the surface properties are incorrectly modeled. These problems can be overcome if the spherical particle model is substituted by a particle model of a more complex geometry. The simplest method to define such particles is to combine several spheres to establish one complex particle; we call such particles *composite particles*. The present section describes Molecular Dynamics simulations of composite particles which are composed of five spheres connected by damped springs [51, 217]. Further similar particle models are discussed in Sect. 2.8.

The particle i consists of four peripheral spheres of radius R_i^p and a central sphere of radius R_i^c see (Fig. 2.13). For simplicity of notation, it is assumed here that all outer spheres of one composite particle are identical. Moreover, for the discussion of single-particle properties, the particle index i is omitted.

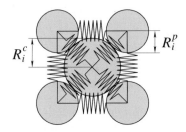

Fig. 2.13. Sketch of a composite particle i that consists of five spheres and eight damped springs

2.6.2 Geometrical Properties

The sphericity of a particle, i.e., its similarity with a sphere, can be varied within certain limits by adjusting the ratio between the radii of the outer and inner spheres. A quantitative measure of the non-sphericity, S, can be defined as a function of the parameters L and R (see Fig. 2.14):

$$S = 1 - \frac{L}{2R} . \tag{2.63}$$

The size L of the enveloping square, i.e., the size of the smallest square the composite particle fits in (see Fig. 2.14) is a measure of the size of the particle. An alternative size definition would be the radius R of the enveloping circle. From geometry follows

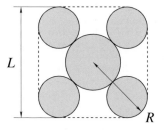

Fig. 2.14. The non-sphericity of a particle depends on the size L of the enveloping square and on the radius R of the enveloping circle

$$L = 2 \max \left(R^{\mathrm{c}}, \frac{R^{\mathrm{c}} + R^{\mathrm{p}}}{\sqrt{2}} + R^{\mathrm{p}} \right) \tag{2.64}$$

$$R = R^{\mathrm{c}} + 2R^{\mathrm{p}} . \tag{2.65}$$

The radius of the central sphere may not be smaller than $(\sqrt{2} - 1)R^{\mathrm{p}}$, otherwise it loses contact with the outer spheres. Figure 2.15 shows the dependency of the non-sphericity S on the ratio R^{p}/R. It assumes its maximum $S_{\mathrm{max}} = 0.255$ for a particle whose convex hull is most similar to a square at a ratio $R^{\mathrm{p}}/R \approx 0.128$ (where $R^{\mathrm{p}}/R^{\mathrm{c}} = 0.17$). In the interval $S \in (0.172, 0.255)$ there are two different particle shapes which correspond to the same value S. It has been shown [51, 217] that both of them reveal similar mechanical behavior. In the limit $R^{\mathrm{p}}/R \to 0$ the particles become spherical and S approaches zero.

Since the space filling of a particle, i.e., the fraction inside its enveloping square that is actually occupied by the spheres, is not a constant but varies with R^{p}/R, the material density of the particles has to be modified by $\rho_0 \to \rho$ when changing $S_0 \to S$ to ensure that the mass of a particle is independent of its shape. For constant L and constant total mass m, the relation

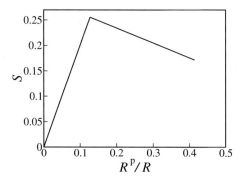

Fig. 2.15. Sphericity as a function of the particle size

$$\frac{\rho(S)}{\rho_0} = \begin{cases} \dfrac{(1-S)^2}{(1-S_0)^2} \left[\dfrac{2S_0^2 - 2S_0 + 1}{2S^2 - 2S + 1} \right] & \text{for } \dfrac{R^{\mathrm{p}}}{R} \le 0.128 \\[2ex] \dfrac{(1-S)^2}{(1-S_0)^2} \left[\dfrac{16S_0^2 + 12\left(\sqrt{2}-2\right)S_0 + 5(3-2\sqrt{2})}{16S^2 + 12\left(\sqrt{2}-2\right)S + 5(3-2\sqrt{2})} \right] & \text{for } \dfrac{R^{\mathrm{p}}}{R} \ge 0.128 \end{cases}$$

$$(2.66)$$

holds true, where the cases distinguish both branches of the function $S(R^{\mathrm{p}}/R)$ as drawn in Fig. 2.15.

2.6.3 Forces

Obviously, when composite particles collide, besides normal forces there are tangential forces due to the particle geometry, i.e., due to the outer particles providing a lever, and therefore a torque. Hence, we bypass the assumption of rather artificial mechanisms necessary for the case of spherical particles (see p. 22). Consequently, for the interaction of the constituting spheres, it is sufficient to describe only the normal forces as there are no tangential forces present between the spheres.

The force acting between two spheres i and j of *different* particles is

$$\vec{F} = \begin{cases} F^{\mathrm{n}} \dfrac{\vec{r}_j - \vec{r}_i}{|\vec{r}_j - \vec{r}_i|} & \text{if } |\vec{r}_j - \vec{r}_i| < R_i + R_j \\[2ex] 0 & \text{otherwise} , \end{cases}$$

$$(2.67)$$

with

$$F^{\mathrm{n}} = \frac{2Y\sqrt{R^{\mathrm{eff}}}}{3\left(1 - \nu^2\right)} \left(\xi^{3/2} + A\sqrt{\xi}\, \frac{\mathrm{d}\xi}{\mathrm{d}t} \right)$$

$$(2.68)$$

$$\xi = R_i + R_j - |\vec{r}_i - \vec{r}_j| ,$$

$$(2.69)$$

i.e., the spheres interact via the modified Hertz law as introduced in Sect. 2.1.1, (2.12). There is an additional force acting between each pair of adjacent spheres that belong to *the same* particle:

$$\vec{F}^{\mathrm{sp}} = \left(k\,\zeta + \gamma^{\mathrm{sp}} m^{\mathrm{eff}}\, \frac{\mathrm{d}\zeta}{\mathrm{d}t} \right) \frac{\vec{r}_j - \vec{r}_i}{|\vec{r}_j - \vec{r}_i|} ,$$

$$(2.70)$$

with

$$\zeta = -|\vec{r}_i - \vec{r}_j| + \begin{cases} R^{\mathrm{c}} + R^{\mathrm{p}} & \text{if either } i \text{ or } j \text{ is the central particle} \\ \sqrt{2}(R^{\mathrm{c}} + R^{\mathrm{p}}) & \text{otherwise} . \end{cases}$$

$$(2.71)$$

This force corresponds to the springs drawn in Fig. 2.13, with spring constant k and damping constant γ^{sp}. In general, the spring constants are different for the central and the peripheral springs.

2.6.4 Implementation

The particle model is implemented in the class `CompositeParticle` which is declared in the file `CompositeParticle.h`:

```
                          CompositeParticle.h
 1  #ifndef _CompositeParticle_h
 2  #define _CompositeParticle_h
 3
 4  #include <vector>
 5  #include "Vector.h"
 6  #include "Sphere.h"
 7
 8  using namespace std;
 9
10  class CompositeParticle {
11    friend istream & operator >> (istream & is, CompositeParticle & p);
12    friend ostream & operator << (ostream & os, const CompositeParticle & p);
13    friend double Distance(const CompositeParticle & p1,
14                           const CompositeParticle & p2,
15                           double lx, double ly);
16    friend void force(CompositeParticle & p1, CompositeParticle & p2,
17                      double lx, double ly);
18  public:
19    Vector & pos(){return particles[0].pos();}
20    Vector pos() const {return particles[0].pos();}
21    double & x(){return particles[0].x();}
22    double x() const {return particles[0].x();}
23    double & y(){return particles[0].y();}
24    double y() const {return particles[0].y();}
25    double & vx(){return particles[0].vx();}
26    double vx() const {return particles[0].vx();}
27    double & vy(){return particles[0].vy();};
28    double vy() const {return particles[0].vy();};
29    const Vector & velocity() const {return particles[0].velocity();}
30    double r() const { return _r;}
31    double m() const { return _m;}
32
33    void read_positions(istream & os);
34    double kinetic_energy() const;
35    void predict(double dt);
36    void correct(double dt);
37
38    void set_force_to_zero()
39    {
40      for(unsigned int i=0;i<particles.size();i++)
41        particles[i].set_force_to_zero();
42    }
43    int ptype() const { return particles[0].ptype();}
44    void boundary_conditions(int n, double timestep, double Time)
45    {
46      for(unsigned int i=0;i<particles.size();i++){
47        particles[i].boundary_conditions(n,timestep,Time);
48      }
49    }
50
51    void periodic_bc(double x_0, double y_0, double lx, double ly);
52  private:
53    vector<Sphere> particles;
54    double k1,k2,gamma1,gamma2,_r,_m;
55    vector<double> restlength1,restlength2;
56    void internal_forces();
57  };
58  #endif
```

The interface of this class (the part declared as `public`) is identical to the interface of the class `Sphere` (see Sect. 2.3.3). The classes differ, however, in their private data members and functions. Since the composite particle consists of spheres, the class `CompositeParticle` contains a list `particles` of its constituting spheres which is, hence, of type `vector<Sphere>`. The spheres are connected by springs of two types. Springs of type 1 connect the outer particles with each other, springs of type 2 connect the outer particles with the central one. So in addition to the array of spheres, four parameters are needed to describe the two spring types with elastic constants `k1` and `k2` and damping constants `gamma1` and `gamma2`. Finally, the lengths of theses springs at rest are stored in the lists `restlength1` for the springs between outer particles and `restlength2` for the springs connecting the central with the peripheral spheres.

As a convention, the first sphere (`particles[0]`) is the central sphere. For modeling walls there are also composite particles which consist only of the central sphere. In this case the definition of `k1` through `gamma2` is superfluous but the impact of this small overhead on the overall performance is negligible.

The actual implementation of the class methods is contained in the file `CompositeParticle.cc`. First the header file which contains the class definition is included.

```
──────── CompositeParticle.cc ────────
1  #include <string>
2  #include "CompositeParticle.h"
──────── to be continued ────────
```

Since the interaction force defines a particle, the implementation starts with this crucial ingredient. In contrast to the sphere implementation there are two separate contributions to the force acting on a particle. First the interparticle forces, acting on spheres of different particles, are computed by the function `force()`. In the main loop this function is called for all pairs of spheres of different particles. Remember that the function `force()` of the class `Sphere` called in line 8 also adds the force to the total force acting on a sphere.

```
──────── CompositeParticle.cc ────────
3   void force(CompositeParticle & p1, CompositeParticle & p2,
4               double lx, double ly)
5   {
6     for(unsigned int i=0;i<p1.particles.size();i++){
7       for(unsigned int k=0;k<p2.particles.size();k++){
8         force(p1.particles[i],p2.particles[k], lx, ly);
9       }
10    }
11  }
──────── to be continued ────────
```

The forces acting between the constituting spheres of a composite particle are computed in the function `internal_forces()`. The rest lengths of the connecting springs are read from the arrays `restlength1` and `restlength2`. The elongation of a spring is the difference between the actual distance of the two spheres involved and the rest length.

```
                          CompositeParticle.cc
12  void CompositeParticle::internal_forces()
13  {
14    for(unsigned int i=1;i<particles.size();i++){
15      int k=i%(particles.size()-1)+1;
16      double dx=particles[i].x()-particles[k].x();
17      double dy=particles[i].y()-particles[k].y();
18      double dvx=particles[i].vx()-particles[k].vx();
19      double dvy=particles[i].vy()-particles[k].vy();
20      double RR=sqrt(dx*dx+dy*dy);
21      double dr=RR-restlength1[i-1];
22      Vector fik;
23
24      fik.x()=-(k1*dr*dx/RR+gamma1*dvx);
25      fik.y()=-(k1*dr*dy/RR+gamma1*dvy);
26      particles[i].add_force(fik);
27      particles[k].add_force((-1)*fik);
28
29      dx=particles[i].x()-particles[0].x();
30      dy=particles[i].y()-particles[0].y();
31      dvx=particles[i].vx()-particles[0].vx();
32      dvy=particles[i].vy()-particles[0].vy();
33      RR=sqrt(dx*dx+dy*dy);
34      dr=RR-restlength2[i-1];
35
36      fik.x()=-(k2*dr*dx/RR+gamma2*dvx);
37      fik.y()=-(k2*dr*dy/RR+gamma2*dvy);
38      particles[i].add_force(fik);
39      particles[0].add_force((-1)*fik);
40    }
41  }
                        to be continued
```

The distance of two particles is defined as the distance of their central particles.

```
                          CompositeParticle.cc
42  double Distance(const CompositeParticle & p1, const CompositeParticle & p2,
43                  double lx, double ly)
44  {
45    double d=Distance(p1.particles[0],p2.particles[0], lx, ly);
46    return d;
47  }
                        to be continued
```

The functions **predict()** and **correct()** implement the predictor and corrector steps of the Gear-algorithm by simply calling the corresponding functions for each of the constituting spheres. At the start of correct, the internal forces are computed, thus, ensuring that it is computed only once per time step.

```
                          CompositeParticle.cc
48  void CompositeParticle::predict(double dt)
49  {
50    for(unsigned int i=0;i<particles.size();i++){
51      particles[i].predict(dt);
52    }
53  }
54
55  void CompositeParticle::correct(double dt)
56  {
57    internal_forces();
58    for(unsigned int i=0;i<particles.size();i++){
59      particles[i].correct(dt);
60    }
61  }
                        to be continued
```

To enforce periodic boundary conditions the function `periodic_bc()` is called. A particle is replaced by its periodic image once the central particle crosses the system's boundaries.

```
———————————————— CompositeParticle.cc ————————————————
62  void CompositeParticle::periodic_bc(double x_0, double y_0,
63                                       double lx, double ly)
64  {
65    for(unsigned int i=particles.size()-1; i>=0;i--){
66      while(particles[0].pos().x()<x_0) particles[i].pos().x()+=lx;
67      while(particles[0].pos().x()>x_0+lx) particles[i].pos().x()-=lx;
68      while(particles[0].pos().y()<y_0) particles[i].pos().y()+=ly;
69      while(particles[0].pos().y()>y_0+ly) particles[i].pos().y()-=ly;
70    }
71  }
———————————————————— to be continued ————————————————————
```

The function `kinetic_energy()` computes the kinetic energy for each of the spheres and returns the sum.[12]

```
———————————————— CompositeParticle.cc ————————————————
72  double CompositeParticle::kinetic_energy() const
73  {
74    double sum=0;
75
76    for(unsigned int i=0;i<particles.size();i++){
77      sum+=particles[i].vx()*particles[i].vx()/2;
78      sum+=particles[i].vy()*particles[i].vy()/2;
79    }
80    return sum;
81  }
———————————————————— to be continued ————————————————————
```

A discussion of the input and output functions remains. An entry of an initialization file for one composite particle has the form

BeginCmpPart: k_1 γ_1 k_2 γ_2
p   *data line for first sphere*

$\vdots$

p   *data line for n.th sphere*
end

The action of the output operator is straightforward, one only has to take note to output the formatting strings along with the data.

```
———————————————— CompositeParticle.cc ————————————————
82  ostream & operator << (ostream & os, const CompositeParticle & p)
83  {
84    os << "BeginCmpPart " << p.k1 << " " << p.gamma1
85       << " " << p.k2 << " " << p.gamma2 << endl;
86    for(unsigned int i=0;i<p.particles.size();i++){
87      os << "p " << p.particles[i];
88    }
89    os << "end\n";
90    return os;
91  }
———————————————————— to be continued ————————————————————
```

In the input function `operator >>`, a string is first read into the variable `type` and checked for the leading "`BeginCmpPart:`". Then the parameters

---

[12] The particle model disregards spinning of the spheres. For a critical discussion of the particle model, see Sect. 2.6.6.

k_1 through γ_2 are read. Next the lines describing spheres are read, i.e., it is checked for the leading "p" and, if found, input the sphere in line 112 using the already defined input function `operator >>` of the class `Sphere`. If there is no leading "p" but "end" instead, the complete entry is read. If `type` did not contain either "p" not "end" the input file is syntactically incorrect. An error message is displayed and the program exits. After the complete entry for the composite particle is read (after finding "end"), the mass and radius of the particle are computed and stored in `_m` and `_r`. At last the rest lengths of the springs connecting the spheres are determined. It is assumed that the springs do not exert forces if the spheres only touch without deforming.

```
                         CompositeParticle.cc
92  istream & operator >> (istream & is, CompositeParticle & p)
93  {
94    p.particles.clear();
95    p.restlength1.clear();
96    p.restlength2.clear();
97    string type;
98
99    is >> type;
100   if(is){
101     if(type!="BeginCmpPart"){
102       cerr << "syntax error in input file: missing BeginCmpPart\n";
103       exit(0);
104     }
105
106     is >> p.k1 >> p.gamma1 >> p.k2 >> p.gamma2;
107
108     do{
109       is >> type;
110       if(type=="p"){
111         Sphere sp;
112         is >> sp;
113         p.particles.push_back(sp);
114       } else if(type!="end"){
115         cerr << "unknown type: '" << type << "'\n";
116         exit(0);
117       }
118     } while((type!="end") && is);
119     if(type!="end"){
120       cerr << "premature end of input file: " << type << "\n";
121       exit(0);
122     }
123   }
124
125   /* Compute total mass */
126   if(p.particles.size()>0){
127     p._r=p.particles[0].r();
128     double Rmax=0;
129     for(unsigned int i=1;i<p.particles.size();i++){
130       if(p.particles[i].r()>Rmax) Rmax=p.particles[i].r();
131     }
132     p._r+=2*Rmax;
133
134     p._m=0;
135     for(unsigned int i=0;i<p.particles.size();i++){
136       p._m+=p.particles[i].m();
137     }
138   }
139   /* Compute rest lengths */
140   if(p.particles.size()>0){
141     p.restlength1.resize(p.particles.size()-1);
142     p.restlength2.resize(p.particles.size()-1);
```

```
143    for(unsigned int i=1;i<p.particles.size();i++){
144        int k=i%(p.particles.size()-1)+1;
145        double r0=p.particles[0].r();
146        double ri=p.particles[i].r();
147        double rk=p.particles[k].r();
148        double rl1 = sqrt((r0+ri)*(r0+ri)+(r0+rk)*(r0+rk));
149        double rl2 = r0+ri;
150        p.restlength1[i-1]=rl1;
151        p.restlength2[i-1]=rl2;
152    }
153  }
154  return is;
155 }
```

The implementation of this more complicated particle model is actually rather simple. This is due to the fact that some earlier described functions of the class Sphere have been used.

2.6.5 Three-Dimensional Composite Particles

The composite particle model can be easily generalized to three dimensions. In analogy to the two-dimensional case in three dimensions, a particle consists of one central sphere and eight identical peripheral spheres which are connected via 20 damped springs. Figure 2.16 shows such a particle. The interaction forces are given again by (2.67) and (2.70).

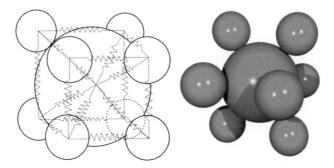

Fig. 2.16. Three-dimensional variant of the composite particle

It turns out that this model can describe the damping properties of a horizontally shaken container filled with granules much more realistically than its two dimensional analogue [248]. Apparently the interlocking of particles which is modeled more realistically in three dimensions is crucial for dissipating energy in this process. Figure 2.17 shows a snapshot of the simulated system.

For simulations using this model, one has to keep in mind that the computational effort is far higher as compared with the two-dimensional variant as well as with three-dimensional spheres since for each particle now, the equations of motion have to be integrated for nine spheres and the corresponding

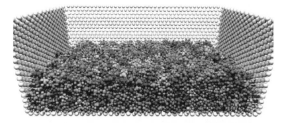

Fig. 2.17. Snapshot of a simulation of 3600 horizontally vibrated three-dimensional composite particles

20 springs. The system shown in Fig. 2.17 is already relatively large and required a considerable computing effort. The use of parallelized algorithms is recommended here (see Sect. 2.10).

2.6.6 Discussion

Using spherical particles, static phenomena can be simulated only under rather artificial assumptions of the tangential forces (see discussion in Sect. 2.3.6). As an example, using the tangential force ansatz (2.18), a heap of spherical particles becomes more shallow with growing number of particles—it dissolves under its own weight. There are several more cases when the simulation results are in obvious contradiction with experimental observations.

The composite particle model has a more complex geometrical structure which allows the transmission of torques during collisions *without* artificial assumptions about the surface properties (roughness) and, therefore, about tangential forces.

However, the more realistic simulation of static phenomena comes at the price of a significantly higher computing effort. Instead of a single sphere, each particle consists of five spheres. Furthermore additional forces have to be computed due to the springs connecting the spheres.

For systems whose static properties are crucial for the system's behavior, the presented particle model is a good compromise between computational expense and realism in that the model captures the main features of the behavior of realistic particles.

As shown in Fig. 2.15, one cannot model an arbitrary non-sphericity with a few spheres. The model can be improved by building the composite particles from a larger number of spheres (e.g., nine) and a correspondingly higher number of springs. In this case, however, the required computational effort grows significantly. As an alternative in Sect. 2.7, a particle model is presented which is composed of triangles and beam springs. This model allows us, in principle, to simulate almost arbitrarily shaped particles, although it requires yet higher computational effort.

There are three simple but interesting generalizations: Throughout this section it was assumed that the spheres of the composite particle do not over-

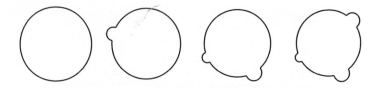

Fig. 2.18. Related composite particles that can be simulated with or without minor modifications of the described program

lap in the absence of external forces. Since the spheres of the same particle do not exert forces directly on each other, this limitation can be abandoned with only small modifications of the program. The second generalization concerns the radii of the particles. Using different sized peripheral spheres, more realistic, irregularly-shaped particles can be simulated. Third, the restriction of four peripheral particles (in two dimensions) can be abandoned. All these generalizations can be immediately adopted in the program, one only has to provide a suitable initialization file. Particle models similar to this generalized model, but where the springs have been replaced by rigid connections, have also been reported in literature [80]. Figure 2.18 shows examples.

There is a drawback to using the composite particles in dynamical simulation. Since the spheres do not exert tangential forces they do not rotate (the particle itself rotates, but the spinning motion of the constituting spheres is disregarded). Therefore, the particles' moment of inertia is systematically underestimated. This may cause inaccurate results in some situations.

2.7 Simulation of Sharp-Edged Particles

2.7.1 Model

The simulation of certain phenomena requires more realistic particle models than the simple sphere model. As a first step in that direction, Sect. 2.6 describes particles which are composed of spheres. Such particle models allow for a significantly more realistic simulation of granular systems, in particular of their static properties, although their potential for approximating sharp-edged particles are rather limited. Composing each of the particles of a feasible number of spheres, e.g., 5 spheres in two dimensions, or 9 spheres in three dimensions, their shape remains still somewhat round (see Figs. 2.13, 2.16, and 2.18) and deviates from the sharp-edged shape that is usually seen in many granular materials.

In this section a particle model is presented that allows for Molecular Dynamics simulations of almost arbitrarily shaped sharp-edged granular particles [52]. Each of the particles is composed of elastic triangles which are connected by dissipatively deformable beam springs. The beams are fixed to the centers of mass of the triangles. This particle model is rather flexible, i.e., a wide variety of particle shapes can be modeled, including concave particles. Figure 2.19 shows a few examples.

Fig. 2.19. Examples of sharp-edged particles composed of different numbers of triangles, connected by deformable beam springs. The model is not limited to convex particles

The connecting beams deform under the action of normal forces (e.g., elongation), tangential forces (shearing), and of bending moments. The interaction between two composite particles can be reduced to the interaction of their constituent triangles. Triangles which belong to the same composite particle interact exclusively via beams. Triangles from different composite particles interact by deforming each other. This deformation can be characterized by an apparent overlapping area of the triangles in contact. The repulsive force between the contacting triangles depends on the size and shape of the overlapping area.

Consider a triangle i being part of the particle k which itself consists of N_k triangles. The force acting on this triangle due to the contact with a triangle j of particle l is denoted by $\vec{F}_{ij}^{kl}$. The internal (beam) force caused by interaction with triangle m of the *same* particle is $\vec{f}_{im}^{k}$. The total number of beams acting on triangle i of particle k is n_i^k. Finally, $\vec{F}^{\text{ext}}(\vec{r}_i^k, \dot{\vec{r}}_i^k)$ is an external force (e.g., gravity) acting on the triangle i. With N being the number of particles, the total force acting on triangle i of particle k is

$$\vec{\mathcal{F}}_i^k = \sum_{\substack{l=1 \\ l \neq k}}^{N} \sum_{j=1}^{N_l} \vec{F}_{ij}^{kl} + \sum_{\substack{m=1 \\ m \neq i}}^{n_i^k} \vec{f}_{im}^{k} + \vec{F}^{\text{ext}}\left(\vec{r}_i^k, \dot{\vec{r}}_i^k\right) . \tag{2.72}$$

The particle $l = k$ is excluded from the first sum since interactions with triangles of the same particle are addressed, by definition, in the second term.

When deformed the beams dissipate energy proportionate to the deformation rate, in the same way as linearly damped springs. In the following, Sect. 2.7.2 considers the interaction forces between contacting triangles; in Sect. 2.7.4 the beam forces are discussed.

2.7.2 Interaction of Colliding Triangles

Triangles i and j (of particle k and l) in contact interact via the force $\vec{F}_{ij}^{kl}$. For simplicity, the upper indices k and l are omitted; it is implied that the interacting triangles belong to different particles. Physically this force is caused by mutual deformations of the particles. This deformation can be quantified by the overlapping area A of the undeformed triangles. The most natural assumption seems to be that the force is proportional to this overlapping area A. There is, however, a serious problem: Consider a triangle which overlaps a surface by the area A (corresponding to the force $\vec{F}$ as drawn in the left hand side of Fig. 2.20.[13] The assumption $F \sim A$ corresponds to $F(h) = \alpha h^2$ where

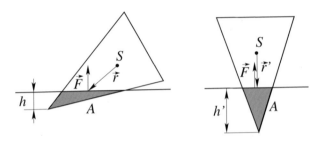

Fig. 2.20. A triangle penetrates a surface (see text for explanation)

α is a constant factor. The elastic energy reads then

$$E_{\text{el}} = \int_0^h F(h)\mathrm{d}h = \frac{\alpha}{3}h^3 = \frac{1}{3}Fh . \tag{2.73}$$

Now the triangle is turned quasistatically in such a way that the area A remains constant (right part of Fig. 2.20). The center of mass point S moves upward ($\vec{r} \rightarrow \vec{r}'$), i.e., the particles gain energy E_r by the action of the repulsive interaction force $\vec{F}$. At the same time the particle loses energy E_φ due to its move against the torque $\vec{r} \times \vec{F}$. For a small rotation $\mathrm{d}\vec{\varphi}$ both energies relate via

$$\mathrm{d}\vec{r} = \mathrm{d}\vec{\varphi} \times \vec{r} \tag{2.74}$$

$$\mathrm{d}E_r = -\vec{F} \cdot (\mathrm{d}\vec{\varphi} \times \vec{r}) = -\left(\vec{r} \times \vec{F}\right) \cdot \mathrm{d}\vec{\varphi} = -\mathrm{d}E_\varphi . \tag{2.75}$$

In other words, if the center of mass of the triangle is lifted by rotating it, the corresponding energy gain due to the action of the repulsive force equals the loss due to the torque $\vec{r} \times \vec{F}$. Therefore, the quasistatic motion does not

---

[13] In all figures of this section the overlapping area drawn is greatly exaggerated. In realistic simulations it is implied that the overlapping area is only a very small fraction of the area of both colliding triangles.

require any extra energy when the area A and, therefore, the force $\vec{F}$ remains constant. From geometry follows that the penetration depth h' in the right figure is larger than the previous depth h. Hence, applying (2.73), the elastic energy in the right figure is larger than the energy for the left triangle. Thus the (completely conservative) system violates the principle of conservation of energy. This is an obvious contradiction since the transition to the new state did not require any work. Thus, the assumption $F \sim A$ is not in agreement with mechanics. Instead, the force must depend also on the shape of the overlapping area A. Following the above arguments, we see that the energy is now conserved.

Let us assume that instead of the force, the elastic energy E_{el} (or the work to deform the triangles) is proportional to the overlapping area. The force acting on a triangle with center of mass $\vec{r}$ reads then

$$\vec{F} = -\frac{\partial E_{el}}{\partial \vec{r}} \, . \tag{2.76}$$

Correspondingly, the torque acting on that triangle is

$$M = -\frac{\partial E_{el}}{\partial \varphi} \, , \tag{2.77}$$

where φ is its orientation angle. In order to determine the forces and torques according to (2.76) and (2.77), we express the overlapping area of the interacting triangles in terms of their center of mass positions and their orientation. This derivation will be given in the remaining part of this section.

Contact Type A

The most frequent type of interaction is sketched in Fig. 2.21 which defines the nomenclature. The gray-shaded overlapping area is a triangle itself. In the following, the coordinate vectors of the points are denoted by the vector arrows, e.g., the coordinate vector of the vertex A_j is $\vec{A}_j$. Although the triangles are two-dimensional objects, all vectors are defined in three dimensions to simplify the notation, e.g.,

$$\vec{A}_j = \begin{pmatrix} [A_j]_x \\ [A_j]_y \\ 0 \end{pmatrix} \, , \tag{2.78}$$

where $[]_{x,y,z}$ denotes the components of a vector in x-, y-, or z-direction. The symbols $\triangle(a, b, c)$ stands at the same time for the triangle spanned by the vertices a, b, c and for the area of that triangle. The same applies for the symbols $\square(a, b, c, d)$ and $\bigcirc(a, b, c, d, e)$. The area of the triangle $\triangle(S_1 S_2 A_j)$ follows from basic geometry:

$$\triangle(S_1 S_2 A_j) = \frac{1}{2} \left[\left(\vec{S}_2 - \vec{A}_j \right) \times \left(\vec{S}_1 - \vec{A}_j \right) \right]_z \, . \tag{2.79}$$

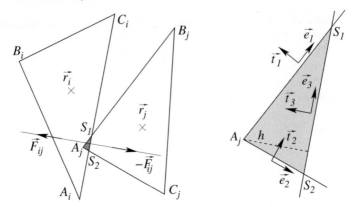

Fig. 2.21. Standard type of a triangle contact. The figure defines the notation. The right sketch shows a magnification of the gray-shaded intersection area with the definition of the unit vectors $\vec{e}_i$, $\vec{t}_i$ ($i = 1, 2, 3$) according to (2.80) and the height h according to (2.82)

To compute the derivatives (2.76) and (2.77) of this expression with respect to the coordinates and orientations of the particles, $\vec{S}_1$ and $\vec{S}_2$ are expressed in terms of the vertex coordinates. We define the unit vectors

$$\vec{e}_1 \equiv \frac{\vec{B}_j - \vec{A}_j}{\left|\vec{B}_j - \vec{A}_j\right|} , \qquad \vec{e}_2 \equiv \frac{\vec{C}_j - \vec{A}_j}{\left|\vec{C}_j - \vec{A}_j\right|} , \qquad \vec{e}_3 \equiv \frac{\vec{C}_i - \vec{A}_i}{\left|\vec{C}_i - \vec{A}_i\right|} \qquad (2.80)$$

and the corresponding perpendicular vectors $\vec{t}_n$ with $n \in \{1, 2, 3\}$:

$$\vec{t}_n \equiv \begin{pmatrix} -\,[e_n]_y \\ [e_n]_x \\ 0 \end{pmatrix} \qquad (2.81)$$

(see Fig. 2.21). Using the height of the overlap triangle with respect to the basis $\overline{S_1 S_2}$,

$$h \equiv \vec{t}_3 \cdot \left(\vec{A}_j - \vec{A}_i\right) , \qquad (2.82)$$

the intersection points are described by

$$\vec{S}_1 = \vec{A}_j - \vec{e}_1 \frac{h}{\vec{t}_3 \cdot \vec{e}_1} , \qquad \vec{S}_2 = \vec{A}_j - \vec{e}_2 \frac{h}{\vec{t}_3 \cdot \vec{e}_2} . \qquad (2.83)$$

The overlapping area (2.79) then becomes

$$\Delta \left(S_1 S_2 A_j\right) = \frac{h^2}{2} \frac{[\vec{e}_2 \times \vec{e}_1]_z}{(\vec{t}_3 \cdot \vec{e}_1)(\vec{t}_3 \cdot \vec{e}_2)} . \qquad (2.84)$$

In the present form this expression does not look very familiar, however, with the base length of the overlapping triangle

$$L \equiv \overline{S_2 S_1} = \left| \vec{S}_2 - \vec{S}_1 \right| = \frac{h \left[\vec{e}_2 \times \vec{e}_1 \right]_z}{(\vec{t}_3 \cdot \vec{e}_1)(\vec{t}_3 \cdot \vec{e}_2)} , \tag{2.85}$$

we obtain the overlapping area

$$\Delta (S_1 S_2 A_j) = \frac{Lh}{2} . \tag{2.86}$$

The advantage of the form (2.84) together with the definition (2.82) over the intuitive form (2.86) is that it allows for the computation of the derivatives (2.76) and (2.77).

The elastic energy is proportional to the overlapping area, $E_{\mathrm{el}} = Y \Delta (S_1 S_2 A_j)$, where Y is the Young modulus of the material (or more strictly its two-dimensional analogue). For the force acting on triangle j with the center of mass position $\vec{r}_j$, we obtain (note that the unit vectors $\vec{e}_n$ and $\vec{t}_n$ are independent of $\vec{r}_j$):

$$\vec{F}_j = -\frac{\partial E_{\mathrm{el}}}{\partial \vec{r}_j} = -Y \frac{[\vec{e}_2 \times \vec{e}_1]_z}{(\vec{t}_3 \cdot \vec{e}_1)(\vec{t}_3 \cdot \vec{e}_2)} \frac{\partial}{\partial \vec{r}_j} \frac{h^2}{2} = -YL \frac{\partial h}{\partial \vec{r}_j} . \tag{2.87}$$

To evaluate the last derivative, the vectors

$$\vec{\xi}_A \equiv \vec{A}_j - \vec{r}_j , \qquad \vec{\xi}_B \equiv \vec{B}_j - \vec{r}_j , \qquad \vec{\xi}_C \equiv \vec{C}_j - \vec{r}_j \tag{2.88}$$

are defined. With this notation, (2.82) transforms into

$$h = \vec{t}_3 \cdot \left(\vec{r}_j + \vec{\xi}_A - \vec{A}_i \right) , \tag{2.89}$$

and thus

$$\frac{\partial h}{\partial \vec{r}_j} = \vec{t}_3 \tag{2.90}$$

since $\vec{A}_i$ and $\vec{\xi}_A$ are independent of $\vec{r}_j$. The force (2.87) becomes then

$$\vec{F}_j = -YL\vec{t}_3 , \qquad \text{and hence} \qquad \vec{F}_i = YL\vec{t}_3 . \tag{2.91}$$

For the derivation of the torque (2.77), the derivative of the overlapping area with respect to the orientation φ_j of particle j is needed. The term $[\vec{e}_2 \times \vec{e}_1]_z$ is independent of the orientation of the triangles, therefore,

$$M_j = -\frac{\partial E_{\mathrm{el}}}{\partial \varphi_j} = -Y \frac{[\vec{e}_2 \times \vec{e}_1]_z}{2} \frac{\partial}{\partial \varphi_j} \frac{h^2}{(\vec{t}_3 \cdot \vec{e}_1)(\vec{t}_3 \cdot \vec{e}_2)}$$

$$= -E_{\mathrm{el}} \left(\frac{2}{h} \frac{\partial h}{\partial \varphi_j} - \frac{1}{\vec{t}_3 \cdot \vec{e}_1} \frac{\partial (\vec{t}_3 \cdot \vec{e}_1)}{\partial \varphi_j} - \frac{1}{\vec{t}_3 \cdot \vec{e}_2} \frac{\partial (\vec{t}_3 \cdot \vec{e}_2)}{\partial \varphi_j} \right)$$

$$= -E_{\mathrm{el}} \left(\frac{2}{h} \frac{\partial (\vec{t}_3 \cdot \vec{\xi}_A)}{\partial \varphi_j} - \frac{1}{\vec{t}_3 \cdot \vec{e}_1} \frac{\partial (\vec{t}_3 \cdot \vec{e}_1)}{\partial \varphi_j} - \frac{1}{\vec{t}_3 \cdot \vec{e}_2} \frac{\partial (\vec{t}_3 \cdot \vec{e}_2)}{\partial \varphi_j} \right) .$$

$$(2.92)$$

For the last line, (2.90) and the orientation-independence of $\vec{A}_i$ and $\vec{r}_j$ have been used. The vector $\vec{t}_3$ is invariant as well. Rotating by an angle of $\delta\varphi_j$, the vector $\vec{\xi}_A$ changes, according to

$$\vec{\xi}'_A = \vec{\xi}_A + \delta\vec{\varphi}_j \times \vec{\xi}_A , \tag{2.93}$$

where the rotation vector $\delta\vec{\varphi}$ points in z-direction with absolute value $\delta\varphi$. Hence

$$\delta\left(\vec{t}_3 \cdot \vec{\xi}_A\right) = \vec{t}_3 \cdot \left(\delta\vec{\varphi} \times \vec{\xi}_A\right) = \left(\vec{t}_3 \times \delta\vec{\varphi}\right) \cdot \vec{\xi}_A = \delta\varphi\vec{e}_3 \cdot \vec{\xi}_A \tag{2.94}$$

and likewise the terms $\vec{t}_3 \cdot \vec{e}_1$ and $\vec{t}_3 \cdot \vec{e}_2$ are treated. The three derivatives in (2.92) then read

$$\frac{\partial\left(\vec{t}_3 \cdot \vec{\xi}_A\right)}{\partial\varphi} = \vec{e}_3 \cdot \vec{\xi}_A , \quad \frac{\partial\left(\vec{t}_3 \cdot \vec{e}_1\right)}{\partial\varphi_j} = \vec{e}_3 \cdot \vec{e}_1 , \quad \frac{\partial\left(\vec{t}_3 \cdot \vec{e}_2\right)}{\partial\varphi_j} = \vec{e}_3 \cdot \vec{e}_2 . \tag{2.95}$$

With (2.82) and (2.83), (2.92) then simplifies to

$$M_j = -E_{el}\frac{\vec{e}_3}{h} \cdot \left(2\vec{\xi}_A - \frac{\vec{e}_1 h}{\vec{t}_3 \cdot \vec{e}_1} - \frac{\vec{e}_2 h}{\vec{t}_3 \cdot \vec{e}_2}\right) = -E_{el}\frac{2\vec{e}_3}{h} \cdot \left(\frac{\vec{S}_1 + \vec{S}_2}{2} - \vec{r}_j\right)$$

$$= -YL\vec{e}_3 \cdot \left(\frac{\vec{S}_1 + \vec{S}_2}{2} - \vec{r}_j\right) . \tag{2.96}$$

For an arbitrary vector $\vec{x}$ the relation $\left[\vec{x} \times \vec{t}_3\right]_z = \vec{x}\vec{e}_3$ holds, therefore, the result can be rewritten in the more intuitive form as follows:

$$M_j = \left[\left(\frac{\vec{S}_1 + \vec{S}_2}{2} - \vec{r}_j\right) \times \vec{F}_j\right]_z , \quad M_i = -\left[\left(\frac{\vec{S}_1 + \vec{S}_2}{2} - \vec{r}_i\right) \times \vec{F}_j\right]_z . \tag{2.97}$$

The term for M_i given above can be derived following the same recipe as for M_j. The term $(\vec{S}_1 + \vec{S}_2)/2$ is the center of the base of the overlapping triangle. Therefore, the interaction force acts on the center of the base of the overlapping triangle.

The vast majority of contacts one encounters in a typical simulation is of the standard type as drawn in Fig. 2.21. The results (2.91) and (2.97) are our final results for the acting forces and torques that apply for this contact type.

The contact sketched at the right hand side in Fig. 2.22 may be reduced to the standard type (left hand side) since the overlapping area is $\Delta\left(A_i B_i C_i\right) - \Delta\left(A_i S_1 S_2\right)$, where $\Delta\left(A_i B_i C_i\right)$ is a constant. Hence, the same result as for the standard case is obtained, but with the opposite sign.

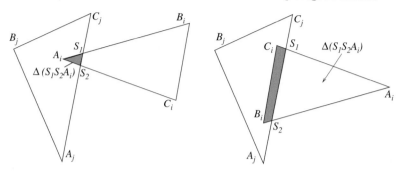

Fig. 2.22. Two forms of contact type A

Contact Type B

The contact type B (Fig. 2.23) comes in two complementary versions too. In the left hand side version the overlapping area is the tetragon $\square\,(A_iS_1A_jS_2)$. On the right hand side the overlapping area is the pentagon $\bigcirc(A_iS_1C_jB_jS_2)$ whose area is $\triangle\,(A_jB_jC_j) - \square\,(A_jS_1A_iS_2)$. Since $\triangle\,(A_jB_jC_j)$ is a constant, this version can be mapped to the one on the left hand side.

The overlapping area of the contact sketched at the left hand side is

$$\square\,(A_iS_1A_jS_2) = \frac{1}{2}\left[\left(\vec{A}_i - \vec{A}_j\right) \times \left(\vec{S}_1 - \vec{S}_2\right)\right]_z . \tag{2.98}$$

Again the unit vectors are defined:

$$\vec{e}_1 \equiv \frac{\vec{B}_j - \vec{A}_j}{\left|\vec{B}_j - \vec{A}_j\right|}, \quad \vec{e}_2 \equiv \frac{\vec{C}_j - \vec{A}_j}{\left|\vec{C}_j - \vec{A}_j\right|}, \quad \vec{e}_3 \equiv \frac{\vec{C}_i - \vec{A}_i}{\left|\vec{C}_i - \vec{A}_i\right|}, \quad \vec{e}_4 \equiv \frac{\vec{B}_i - \vec{A}_i}{\left|\vec{B}_i - \vec{A}_i\right|}, \tag{2.99}$$

with the according unit vectors $\vec{t}_n$, $n = 1, 2, 3, 4$ (see (2.81)). With the coordinates for the intersections S_1 and S_2,

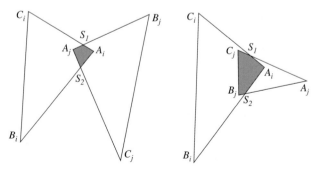

Fig. 2.23. Contact type B

$$
\begin{aligned}
\vec{S}_1 &= A_j + \vec{e}_1 \frac{\vec{t}_3 \cdot \left(\vec{A}_i - \vec{A}_j\right)}{\vec{t}_3 \cdot \vec{e}_1} = \vec{A}_i + \vec{e}_3 \frac{\vec{t}_1 \cdot \left(\vec{A}_j - \vec{A}_i\right)}{\vec{t}_1 \cdot \vec{e}_3} \\
\vec{S}_2 &= \vec{A}_j + \vec{e}_2 \frac{\vec{t}_4 \cdot \left(\vec{A}_i - \vec{A}_j\right)}{\vec{t}_4 \cdot \vec{e}_2} = A_i + \vec{e}_4 \frac{\vec{t}_2 \cdot \left(\vec{A}_j - \vec{A}_i\right)}{\vec{t}_2 \cdot \vec{e}_4} \quad ,
\end{aligned}
\tag{2.100}
$$

and $\Delta\vec{A} \equiv \vec{A}_i - \vec{A}_j$, (2.98) reads

$$
\Box\left(A_i S_1 A_j S_2\right) = \frac{1}{2}\left[\Delta\vec{A} \times \left(\vec{e}_1 \frac{\vec{t}_3 \cdot \Delta\vec{A}}{\vec{t}_3 \cdot \vec{e}_1} - \vec{e}_2 \frac{\vec{t}_4 \cdot \Delta\vec{A}}{\vec{t}_4 \cdot \vec{e}_2}\right)\right]_z .
\tag{2.101}
$$

With $\left|\Delta\vec{A} \times \vec{e}_n\right| = -\vec{t}_n \cdot \Delta\vec{A}$ and $E_{el} = Y\Box\left(A_i S_1 A_j S_2\right)$ we obtain finally

$$
E_{el} = -Y\left[\frac{\left(\vec{t}_1 \cdot \Delta\vec{A}\right)\left(\vec{t}_3 \cdot \Delta\vec{A}\right)}{2\vec{t}_3 \cdot \vec{e}_1} - \frac{\left(\vec{t}_2 \cdot \Delta\vec{A}\right)\left(\vec{t}_4 \cdot \Delta\vec{A}\right)}{2\vec{t}_4 \cdot \vec{e}_2}\right] .
\tag{2.102}
$$

In the same way as for contact type A,

$$
\frac{\partial \vec{t}_n \cdot \Delta\vec{A}}{\partial \vec{r}_j} = -\vec{t}_n
\tag{2.103}
$$

is derived. All unit vectors $\vec{t}_n$ and $\vec{e}_n$ are independent of $\vec{r}_i$ and $\vec{r}_j$, hence,

$$
\vec{F}_j = -\frac{\partial E_{el}}{\partial \vec{r}_j} = -Y\left[\frac{\vec{t}_1\left(\vec{t}_3 \cdot \Delta\vec{A}\right) + \left(\vec{t}_1 \cdot \Delta\vec{A}\right)\vec{t}_3}{2\vec{t}_3 \cdot \vec{e}_1} - \frac{\vec{t}_2\left(\vec{t}_4 \cdot \Delta\vec{A}\right) + \left(\vec{t}_2 \cdot \Delta\vec{A}\right)\vec{t}_4}{2\vec{t}_4 \cdot \vec{e}_2}\right].
\tag{2.104}
$$

To further simplify this expression, the transposition operator $\hat{T}$ is defined, which transforms each vector into its perpendicular vector,

$$
\hat{T}\vec{e}_n = \vec{t}_n
\tag{2.105}
$$

(see (2.81) for the component-wise definition of $\hat{T}$). We rewrite the terms in (2.104) using (2.100):

$$
\begin{aligned}
\vec{e}_1\left(\vec{t}_3 \cdot \Delta\vec{A}\right) &= \left(\vec{S}_1 - \vec{A}_j\right)\left(\vec{t}_3 \cdot \vec{e}_1\right) \\
\vec{e}_3\left(\vec{t}_1 \cdot \Delta\vec{A}\right) &= -\left(\vec{S}_1 - \vec{A}_i\right)\left(\vec{t}_1 \cdot \vec{e}_3\right) = \left(\vec{S}_1 - \vec{A}_i\right)\left(\vec{t}_3 \cdot \vec{e}_1\right) \\
\vec{e}_2\left(\vec{t}_4 \cdot \Delta\vec{A}\right) &= \left(\vec{S}_2 - \vec{A}_j\right)\left(\vec{t}_4 \cdot \vec{e}_2\right) \\
\vec{e}_4\left(\vec{t}_2 \cdot \Delta\vec{A}\right) &= -\left(\vec{S}_2 - \vec{A}_i\right)\left(\vec{t}_2 \cdot \vec{e}_4\right) = \left(\vec{S}_2 - \vec{A}_i\right)\left(\vec{t}_4 \cdot \vec{e}_2\right) .
\end{aligned}
\tag{2.106}
$$

The interaction force is then

$$\vec{F}_j = -\frac{Y}{2}\hat{T}\left[\frac{\vec{e}_1\left(\vec{t}_3 \cdot \Delta\vec{A}\right) + \vec{e}_3\left(\vec{t}_1 \cdot \Delta\vec{A}\right)}{\vec{t}_3 \cdot \vec{e}_1} - \frac{\vec{e}_2\left(\vec{t}_4 \cdot \Delta\vec{A}\right) + \vec{e}_4\left(\vec{t}_2 \cdot \Delta\vec{A}\right)}{\vec{t}_4 \cdot \vec{e}_2}\right]$$

$$= -\frac{Y}{2}\hat{T}\left(\vec{S}_1 - \vec{A}_j + \vec{S}_1 - \vec{A}_i - \vec{S}_2 + \vec{A}_j - \vec{S}_2 + \vec{A}_i\right)$$

$$= Y\hat{T}\left(\vec{S}_2 - \vec{S}_1\right) . \tag{2.107}$$

Thus the force acting on particle j is proportional to the length of the diagonal $\overline{S_1 S_2}$ and points in the perpendicular direction.

To compute the torque M, the overlapping area is derived with respect to the orientation of the particles. We start with particle j and note that $\vec{e}_3$, $\vec{e}_4$, $\vec{t}_3$ and $\vec{t}_4$ are invariant with respect to the orientation of particle j. Similarly as for contact type A, the derivatives of the terms in (2.102) are evaluated using the definitions (2.88) and (2.93):

$$\frac{\partial\left(\vec{t}_1 \cdot \Delta\vec{A}\right)}{\partial\varphi_j} = -\vec{e}_1 \cdot \left(\vec{A}_i - \vec{r}_j\right) , \qquad \frac{\partial\left(\vec{t}_2 \cdot \Delta\vec{A}\right)}{\partial\varphi_j} = -\vec{e}_2 \cdot \left(\vec{A}_i - \vec{r}_j\right) ,$$

$$\frac{\partial\left(\vec{t}_3 \cdot \Delta\vec{A}\right)}{\partial\varphi_j} = -\vec{e}_3 \cdot \left(\vec{A}_j - \vec{r}_j\right) , \qquad \frac{\partial\left(\vec{t}_4 \cdot \Delta\vec{A}\right)}{\partial\varphi_j} = -\vec{e}_4 \cdot \left(\vec{A}_j - \vec{r}_j\right) ,$$

$$\frac{\partial\left(\vec{t}_3 \cdot \vec{e}_1\right)}{\partial\varphi_j} = \vec{e}_3 \cdot \vec{e}_1 , \qquad \frac{\partial\left(\vec{t}_4 \cdot \vec{e}_2\right)}{\partial\varphi_j} = \vec{e}_4 \cdot \vec{e}_2 . \tag{2.108}$$

With these expressions, (2.102) and the relations $\vec{t}_1 \cdot \vec{e}_3 = -\vec{t}_3 \cdot \vec{e}_1$ and $\vec{t}_2 \cdot \vec{e}_4 = -\vec{t}_4 \cdot \vec{e}_2$, the torque reads

$$M_j = -\frac{\partial E_{\text{el}}}{\partial\varphi_j}$$

$$= -\frac{Y}{2}\left\{\frac{\left[\vec{e}_1 \cdot \left(\vec{A}_i - \vec{r}_j\right)\right]\left(\vec{t}_3 \cdot \Delta\vec{A}\right)}{\vec{t}_3 \cdot \vec{e}_1} + \frac{\left(\vec{t}_1 \cdot \Delta\vec{A}\right)\left[\vec{e}_3 \cdot \left(\vec{A}_j - \vec{r}_j\right)\right]}{\vec{t}_3 \cdot \vec{e}_1}\right.$$

$$+ \frac{\left(\vec{t}_1 \cdot \Delta\vec{A}\right)\left(\vec{t}_3 \cdot \Delta\vec{A}\right)\left(\vec{e}_3 \cdot \vec{e}_1\right)}{\left(\vec{t}_3 \cdot \vec{e}_1\right)^2} - \frac{\left[\vec{e}_2 \cdot \left(\vec{A}_i - \vec{r}_j\right)\right]\left(\vec{t}_4 \cdot \Delta\vec{A}\right)}{\vec{t}_4 \cdot \vec{e}_2}$$

$$\left. - \frac{\left(\vec{t}_2 \cdot \Delta\vec{A}\right)\left[\vec{e}_4 \cdot \left(\vec{A}_j - \vec{r}_j\right)\right]}{\vec{t}_4 \cdot \vec{e}_2} - \frac{\left(\vec{t}_2 \cdot \Delta\vec{A}\right)\left(\vec{t}_4 \cdot \Delta\vec{A}\right)\left(\vec{e}_4 \cdot \vec{e}_2\right)}{\left(\vec{t}_4 \cdot \vec{e}_2\right)^2}\right\} \tag{2.109}$$

which simplifies by means of (2.106) after some algebraic manipulation as

$$
\begin{aligned}
M_j = -\frac{Y}{2} &\left[\left(\vec{A}_i - \vec{r}_j \right) \cdot \left(\vec{S}_1 - \vec{A}_j \right) + \left(\vec{S}_1 - \vec{A}_i \right) \cdot \left(\vec{A}_j - \vec{r}_j \right) \right. \\
&+ \left(\vec{S}_1 - \vec{A}_i \right) \cdot \left(\vec{S}_1 - \vec{A}_j \right) - \left(\vec{A}_i - \vec{r}_j \right) \cdot \left(\vec{S}_2 - \vec{A}_j \right) \\
&\left. - \left(\vec{S}_2 - \vec{A}_i \right) \cdot \left(\vec{A}_j - \vec{r}_j \right) - \left(\vec{S}_2 - \vec{A}_i \right) \cdot \left(\vec{S}_2 - \vec{A}_j \right) \right] \\
= Y &\left(\vec{S}_2 - \vec{S}_1 \right) \cdot \left(\frac{\vec{S}_1 + \vec{S}_2}{2} - \vec{r}_j \right) .
\end{aligned}
\tag{2.110}
$$

For vectors $\vec{u}$ and $\vec{v}$ on the xy-plane, $[\vec{u} \times \vec{v}]_z = -\vec{u} \cdot (\hat{T}\vec{v})$ and $\hat{T}\hat{T} = -1$. Therefore, the result can be rewritten by means of the force formula (2.107):

$$
\begin{aligned}
M_j &= - \left[\hat{T}\hat{T}Y \left(\vec{S}_2 - \vec{S}_1 \right) \right] \cdot \left(\frac{\vec{S}_1 + \vec{S}_2}{2} - \vec{r}_j \right) \\
&= - \left(\frac{\vec{S}_1 + \vec{S}_2}{2} - \vec{r}_j \right) \cdot \left(\hat{T}\vec{F}_j \right) = \left[\left(\frac{\vec{S}_1 + \vec{S}_2}{2} - \vec{r}_j \right) \times \vec{F}_j \right]_z .
\end{aligned}
\tag{2.111}
$$

Hence, the interaction force acts on the middle point $(\vec{S}_1 + \vec{S}_2)/2$ of the diagonal $\overline{S_1 S_2}$. Therefore, the torque acting on particle i reads

$$
M_i = \left[\left(\frac{\vec{S}_1 + \vec{S}_2}{2} - \vec{r}_i \right) \times \vec{F}_i \right]_z = -Y \left(\vec{S}_2 - \vec{S}_1 \right) \cdot \left(\frac{\vec{S}_1 + \vec{S}_2}{2} - \vec{r}_i \right) .
$$

$$\tag{2.112}$$

The results (2.107) and (2.111) which have been derived for the contact, as sketched in the left hand side of Fig. 2.23, apply to the case drawn at the right hand side as well. Here the overlapping area is the pentagon $\bigcirc (A_i S_1 C_j B_j S_2)$. This area can be expressed as the difference between the triangle area $\triangle (A_j C_j B_j)$ and the (concave) tetragon $\square (A_i S_2 A_j S_1)$. Since the triangle area is constant, the forces and torques agree with (2.107) and (2.111) but with the opposite sign.

Contact Type C

Contacts of type C differ from types A and B in that the vertices of one particle move completely through the other particle and reemerge at its far side (Fig. 2.24).

At first glance, such contacts look quite improbable but let us discuss one common scenario for which this situation may actually emerge. Figure 2.25 (left) shows a contact of standard type A. Assume now that the upper triangle moves as sketched in the figures. Thus, inevitably, at a certain moment the lower vertex of the upper triangle reaches the lower edge of the lower triangle. Further motion in the same direction leads to a contact of type C. The contact

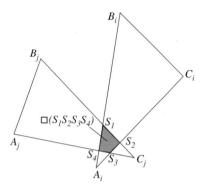

Fig. 2.24. Contact type C

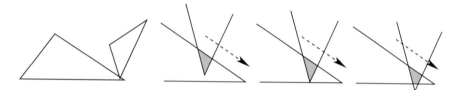

Fig. 2.25. Emergence of a type C contact. A type A contact (*left*) turns into type C when the upper triangle moves along the edge of the *lower* triangle

type C is an abstraction – in reality the particles would deform each other but they would, of course, maintain their physical integrity.

The basic postulate for the derivation of the interaction forces and torques of contact types A and B was the assumption that the elastic energy is proportional to the overlapping area. For case C, this assumption leads to an artifact: Suppose a triangle penetrates another as sketched in Fig. 2.26 (left). The force

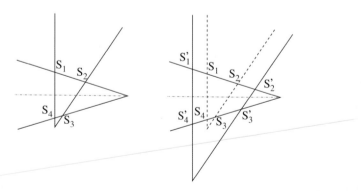

Fig. 2.26. The assumption that the overlapping area is proportional to the elastic energy fails for type C contacts. For the sketched situation, the more the triangles penetrate each other, the smaller is the restoring force (see text for explanation)

acting on the upper particle is $Y \overline{S_1 S_2}\, \vec{t}_{12} - Y \overline{S_3 S_4}\, \vec{t}_{34}$ with the vectors $\vec{t}$ normal to $\overline{S_1 S_2}$ and $\overline{S_3 S_4}$. For slender triangles we can approximate the absolute value of the force as $Y\left(\overline{S_1 S_2} - \overline{S_3 S_4}\right)$. When the upper triangle penetrates further (right sketch), the overlapping area $\square(S_1' S_2' S_3' S_4') > \square(S_1 S_2 S_3 S_4)$ leads to the restoring force $F' \sim \left(\overline{S_1' S_2'} - \overline{S_3' S_4'}\right)$. From geometry follows $\overline{S_1' S_1} = \overline{S_4' S_4}$, $\overline{S_2' S_2} < \overline{S_3' S_3}$. Consequently, the contact force grows slower than prescribed by Hooke's law, i.e., the resistance against penetration decreases with increasing penetration depth. We even have the absurd situation that very slender triangles do not resist further penetration after a certain penetration depth is achieved. This behaviour is physically incorrect.

The problem can be resolved by considering the overlapping area as drawn in Fig. 2.27, i.e., the gray drawn area consists of the true overlapping area $\square(S_1 S_2 S_3 S_4)$ and the smaller of the cut-off triangles, either $\Delta(S_3 A_i S_4)$ or $\Delta(S_2 C_j S_3)$. Indeed the *smaller* of the triangles has to be added, as both alternatives—(a) adding the areas of both triangles and (b) adding the larger triangle—lead to a violation of the conservation of energy. In fact, it is enough to show that for case (b), case (a) is then implied: Consider a type C contact which turns into type A as sketched in Fig. 2.28. When moving the left triangle as indicated by the arrow, the gray drawn area (which is assumed to be

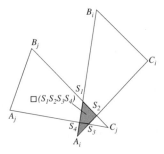

Fig. 2.27. For type C contacts, the energy is assumed to be proportional to the sum of the overlapping area $\square(S_1 S_2 S_3 S_4)$ and the area of the smaller cut-off triangles ($\Delta(S_3 A_i S_4)$ here)

Fig. 2.28. The gray area, which is proportional to the elastic energy, consists of the true overlap and the larger of the cut-off triangles. When a type C contact turns into type A due to a motion along the arrow, the energy changes non-continuously which implies the appearance of an infinite repulsive force

proportional to the elastic energy) decreases continuously until the vertex of the left triangle reaches the edge of the right one. At this moment the contact turns from type C to type A and, correspondingly, the energy changes non-continuously. This jump of the elastic energy goes along with an infinite repulsive force.

If the energy is assumed to be proportional to the total area of the true overlap and the smaller of the cut-off triangles, as proposed above, the non-physical jump of the energy is avoided. The same situation as drawn in Fig. 2.28 is shown again in Fig. 2.29. Initially the elastic energy is proportional to the sum of the overlapping area and the cut-off area of the right triangle, $E_{\rm el} \sim (\Box + \Delta_r)$. As soon as the other cut-off part Δ_r exceeds Δ_l (middle figure), $E_{\rm el} \sim (\Box + \Delta_l)$ does not reveal any jump of the elastic energy. When later (right figure) the contact turns from type C to type A, the energy varies continuously as well.

Therefore, with the assumption $E_{\rm el} \sim \min(\Box + \Delta_r, \Box + \Delta_l)$ type C contacts can be resolved in a physically consistent way. Since the overlapping area is reduced to a triangle, the interaction forces and torques can be computed in perfect analogy to contacts of type A.

Contact Types D and E

There are two more contact types, as sketched in Fig. 2.30. Type D can be addressed in the same way as type C. Again, it is assumed that the energy is proportional to the total of the true overlapping area and the smaller of the cut-off triangles (Fig. 2.31, left).

Type E contacts reveal four cut-off triangles. If we, however, notionally cut the particles along the line $B_i - B_j$ (dashed line), both parts are of type D. Therefore the area which is proportional to the energy is the sum of the overlapping area, the minimum of the areas $\Delta\,(A_j S_2 S_3)$ and $\Delta\,(C_i S_3 S_4)$ and the minimum of $\Delta\,(A_i S_5 S_6)$ and $\Delta\,(C_j S_6 S_1)$.

Thus, both types D and E may be mapped to types A and B.

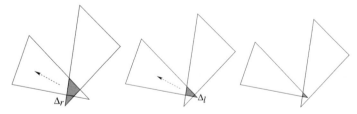

Fig. 2.29. Same situation as in Fig. 2.28. The energy is assumed to be proportional to the sum of the overlapping area and the area of the smaller cut-off triangle. Here no diverging forces appear

Contact Type F

There is one more type which is mentioned only for completeness: one of the triangles may be completely inside the other one. According to our presumptions, the overlapping area must be much smaller than each of the triangles. Hence, type F contacts should never occur. If such contacts occurred in a simulation, the parameters of the simulation must have been chosen inappropriately for modeling a granular system of given material parameters. A Molecular Dynamics program should, hence, detect type F contacts and abort the simulation on occurrence.

2.7.3 Contact Classification

To evaluate the forces and torques for contacting triangles, the contacts must first be classified into types A–F. For each pair of triangles i and j all pairs of edges are checked for intersections using equations of the type (2.83). Then it can be assessed if a vertex of either triangle is inside the other one. From the numbers of intersections and inner vertices of each triangle, we classify the contact as follows:

- One inner vertex, two intersections: type A (Fig. 2.22, left)

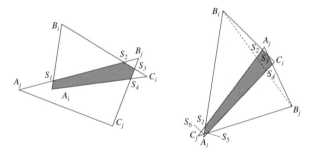

Fig. 2.30. Contact types D (*left*) and E (*right*)

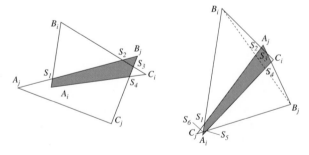

Fig. 2.31. For contacts of type D and E, the elastic energy is proportional to the gray area (see text for explanation)

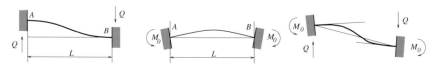

Fig. 2.32. Shearing (*left*) and bending (*center*) of beams. Shearing is the result of an acting force pair (Q) while bending is due to the application of a bending moment (M_0). A general deformation is a superposition of both bending and shearing (*right*)

- Two inner vertices in the same triangle, two intersections: type A (Fig. 2.22, right)
- One inner vertex in each triangle, two intersections: type B (Fig. 2.23, left)
- One inner vertex in one triangle, two in the other, two intersections: type B (Fig. 2.23, right)
- Four intersections, no inner vertices: type C (Fig. 2.24), can be mapped to type A
- One inner vertex, four intersections: type D (Fig. 2.30, left), can be mapped to type C
- No inner vertices, six intersections: type E (Fig. 2.30, right), can be mapped to either type A or B
- Three inner vertices in one triangle, no intersections: type F: program aborts

2.7.4 Beam Forces

The triangles that constitute a particle are connected by beams of elasticity[14] E, length L, and moment of area I. The total deformation of the beam is a superposition of three different mechanisms:

- Elongation
- Shearing (see Fig. 2.32, left)
- Bending (see Fig. 2.32, center)

The total beam deformation (Fig. 2.32, right) is a superposition of all three deformations. Elongation by ΔL is the simplest deformation, where the resulting restoring force is

$$\left| \vec{F}_{elong} \right| = E \frac{\Delta L}{L} \ . \tag{2.113}$$

Shearing is caused by the shear forces Q and bending is due to the global bending moment M_0. Since the system as a whole will not move accelerated, it is required that both shear forces are of the same magnitude Q, pointing in opposite direction. We restrict ourselves to the case of small deformations here to simplify the calculation. The x-axis is defined along the undeformed

---

[14] The elasticity of a spring is the same as its Young modulus. To avoid confusion, however, the symbol Y here is used for the bulk elasticity of the triangles and E for the spring elasticity.

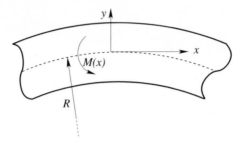

Fig. 2.33. A segment of a bent beam. The beam may be approximated locally by an arc of radius R. For an unbent beam $R \to \pm\infty$

beam (see Fig. 2.33). The (local) bending moment, caused by the external forces and torques, is expressed as

$$M(x) = Qx - Q(L - x) + M_0 = M_0 + Q(2x - L) , \tag{2.114}$$

i.e., the force Q on particle A acts by the lever x, the force $-Q$ on particle B acts by the lever $L - x$, and there is an additional global (external) bending moment M_0. This torque $M(x)$ causes a local curvature of radius R:

$$\frac{1}{R} = \frac{M(x)}{EI} , \tag{2.115}$$

i.e., the larger the bending moment, the more the beam is bent (the smaller is R). For small deformations,

$$\frac{1}{R} = \frac{\mathrm{d}^2 y}{\mathrm{d}x^2} = y'' . \tag{2.116}$$

For a beam bent downward (see Figs. 2.32 and 2.33) whose center of curvature is below the abscissa, the curvature is considered negative. Combining (2.114)–(2.116) the final equation governing the shape of the beam is obtained:

$$y'' = \frac{M_0}{EI} + \frac{Q}{EI}(2x - L) . \tag{2.117}$$

The general solution of this equation is

$$y(x) = C_0 + C_1 x + \frac{M_0 x^2}{2EI} + \frac{Q}{EI}\left(\frac{x^3}{3} - \frac{Lx^2}{2}\right) , \tag{2.118}$$

with integration constants C_0 and C_1. To simplify the discussion of the boundary conditions, in Fig. 2.34 we redraw the right picture in Fig. 2.32 where the system is rotated such that the beam ends are at $y = 0$. The rotation in the xy-plane does not change the bending moment $M(x)$ since its vector points in the z-direction. For small deformations (2.116) remains valid as well, thus (2.117) is invariant to small rotations. The boundary conditions are now

$$y(0) = 0 , \qquad y(L) = 0 , \qquad y'(0) = y'_0 , \qquad y'(L) = y'_L . \qquad (2.119)$$

Inserting (2.119) into (2.118) yields

$$C_0 = 0 , \qquad C_1 = \frac{QL^2}{6EI} - \frac{M_0 L}{2EI} \qquad\qquad (2.120)$$

and the final solution for the shape of the beam reads

$$y(x) = \frac{M_0}{2EI}(x^2 - Lx) + \frac{Q}{6EI}\left(2x^3 - 3Lx^2 + L^2 x\right) . \qquad (2.121)$$

From (2.121) follow the orientations y'_0 and y'_L as functions of the force Q and the bending moment M_0:

$$y'_0 = \frac{QL^2}{6EI} - \frac{M_0 L}{2EI} , \qquad y'_L = \frac{QL^2}{6EI} + \frac{M_0 L}{2EI} . \qquad (2.122)$$

Up to now we assumed Q and M_0 are known values which determine the shape of the beam. For simulations, however, one is interested in the reverse case, i.e., the beam ends are oriented in directions y'_0 and y'_L and the corresponding force and bending moment have to be determined. The system of equations (2.122) can be solved to obtain the force and moment for the *elastic* deformation of the beam, $Q^{\mathrm{el}} \equiv Q$ and $M_0^{\mathrm{el}} \equiv M_0$:

$$Q^{\mathrm{el}} = 3EI\frac{y'_L + y'_0}{L^2} \qquad\qquad (2.123)$$

$$M_0^{\mathrm{el}} = EI\frac{y'_L - y'_0}{L} . \qquad\qquad (2.124)$$

Assuming damped motion of the beams, proportionate with the deformation rate (linear damping), the dissipative forces and the torque due to elongation, shearing, and bending becomes

$$\vec{F}^{\mathrm{diss}}_{\mathrm{elong}} = -\gamma \Delta \dot{L} \vec{b} \qquad\qquad (2.125)$$

$$Q^{\mathrm{diss}} = \frac{3\gamma I}{L^2} \left(\dot{y}'_L + \dot{y}'_0\right) \qquad\qquad (2.126)$$

$$M_0^{\mathrm{diss}} = \frac{\gamma I}{L} \left(\dot{y}'_L - \dot{y}'_0\right) , \qquad\qquad (2.127)$$

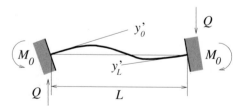

Fig. 2.34. A typical deformation of a beam. The beam is both bent and sheared

with γ being the dissipative parameter of the beam and $\vec{b}$ is the unit vector pointing between the ends of the beam.

Given the coordinates of the triangles which are connected by the beam, the orientation of the beam, described by y_0' and y_L', can be evaluated particularly efficiently if the beams are perpendicular to the adjacent edges of the triangles (characterized by the unit vectors $\vec{n}_A$ and $\vec{n}_B$).

The implementation is restricted to particles whose adjacent triangles are shaped such that the line connecting their center of mass points is perpendicular to the direction of adjacent edges in the relaxed case, i.e., no forces nor torques apply. Obviously this condition restricts the admissible particle shapes, but as shown in Fig. 2.19 it still provides enough freedom for constructing particles. The generalization to arbitrary shapes is straightforward but requires extra numerical effort.

For small deformation we obtain y_0' and y_L' from the unit vector $\vec{b}$, characterizing the spatial beam orientation, and the vectors $\vec{n}_A$ and $\vec{n}_B$ that define the orientation of the triangles:

$$\left[\vec{b} \times \vec{n}_A\right]_z = \sin\varphi_A \approx \tan\varphi_A = y_0' , \qquad -\left[\vec{b} \times \vec{n}_B\right]_z \approx y_L' , \qquad (2.128)$$

with $\varphi_{A/B}$ being the angle between $\vec{b}$ and $\vec{n}_A$ or $\vec{n}_B$, respectively. Since $\dot{\vec{n}}_{A/B} = \vec{\omega} \times \vec{n}_{A/B}$ the deformation rates read

$$\dot{y}_0' = -\omega_A\left(\vec{b} \cdot \vec{n}_A\right) , \qquad \dot{y}_L' = \omega_B\left(\vec{b} \cdot \vec{n}_B\right) , \qquad (2.129)$$

where ω_A and ω_B are the angular velocities of the particles.

With (2.128) and (2.129), the total bending moment $M_0 \equiv M_0^{\text{el}} + M_0^{\text{diss}}$ and the total force $Q \equiv Q^{\text{el}} + Q^{\text{diss}}$ can be computed.

Using the contact forces described in Sect. 2.7.2 and the contact forces derived here, a Molecular Dynamics simulation can be performed. To this end the bending moment is added to the sum of torques due to triangular contacts for both triangles. When adding the force one has to take note to add $Q+Q^{\text{diss}}$ to the force of triangle A but $-Q - Q^{\text{diss}}$ to the force of triangle B.

2.7.5 Examples

Collision of Two Particles

As a first example we present a simulation of two colliding quadratic particles which are composed of four identical triangles each, as sketched in Fig. 2.19 (left). This simple example reveals already complex system behavior, caused by the geometrical sharp-edged shape of the particles. Originally we studied this system to test the algorithm and its implementation.

Figure 2.35 shows three series of snapshots of a moving particle impinging on another identical particle at rest, for different impact velocity (100 cm/sec,

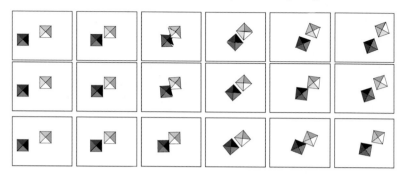

Fig. 2.35. Snapshots of colliding particles for large (*top row*), intermediate (*middle*) and small (*bottom row*) impact velocity (see text for explanation). The time scale is different in the series and the snapshots are not taken in equidistant intervals. Different gray shading for the triangles has been used to visualize the particle rotation. For this example the beams have been chosen to be extraordinarily weak to visualize the particle deformation

50 cm/sec, and 10 cm/sec). Obviously, the top row (large velocity) differs qualitatively in that there occurs only one contact of particle edges, whereas in the other cases there are two contacts in short succession (fifth picture of the series). An animation of the collisions can be found at the web site.

The corresponding traces of the particles (Fig. 2.36) differ qualitatively; for large impact rate the traces form a small angle whereas for the other cases a much wider angle is observed. Interestingly, the traces do not differ much for impact rates of 50 cm/sec and 10 cm/sec. For high impact rate the angular velocity is much larger than for the other cases which is also visible in the figure. This behavior can be explained by the double contact: In all three cases there is a large momentum transferred at the first contact which causes rapid rotation. For a high impact velocity this rotation stays constant after the impact. For a lower impact rate there occurs a second contact where a

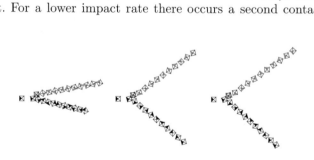

Fig. 2.36. Traces of the particles corresponding to the time series in Fig. 2.35. Impact velocity: *Left* – 100 cm/sec, *middle* – 50 cm/sec, *right* – 10 cm/sec. The first case differs from the others due to the double contact for low impact rate

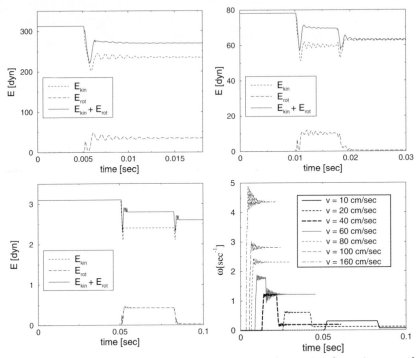

Fig. 2.37. Kinetic energies of the translational and rotational motion together with the total energy of both particles as a function of time. *Top*: high (*left*) and intermediate (*right*) impact rate, *bottom left*: low impact rate. The fourth plot shows the angular velocity of the initially resting particle as a function of time

torque is directed just in the opposite direction. Therefore, after the second collision the particles rotate at a much lower rate.

This effect can be assessed quantitatively from the translational and rotational energies of the initially resting particle, as shown in Fig. 2.37 (bottom right). Initially the total energy is carried by the translational motion of the left particle. This energy is partly transferred to the other particle in the course of the collision, where a certain part of the energy is transformed into rotational energy. Another part of the energy is dissipated by the springs. The fraction of the total rotational energy depends strongly on the impact velocity due to the described double-contact scenario for lower impact rate.

At high impact velocity one notes a sudden change of the angular motion at the time of contact of the particles, and a short damped oscillation which results from excitations of internal degrees of freedom (i.e., deformed dissipative springs). After the collision the particles move with constant angular and translational velocities. For intermediate and low impact velocity the behavior is qualitatively different: first the particles contact each other just as in the case of high impact rate resulting in a certain rotational energy. Shortly after that, however, the particles contact each other for a second time at different edges due to their rotation. After the full collision (consisting of two contacts) the angular velocities are small.

Although the total energy for intermediate and low impact rate is very different, the qualitative behavior is essentially the same. For the given collision geometry, there is a certain marginal impact velocity where the collision changes its nature qualitatively. This can be seen in Fig. 2.37 (bottom right), where the evolution of the angular velocity of the initially resting particle is illustrated for different impact rates (see [52] for details).

Hence, even for the simplest numerical experiment—the collision of two particles in an otherwise empty system—a complex and strictly nonlinear behavior is observed which cannot be found (at least not in such a striking way) when simulating spheres (Sect. 2.3) or particles composed of spheres (Sect. 2.6). This interesting and realistic behavior is a direct consequence of the sharp-edged shape of the simulated grains.

Clogging Hopper

As a second example for the application of the presented particle model, we discuss briefly the flow of a granular material in a hopper. This problem is of interest not only due to its technological importance but also due to interesting phenomena like arc formation and clogging which are the subject of many theoretical and experimental works. Such systems have been frequently simulated numerically using various methods, among those also Molecular Dynamics simulations in two and three dimensions (e.g., [24, 141, 150, 238, 241]).

Here we do not wish to report any quantitative result of hopper simulation, but only demonstrate that using our model for simulating this interesting system yields qualitatively correct results. For a detailed discussion see [54].

Figure 2.38 shows snapshots of a system of $N = 400$ complex particles

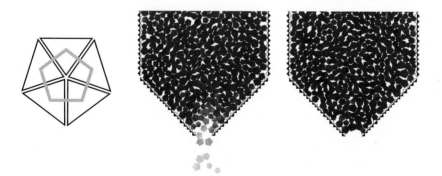

Fig. 2.38. Snapshots of the simulation of hopper flow. The particle model is sketched at the *left*. The system consists of $N = 400$ pentagons built up from five triangles each. The gray value encodes the particle velocities

(animations can be found on the web site of the book). Each grain consists of five triangles forming a pentagon as sketched in Fig. 2.38. The hopper flow is nonuniform. In the animations density waves appear which have been found experimentally (e.g., [24, 233, 265]) and in simulations using spherical particles (e.g., [238, 241]). The right part of the figure shows the same system a few moments later: the flow breaks down, the hopper clogs. We have not been able to reproduce complete clogging in simulation using spheres with the force law given by (2.14) and (2.18) (and similar ones). We conclude that the realistic description of the system is a consequence of the sharp-edged particle geometry.

2.7.6 Fragmentation of Sharp-Edged Particles

By introducing a threshold value for the deformation above which the beams break, one can also study plastic deformation, fragmentation, and grinding with the presented model. Similar models have been discussed in literature [108, 223, 275]. However, if the number of triangles in the system is fixed, the effective material properties change after fragmentation events. The number of triangles per particle is reduced due to fragmentation. Consequently the particles grow harder in the coarse of time (see Sect. 2.9 for a more detailed discussion of this effect). This artifact of the method has to be compensated by adjusting the elastic and dissipative material characteristics of the triangles and beams.

2.8 Further Particle Models

In literature a variety of further particle models has been proposed for Molecular Dynamics of granular materials. In this section we briefly describe a few of them. The central problem for choosing the appropriate model is finding a proper compromise between numerical effort and realistic system behavior, in order to simulate a system at good precision and simultaneously to perform the simulation over a sufficiently long (real) time interval in acceptable computer time.

Several authors [194, 277, 278, 302] proposed a particle model whose surface is defined by simple or generalized ellipsoids, i.e., in local particle coordinates:

$$f_i\left(x_i, y_i\right) = \left(\frac{|x_i|}{a_i}\right)^{n_i} + \left(\frac{|y_i|}{b_i}\right)^{n_i} - 1 = 0 . \tag{2.130}$$

Choosing n_i between 2 and ∞ the particle's shape can be varied continuously between and ellipse/ellipsoid and a square/parallelepiped. In each simulation time step the contacts of all pairs of particles have to be determined, which for this model requires numerical root finding of a polynomial of high order. The higher the order n_i of the generalized ellipsoid, i.e., the better the particles approximate a parallelepiped, the more costly is the numerical root finding. The model is limited to convex particles.

Gallas and Sokołowski [85] defined a composite particle model which consists of two spheres. Similar models have been proposed by Walton and Brown [288] and other authors. The difference from the model type described in Sect. 2.6 in these models is that the particles are rigidly attached to each other.

Hogue and Newland [120–122] studied the flow of granular material in two dimensions on an inclined plane and in an outflowing hopper using a convex polygon particle of up to 24 edges. To decide whether two particles are in contact with each other, each pair of faces of different particles had to be checked for intersections. The particle interaction was determined according to a model [269, 290] that takes into account sliding motion according to Coulomb's law.

Tillemans and Herrmann [275, 276] had also proposed a two dimensional convex polygon particle model. When two particles i and j deform each other, they experience a force

$$\vec{F} = F^{\mathrm{n}} \, \vec{e}^{\,\mathrm{n}} + F^{\mathrm{t}} \, \vec{e}^{\,\mathrm{t}} \, , \tag{2.131}$$

with

$$\begin{aligned} F^{\mathrm{n}} &= -\frac{YA}{L_c} - m^{\mathrm{eff}} \, \gamma^{\mathrm{n}} \left(\dot{\vec{r}}_i - \dot{\vec{r}}_j \right) \cdot \vec{e}^{\,\mathrm{n}} \\ F^{\mathrm{t}} &= -\min \left(m^{\mathrm{eff}} \, \gamma^{\mathrm{t}} \left| \vec{v}_{\mathrm{rel}} \cdot \vec{e}^{\,\mathrm{t}} \right|, \mu \left| F^{\mathrm{n}} \right| \right) \, , \end{aligned} \tag{2.132}$$

where A is the compression (intersection area) of the contacting polygons and L_c is the characteristic size of the particles. Since the particle interaction force is proportional to the intersection area, for γ^{n} the conservation of energy is violated (see Sect. 2.7.2). The model was applied to simulations of shear cells, earthquakes, as well as to arc formation and clogging in hoppers [275]. A similar model was used by Handley [108] who studied the fragmentation of brittle granular material.

Potapov et al. [221, 223] defined a model similar to the composite triangle particles to study fragmentation as well: At the beginning of the simulation they partitioned the macroscopic two-dimensional particle into small equilateral triangles (elements) which are thought to be perfectly rigid. When loading the particle, forces act between the elements. Two types of forces are distinguished: Adhesive forces act between elements which are *not* separated by a fracture, such elements can, therefore, be loaded in each direction—they cling to each other via the adhesive force. If the adhesive force surpasses a threshold value, a fracture forms. In this case there will be no adhesive force between these elements anymore. Collisional forces act on elements at the surface of the particles when contacting other particles and also between elements which are separated by a fracture. Apart from the considerable numerical effort required for the simulation of such particles, we are faced with problems when considering repeated fractures: Since fractures only occur between the elements of a particle, larger particles reveal different material properties than small ones. Thus, smaller particles are stiffer than larger particles, which would affect the behavior of fragmentation processes.

In [113] the particle surfaces are described by two-dimensional splines. For the detection of contacts a three-step algorithm was applied: In the first step, the set of particles is selected whose center of mass points are close enough to allow for a collision (see also [200, 301] for ellipsoidal particles). In the second step the set is further reduced by checking whether the convex hulls of the particles intersect. For pairs of particles with intersecting hulls, an investigation of the intersection of the splines that constitute the particle surfaces is performed to determine the interaction forces. Similar algorithms for contact detection are described in [192].

There are many more particle models that can be found in literature. Here we could describe only a small selection.

2.9 Particle Fragmentation

2.9.1 Modeling of Fragmentation

Comminution belongs to the most important technological processes in particle engineering. To simulate such processes the Molecular Dynamics method has to be expanded to take into account the fragmentation of particles.

Particle fragmentation is a complex process and is a subject of active research. Even the fracture of a single particle is a highly complicated problem (see, e.g., [71, 154, 188]). In the framework of a Molecular Dynamics program, therefore, one has to restrict to simpler model assumptions.

In a simple approach each granular particle is constructed from many simulated (elementary) particles which are linked to each other, similar to the composite particles presented in Sect. 2.6. Fragmentation of such particles is simulated by cutting the links between the elementary particles given certain conditions, e.g., if the stress at this link exceeds a certain threshold [28, 147, 219, 220, 222]. The results of such simulations frequently agree well with experiments, in particular if the statistical properties of fracture mechanics are taken into account, i.e., the fracture probability (see Sect. 2.9.3) and the size distribution of the fragments (see Sect. 2.9.4).

On the other hand, this model is problematic for the description of granular materials of large particle size dispersion: to describe fragmentation realistically, even the smallest of the particles must consist of many elementary particles. To simulate identical mechanical properties of small and large particles, large ones then consist of a *very* large number of elementary particles. If we assume a dispersion of 10 of the radii and assume that a small particle consists of 10 elementary particles, then the large particle already consists of 10^4 elements. Simulating a material where single grains consist of thousands of elements is presently not feasible. The other problem of this model arises from the request that the particle material properties are invariant, i.e., independent of the number of fragmentation events that have taken place.

We now present a model for particle fragmentation which has been applied to the simulation of milling processes [49]. At each simulation time step, for each particle it is checked whether the conditions for fragmentation are fulfilled. Fragmentation processes do not occur deterministically but according to a probability distribution as a function of the local pressure (see Sect. 2.9.3). If fragmentation occurs, the sizes of the resulting fragments obey a probability distribution too (see Sect. 2.9.4). Here only fractures with two fragments are considered; the generalization is, however, simple.

2.9.2 Molecular Dynamics of Fragmenting Particles

In the comminution simulations spherical particles are used, hence, as a model assumption, the fragments are spheres too. Moreover, conservation of the total mass is assumed, i.e., the radii of the fragments fulfill the condition

$$R_{f_1}^d + R_{f_2}^d = R^d \; , \tag{2.133}$$

where R is the radius of the original particle and d is the system dimension. Satisfying both conditions simultaneously is problematic since the fragment spheres require always more space than the original particle. In general, they cannot fit into the same cavity as the original particle. Hence, it is impossible to replace the original particle by its fragments in an uncompressed state. Any considerable compression corresponds, however, to an unrealistically strong repulsive normal force. An exceptional large deformation energy would be the consequence. We bypass this problem by temporally modifying the interaction between the fragments.

Assume a particle has been selected for fragmentation due to its load and the fracture probability (see Sect. 2.9.3). The fragment sizes R_{f_1} and R_{f_2} are determined according to the fracture size distribution (see Sect. 2.9.4) with the side condition (2.133). The original particle is eliminated from the simulation and the new positions of the fragments are determined randomly with the condition

$$\Delta r \equiv \left| \vec{r}_{f_1}^0 - \vec{r}_{f_2}^0 \right| = 2R - R_{f_1} - R_{f_2} < R \; , \tag{2.134}$$

i.e., the fragment particles are completely contained inside the surface of the original particle and compress each other. From now on the fragment particles interact in the common way with all other particles. The interaction of the fragments with each other is special: Up to the moment of their first complete separation the normal force $F_{f_1 f_2}^n$, e.g., (2.12), is temporarily replaced by

$$\mathcal{F}_{f_1 f_2}^n \left(\vec{r}_{f_1}, \vec{r}_{f_2}, \dot{\vec{r}}_{f_1}, \dot{\vec{r}}_{f_2} \right) =$$

$$= \begin{cases} F^n \left(\left| \vec{r}_{f_1} - \vec{r}_{f_2} \right| - \Delta r, \dot{\vec{r}}_{f_1} - \dot{\vec{r}}_{f_2} \right) & \text{for } \left| \vec{r}_{f_1} - \vec{r}_{f_2} \right| < \Delta r \\ F_{\text{frag}} = \text{const} & \text{otherwise} \, . \end{cases} \tag{2.135}$$

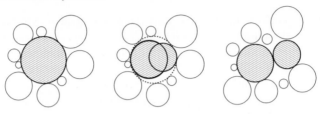

Fig. 2.39. Sketch of the particle fragmentation procedure. Immediately after the fragmentation the particles compress each other due to mismatch of geometry of the spherical fragments (*middle*). A small repulsive force helps to overcome this non-physical transient situation

At the beginning of each time step the parameter Δr is reset to the current distance of the particle. If the particles approach each other, at the end of the predict step the expression $|\vec{r}_{f_1} - \vec{r}_{f_2}| - \Delta r$ (strictly speaking $|\vec{r}^{\,p}_{f_1} - \vec{r}^{\,p}_{f_2}| - \Delta r$, i.e., using the values after the predict step) is negative and the first line of (2.135) applies. This means that the fragments resist further compression like standard particles and experience a small constant repelling force F_{frag} which drives them apart. Starting at the instant of first complete separation of the fragments, i.e., when $\Delta r \geq R_{f_1} + R_{f_2}$ for the first time, they also interact in the standard way, as described by $F_{f_1 f_2}$ (e.g., (2.12)).

The constant repelling force F_{frag} is chosen to be sufficiently small in order to suppress non-physical energy gain of the system; this means that the energy, which is unavoidably pumped into the system due to this force, must be negligible in comparison with the mean kinetic energy of the grains. By carefully adjusting F_{frag} we do not only suppress the elastic energy gain but also assure that the time period requiring a modified interaction force is comparatively short. The procedure of fragmentation is sketched in Fig. 2.39.

The Molecular Dynamics simulation scheme described on p. 28 is, hence, modified to account for particle fragmentation:

1. Initialization
2. Predict the positions and time derivatives of all N particles.
3. Compute the forces $\vec{F}_i$ acting on all particles. For fragment particles which are not separated so far, apply the modified force $\mathcal{F}^n$. Store the maximum normal interaction force F_i^{max} which is exerted on particle i by any of the neighbor particles.
4. Correct the positions and time derivatives of all particles.
5. For each particle i check whether it breaks according to the fragmentation probability as a function of F_i^{max} (Sect. 2.9.3). If the particle breaks:
   (a) Determine the radii R_{f_1} and R_{f_2} of the fragments of each breaking particle according to the fragment size distribution (Sect. 2.9.4).
   (b) Replace each broken particle by its fragments according to the explained scheme. Increase the number of particles N correspondingly.
6. Extract data.
7. Terminate program.

The described algorithm is flexible enough to include multiple fracture events, i.e., a fragment itself may break even before its first complete separation from its original twin fragment. An animated sequence of fragmentation can be found at the web site.

Using the fragmentation probability from Sect. 2.9.3 and the fragment size distribution described in Sect. 2.9.4, the algorithm was applied to the simulation of autogenous dry comminution [49]. In the simulation the introduced non-physical transient situations in which the fragments resulting from a broken particle deform each other were always rather short-lived.

2.9.3 Fragmentation Probability

The precondition for the fragmentation of particles is a sufficient mechanical load. In experiments particles of identical size and material vary with respect to their tensile strengths. That means when compressing particles of a sample by applying a certain force, only a fraction of the sample actually breaks. The reason for this different behavior can be explained by the existence of flaws which provide sites for stress concentration and the initiation of cracks which, subsequently, propagate through the material. These flaws are distributed throughout the whole volume, but several investigations have revealed that surface flaws activated by high tensile stress play the dominant role for the initiation of a crack [294]. Hence, the fragmentation probability is intimately related to the statistical distribution of surface flaws and the resulting stress distribution over the particle's surface.

The fracture probability P of a single particle can be derived employing statistical reasoning [293]. A particle, subjected to a force which stores the specific elastic energy $W_m = W/m$, breaks with probability

$$P\left(R, W_m\right) \propto 1 - \exp\left(-cR^2W_m^z\right) , \tag{2.136}$$

where c and z are material constants. This probability was found empirically by Weibull [291, 292] for the fragmentation of repeatedly loaded brittle rods. For a more general mathematical derivation see [47, 303]. The simple law (2.136) is in good agreement with experimental data. The constants c and z can be extracted from measurements when plotting $\log\left[-\log(1 - P)\right]$ vs. $\log W_m$. Typical values of z are in the range of 1.5–2.5. In agreement with experimental results, the term R^2 in the exponent (for fixed W_m) in general predicts that under the same load large particles break with higher probability than small ones. This fact is rooted in the diminished probability to find a sufficiently large flaw on a smaller surface.

With the assumption of linear elastic material behavior, the Hertz law (2.10) [119, 149] can be employed to derive a relation between the elastic energy W stored in a sphere and the exerted repulsive contact force F:

$$W = \frac{2}{5}\xi F \sim \left(\frac{1}{R}F^5\right)^{1/3} \quad \text{for} \quad \xi > 0, \tag{2.137}$$

where $\xi \equiv R_1 + R_2 - |\vec{r}_1 - \vec{r}_2|$ quantifies the deformation.

Consider the situation sketched in Fig. 2.40: a particle placed in the center is compressed by four contacting particles. The highest compressive stress occurs close to the contact points, a fact well known to engineers [29] and also found in numerical simulations of the particle material [146, 147]. Hence, the fracture probability of the center particle is linked to the probability of finding a flaw of sufficient size in the very vicinity of any of the four contact regions. Thus, the action of different compressing particles can be regarded as independent stress sources. That means we consider the elastic energy stored by each compressing particle separately, and calculate the related fracture probability according to (2.136) and (2.137) as a function of the particle's size and the maximum of all acting normal forces F^n:

$$P\left(R, F^n\right) \propto 1 - \exp\left[-\tilde{c}R^{2-z/3}\left(F^n\right)^{5\,z/3}\right] . \tag{2.138}$$

The constant $\tilde{c}$ is related to the empirical constant c in (2.136); it is found by inserting (2.137) into (2.136) and expressing m in terms of the particle radius ($m \propto R^3$).

2.9.4 Fragment Size Distribution

Once a flaw has been activated it forms an initial crack which rapidly propagates through the particle. Typical crack velocities in the range of 1500 m/sec have been measured [294]. In rock, the ratio of the crack propagation velocity and the speed of sound is about 0.4 [151]. The formation of a variety of differently sized fragments occurs by virtue of branching cascades. Branching is governed by a balance between the energy release rate G and the crack resistivity B [294], the latter accounting for the creation of new surfaces. Because of the extremely short duration of the rupture process one can safely neglect external energetic contributions to this energy balance. The formation of a crack releases stresses in the material and, hence, energy. The energy release

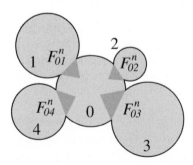

Fig. 2.40. The central particle 0 is being compressed by four neighbors. The four independent stress zones are drawn in dark gray

rate G depends on the crack velocity and the crack length, whereas the crack resistivity B depends on the crack velocity and the number of branches. Due to a maximal crack velocity, i.e., the speed of sound, the balance between both terms requires the formation of new cracks by branching. A crack may release more elastic energy than it can consume. Thus, the crack branches and, therefore, more energy per time unit is consumed by plastic deformation. The branching continues until a balance between energy consumption and energy release is achieved. From these considerations it can be shown [294] that the fragment size distribution (cumulative mass distribution) Q is a function of the product $R_f W_m$, where R_f denotes the fragment size, i.e., $Q = Q(R_f W_m)$.

Experimental evidence [293] for a scaling law $Q \sim R_f^\beta$ is in agreement with the empirical Schuhmann law [259]

$$Q \propto \left(\frac{R_f}{k}\right)^\beta , \tag{2.139}$$

which can be regarded as an approximation of the Rosin–Rammler law [244]:

$$Q \propto 1 - \exp\left[-\left(\frac{R_f}{k}\right)^\beta\right] . \tag{2.140}$$

The variable k has the dimension of length and, thus, can only be identified with the size of the original particle. A rigorous derivation of (2.140) (compatible with $\beta = 1$) is given in [89, 90].

Equations (2.139) and (2.140) describe the distribution of fragments only for certain materials and only below a certain size [89]. For the sizes of the largest fragments an equally well-defined quantitative statement does not exist. There exists a well elaborated theory on fracture mechanics, e.g., [136, 280], where many more material specific quantitative laws for the fracture of materials can be found.

2.10 High Performance Computers

The algorithms and programs described so far may be applied to Molecular Dynamics simulations of medium size systems consisting of up to a few thousand particles using personal computers or workstations. The limited computational power of such computers does not, however, allow for simulations of larger systems or of long-term behavior of granular systems. For these cases it is advisable to apply more powerful computers.

Presently there are essentially two types of high-performance computers: vector computers and parallel computers, where the recent development tends to the latter type. In the following two sections, their principles and their application to Molecular Dynamics simulations of granular matter are described.

2.10.1 Vectorization

Principle of Vector Computers

The idea of vectorization and pipelining is introduced here only very briefly. Much more detailed descriptions of modern vector machines can be found in the literature, e.g., [48, 127, 279].

Vector computers as used since the 1970s are conceptually simple parallel computers that can be classified as SIMD machines (single instruction, multiple data). They are very efficient for processing vectorial data, i.e., data-parallel operations. The C++ notation for the element-wise multiplication of two N-dimensional vectors, V_1 and V_2, reads

```
for(int i=0; i<N; i++) { V[i] = V1[i] * V2[i]; }
```

A serial computer would perform the following sequence of operations:

1. Load $V_1[1]$ from the RAM into CPU register I
2. Load $V_2[1]$ from the RAM into CPU register II
3. Compute $V_1[1] * V_2[1]$, the result is contained in CPU register III
4. Store the result $V[1]$ into the RAM
5. Load $V_1[2]$ from the RAM into CPU register I
6. Load $V_2[2]$ from the RAM into CPU register II
7. Compute $V_1[2] * V_2[2]$, the result is contained in CPU register III
8. Store the result $V[2]$ into the RAM
9. Etc.

Aside from these computations, the addresses, the factors, and the result have to be determined by the computer. Any of the multiplications and the according memory access operations require M processor cycles. Hence, the described program requires approximately a time period of

$$T_1 = M N T , \tag{2.141}$$

where T is the cycle time of the processor.

The underlying idea of vector computers is *pipelining*: Assume an operation (e.g., a floating point multiplication) may be split into P independent sub-operations. The first pair of operands is loaded and the first sub-operation of the multiplication is performed. The intermediate result is transferred to another processor unit, while the first unit proceeds with the first sub-operation of the multiplication of the second pair of operands. The *depth* P of the pipeline is the number of computational units which have to be passed to obtain the result from a pair of operands, i.e., after P operations the first pair of operands has been multiplied. There exist vector computers of pipeline depth as large as $P = 1024$. If the above program was processed by a vector computer with $P = 3$, the computer would perform:

1. Load $V_1[1]$ into CPU register I

2. Load $V_2[1]$ into CPU register II
3. - Perform the first step of the computation $V_1[1] * V_2[1]$
   - Load $V_1[2]$ into CPU register I
   - Load $V_2[2]$ into CPU register II
4. ◇ Perform the first step of the computation $V_1[2] * V_2[2]$
   ◇ Perform the second step of the computation $V_1[1] * V_2[1]$
   - Load $V_1[3]$ into CPU register I
   - Load $V_2[3]$ into CPU register II
5. ◇ Perform the first step of the computation $V_1[3] * V_2[3]$
   ◇ Perform the second step of the computation $V_1[2] * V_2[2]$
   ◇ Perform the third step of the computation $V_1[1] * V_2[1]$
   - Load $V_1[4]$ into CPU register I
   - Load $V_2[4]$ into CPU register II
6. - Store the first result $V[1]$
   ◇ Perform the first step of the computation $V_1[4] * V_2[4]$
   ◇ Perform the second step of the computation $V_1[3] * V_2[3]$
   ◇ Perform the third step of the computation $V_1[2] * V_2[2]$
   - Load $V_1[5]$ into CPU register I
   - Load $V_2[5]$ into CPU register II
7. Etc.

The symbol ◇ labels processes which are performed by parallel operating units.

Since P pairs of operands are processed in parallel in different units, in the limit of infinite vector length ($N \to \infty$), a vector computer is P times as fast as an equivalent serial computer, i.e.,

$$T_P \to T_1/P .\tag{2.142}$$

As soon as the pipeline is filled, at each processor time clock T one result is computed. Therefore,

$$T_P = [P + (N - 1)] \, T .\tag{2.143}$$

For $N > P$ the speed-up factor of a vector computer of pipeline depth P is

$$S_P(N) = \frac{T_1}{T_P} = \frac{P \, N}{P + (N - 1)} = \frac{P}{\frac{P}{N} + \frac{N-1}{N}} ,\tag{2.144}$$

i.e., for very long vectors,

$$\lim_{N \to \infty} S_P(N) = P .\tag{2.145}$$

Each program consists of vectorizable and non-vectorizable operations. If their ratio is given by f ($0 \le f \le 1$), the total speedup factor reads

$$S = \frac{1}{(1 - f) + f/P}.\tag{2.146}$$

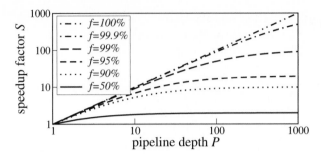

Fig. 2.41. According to Amdahl's law (2.146) it is only worthwhile to apply a vector computer if a major part of the program can be vectorized

Equation (2.146) is called Amdahl's law [6] (see Fig. 2.41).

Assume $f = 90\%$ of a program can be vectorized. Then a vector computer of pipeline depth $P = 64$ can develop only $S = 8.8$ times the performance of an equivalent serial computer, i.e., 7.3 times below the theoretical maximum of $S = 64$. Although there are many additional tricks to increase the performance of vector computers, even for rather short vector length (e.g., [48]), this example shows that the ratio of vectorizable vs. non-vectorizable code sensitively determines the efficiency of vector computers. There are various techniques for the vectorization of code (e.g., [145]) as well as vectorizing compilers which assist the programmer, although in many cases they may not deliver satisfying results.

Vectorized Force Summation Algorithm

The algorithm for Molecular Dynamics of granular particles presented here is completely vectorized [50]. Except for the initialization, the algorithm is of complexity $\mathcal{O}(N)$. For its application it is required that

1. There is only short-range interaction of the particles
2. The particles deform only slightly
3. The grain-size dispersion is small
4. The particle-number density is large enough

The algorithm uses the lattice method for the computation of the forces that is described in Sect. 2.4.4. A rectangular lattice of $L_{\max}$ quadratic sites covers the simulation area. (For periodic boundary condition the lattice covers one period.) The size d of the sites has to be chosen such that no site can contain the centers of more than one particle, hence for undeformable particles $d = \sqrt{2}R_{\min}$ where $R_{\min}$ is the radius of the smallest particle. For deformable particles d has to be chosen so that it is smaller respectively (Fig. 2.42).

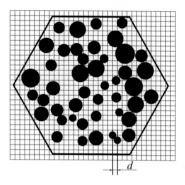

Fig. 2.42. The size of the lattice sites is determined by the minimal distance of the centers of adjacent particles and by the request that any cell can be occupied by at most one particle

The algorithm needs a number of vectors:[15] The L-vector $\vec{\mathbf{I}}$ describes the occupation of the lattice sites, i.e.,

$$
\mathbf{I}_i \equiv \begin{cases} k & \text{if site } i \text{ contains the center of particle } k \\ -1 & \text{else}, \end{cases} \tag{2.147}
$$

with $i = 0, \ldots, L_{\max} - 1$. In other words it maps the index i of the lattice site to the particle index k. For simpler notation, the L-vector $\vec{\mathbf{T}}$ is defined:

$$
\mathbf{T}_i \equiv \begin{cases} 1 & \text{if site } i \text{ contains a particle}, \\ 0 & \text{otherwise}. \end{cases} \tag{2.148}
$$

By means of $\vec{\mathbf{I}}$, a new set of L-vectors $\vec{\mathbf{x}}, \vec{\mathbf{y}}, \vec{\mathbf{F}}_x, \vec{\mathbf{F}}_y, \vec{\mathbf{R}}$, can be defined that map the scalar particle properties, such as x-position, force in x-direction etc., to the lattice site where the particle is located, e.g.,

$$
\vec{\mathbf{x}}_i = \begin{cases} x_k & \text{if site } i \text{ contains the center of particle } k \\ (\textit{undefined}) & \text{otherwise}. \end{cases} \tag{2.149}
$$

According to the precondition that each site is occupied by at most one particle, this mapping is unique.

For the computation of the interaction forces, for each particle only such interaction partners are considered, which are located in neighboring boxes up

---

[15] In the remainder of the section there are no three-dimensional (physical) vectors, such as position, etc. To distinguish vectors that refer to the lattice sites from vectors referring to particles, the former are typeset in bold face, e.g., $\vec{\mathbf{z}} \equiv (\mathbf{z}_0, \mathbf{z}_1, \ldots, \mathbf{z}_{L_{\max}-1})^{\mathrm{T}}$ to discriminate from $\vec{z} \equiv (z_0, z_1, \ldots, z_{N-1})^{\mathrm{T}}$. Furthermore, since vectors referring to particles are of dimension N (number of particles) we call them N-vectors while the vectors referring to lattice sites are called L-vectors (of dimension $L_{\max}$).

15	14	13	12	11
16	4	3	2	10
17	5	0	1	9
18	6	7	8	24
19	20	21	22	23

Fig. 2.43. The neighboring sites of site 0 are enumerated by a mask. For the computation of the interaction forces, this mask is successively moved over all lattice sites

to a certain distance according to the precondition that there are exclusively short-range interaction forces.

For each particle which is located at lattice site (i, j), only interactions with particles that are located at sites $(k, l), (k \in [i-2, i+2], l \in [j-2, j+2], (k, l) \neq (i, j))$ are considered. The neighborhood of a lattice site (i, j) is described by a mask as sketched in Fig. 2.43.

For the computation of the forces the center of this mask (position 0) is moved successively over all lattice sites (i, j). If this site is occupied, we compute the forces between this particle and those particles located at the sites covered by the mask.

The values i ($i \in [1, 24]$) that enumerate the positions of the mask can be used to refer to each site in the neighborhood of the cell in the center of the mask (position 0). This way the distances between adjacent particles as well as all other values which are necessary for the computation of the forces acting on the particle at position (i, j) can be expressed by 24 L-vectors each. These L-vectors are indexed by the number of the lattice site. For the distances of the particles these L-vectors read

$$\mathbf{D}_i = \sqrt{\Delta \mathbf{x}_i \Delta \mathbf{x}_i + \Delta \mathbf{y}_i \Delta \mathbf{y}_i} \qquad i \in [1, 24] \tag{2.150}$$

with

$$\Delta \mathbf{x}_i \equiv \mathbf{x} - \mathbf{x}\{i\} \qquad \text{and} \qquad \Delta \mathbf{y}_i \equiv \mathbf{y} - \mathbf{y}\{i\} . \tag{2.151}$$

where the L-vector $\mathbf{x}\{i\}$ contains the same elements as $\mathbf{x}$ but shifted by i. The square root in (2.150) is meant element-wise with respect to the L-vectors $\Delta \mathbf{x}$ and $\Delta \mathbf{y}$.

We define now a set of L-vectors $\mathbf{C}_i$, $i \in [1, 24]$, which contains the information on whether particles in adjacent lattice sites touch each other, i.e., whether they exert forces on each other:

$$\mathbf{C}_i = \begin{cases} 1 & \text{in case the particles touch} \\ 0 & \text{otherwise ,} \end{cases} \qquad i \in [1, 24] . \tag{2.152}$$

The normal and shear forces can be computed by means of these variables:

$$\mathbf{F}_i^n = \mathbf{T}\ \mathbf{T}\{i\}\ \mathbf{C}_i\ \mathcal{F}^n(0, i)\mathbf{F}_i^t\ = \mathbf{T}\ \mathbf{T}\{i\}\ \mathbf{C}_i\ \mathcal{F}^t(0, i)\ , \tag{2.153}$$

where $\mathcal{F}^n(0, i)$ and $\mathcal{F}^t(0, i)$ are the forces in the normal and tangential direction, respectively, between particles that are possibly located at position 0 of the mask and at position i $(i \in [1, 24])$.

The total force which acts upon each particle is then represented by

$$\overline{\mathbf{F}}_x = \sum_{i=1}^{24} \frac{1}{\mathbf{D}_i}\ \left(\varDelta\mathbf{x}_i\mathbf{F}_i^n - \varDelta\mathbf{y}_i\mathbf{F}_i^t\right)$$

$$\overline{\mathbf{F}}_y = \sum_{i=1}^{24} \frac{1}{\mathbf{D}_i}\ \left(\varDelta\mathbf{y}_i\mathbf{F}_i^n + \varDelta\mathbf{x}_i\mathbf{F}_i^t\right) \tag{2.154}$$

$$\overline{\mathbf{M}} = \sum_{i=1}^{24} \mathbf{R}\ \mathbf{F}_i^t\ ,$$

where again all operations are meant element-wise.

The simulation algorithm is then described by the following steps:

1. *Initialization:* The radii, positions and velocities of the particles are initialized.
2. *Predictor:* The positions, velocities, and higher order time derivatives are determined by means of the predictor–corrector algorithm (see Sect. 2.1.5).
3. *Computation of the lattice vectors:* All particle properties that are necessary for the computation of the interaction forces are mapped to the corresponding lattice site where the particle resides.
4. *Computation of the forces:* Using the lattice vectors the forces acting on each particle are computed according to (2.150)–(2.154).
5. *Corrector:* The positions, velocities and higher-order time derivatives which have been predicted in step 2 are corrected according to the predictor–corrector algorithm (see Sect. 2.1.5).
6. *Time step:* The system time is incremented, $t := t + \varDelta t$, and the algorithm proceeds with the next iteration at step 2.

All operations of the Molecular Dynamics algorithm have been formulated in terms of vector operations, hence, the algorithm is completely vectorized, except for the initialization.

For the described algorithms, the neighborhood of the site (i, j) is defined as all sites (k, l), where $k \in [i - 2, i + 2], l \in [j - 2, j + 2]$, and $(k, l) \neq (i, j)$. The definition of neighborhood restricts the allowed dispersion of the particle radii. To exclude interaction of particles which are not in the neighboring sites requires $R_{max} < d$, with d being the size of a lattice cell. Thus from $\sqrt{2}R_{min} < d < R_{max}$ the admitted dispersion is $R_{max}/R_{min} < \sqrt{2}$ if particle deformation is not taken into account. If the particles do not deform by more than 10% of their radii $(R_i + R_j - |\vec{r}_i - \vec{r}_j| < 0.1 \cdot \min(R_i, R_j))$, which holds

Table 2.1. Estimate of the computer time consumption for the vectorized algorithm. L_{max} is the number of lattice sites

	N_{op}	n	Time steps
Predictor	59	N	$59(N + P)$
Index	12	N	$12(N + P)$
Forces	1656	L_{max}	$1656(L_{max} + P)$
Corrector	54	N	$54(N + P)$

true for realistic granular materials in general, the presented algorithm is correct provided the ratio of the radii of the largest and smallest particle does not exceed $0.9\sqrt{2} \approx 1.27$. The admissible dispersion can be increased by defining a larger neighborhood of lattice sites.

Efficiency of the Algorithm

For the assessment of the efficiency, we determine the number of operations N_{op} which have to be performed in any of the steps in the algorithm from the previous section. depending on the corresponding vector sizes n ($n = N$ for N-vectors and $n = L_{max}$ for L-vectors). For this estimate *chaining* and *multi-pipelining*[16] is disregarded. A pipeline depth P is assumed and any elementary vector operation needs $n + P$ time steps.

Table 2.1 estimates the processing time for each step of the algorithm.

Each iteration step lasts, therefore, $t = 125\ (N + P) + 1656\ (L_{max} + P)$ computer time steps. The performance depends sensitively on the number of lattice sites but hardly on the number of particles. Figure 2.44 displays the computer time consumption per particle as a function of the number of particles for varying particle number density, $\rho = N/L_{max}$.

Obviously, the particle number density determines the efficiency of the algorithm. This result motivates the precondition of large particle number density in the beginning of this section.

The completely vectorized algorithm that we have presented here for Molecular Dynamics of short-range interacting particles is of complexity $\mathcal{O}(N)$. It performs efficiently for large particle number density and small particle size dispersion. The vectorized algorithm has the same algorithmic complexity as the serial algorithm, however, the prefactor is far smaller due to pipelining. The basic idea of the lattice algorithm may be generalized to larger particle size dispersion; however, the interaction mask has to be enlarged appropriately, e.g., from 5×5 to 7×7.

---

[16] Chaining: Under certain preconditions, modern vector computers are able to start a vector operation even before the preceding vector operation is completed, i.e., the pipeline still contains data from the preceding operation. Multi-pipelining: Some computers are able to perform complex vector operations by running several pipelines in parallel.

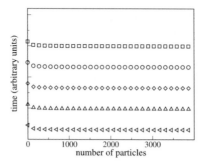

Fig. 2.44. Computer time consumption per particle over the number of particles for varying density ($\square : \rho = 1/12$, $\circ : \rho = 1/10$, $\diamond : \rho = 1/8$, $\triangle : \rho = 1/6$, $\triangleleft : \rho = 1/4$)

The algorithm can be generalized to non-spherical particles (see, e.g., Sect. 2.6) since all particle interaction forces can be expressed in terms of vector operations. The algorithm has been applied to Molecular Dynamics of granular pipe flow [50]. Similar algorithms have been proposed in [77, 78].

2.10.2 Parallelization

Idea of Parallelization

The basic idea of parallel computers is neither new nor particularly surprising: as early as 1842 Luigi Menabrea published an article on the *"Analytical Engine"* which had been developed by Charles Babbage. He proposed computing numerical tables by means of parallel working mechanical calculators [187].

The idea of parallel computers is to break down a mathematical problem in parts which may be processed in parallel, i.e., at the same time and independently of each other. The method for decomposing the problem is decisive for the efficiency of the parallelized program. One of the main difficulties is to keep the load between the parallel working processors balanced (idle processors waste time) and to keep the necessary communication between the processors small.

Parallel computers are frequently classified with respect to instruction and data flow. According to this scheme (which can be found in any book on modern computer architecture) the classical serial Von-Neumann computer belongs to the class *single instruction single data (SISD)*. In this section we deal with algorithms for computers of the class *multiple instruction multiple data (MIMD)*. Such computers consist of a certain number of processors which are equipped with local memory and which are able to exchange information with each other. All processors execute the same program, however, controlled by an individual processor ID; in general, the actual execution of the program is different for each processor.

Starting from a serial program we present two strategies for the parallelization of Molecular Dynamics of granular materials. The serial algorithm is sketched in Fig. 2.45. It is mainly the decomposition of the data which distinguishes strategies of parallelization. All parallel algorithms described here are based on *message passing*. The algorithms have been implemented using the library MPI. A good introduction to MPI can be found in [204, 306].

Data Replication Model

The idea of the data replication model is to store all information on all particles in the individual memory of each processor. Then each processor may compute the trajectories of an arbitrarily chosen subset of particles.

Considering Fig. 2.45, if each processor is assigned an equal number of particles, we can obtain optimal load balancing for the predictor and corrector as well as for the computation of the external forces and the particle motion due to boundary conditions (not drawn in the sketch). Since these computations are independent for each particle, the distribution of the particles, i.e., the set of their indices, is arbitrary.

This principle does not apply to the computation of the particle–particle forces. Here each processor is not responsible for a certain number of particles, but instead the entries of the interaction lists (e.g., Verlet lists) are distributed such that each processor is responsible for the same number of particle contacts. Approximate load balancing is given if the number of particle contacts is approximately the same for all parts of the interaction lists.

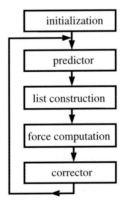

Fig. 2.45. Serial program for Molecular Dynamics of granular materials using Verlet lists. For simplification data output and motion of particles due to boundary conditions have been dropped from the sketch. The list construction contains also house-keeping steps like checking if the list is still up to date. The actual list construction is not necessarily performed in each time step

The computation of the interaction lists themselves can be parallelized as well: each processor generates the lists for the same number of particles in arbitrary distribution.

Hence, except for the initialization and the data output, all parts of the Molecular Dynamics program may be parallelized. As sketched in Fig. 2.46 the data replication model requires substantial communication amongst the processors. Communication is needed three times per step:

1. After predicting the particle positions, velocities, and higher-order time derivatives, for spherical particles the coordinates $\vec{r}_i$, the velocities $\vec{v}_i$, and the angular velocities of all particles are broadcast to all processors. For more complicated particles this also concerns additional data (all particle characteristics which are subject to Newton's equation of motion).
2. Whenever new interaction lists are computed, they must be broadcast to all processors.
3. The particle interaction forces have to be summed up to obtain the forces that act on each individual particle. Each processor must be sent the forces that act on the particles for which it is responsible.

The parallelized program is sketched in Fig. 2.46.

The serious communication requirements restricts the application of the algorithm to computers with highly optimized communication channels between the processors. The algorithm was implemented on a Cray T3E resulting in a speed-up of 70%, i.e., using 16 processors the program runs about 11 times faster than on a single processor.

The communication effort increases quickly with increasing number of processors, hence the program becomes inefficient. Thus, the data replication model is unsuitable for computers with many processors using less efficient communication channels, such as most clusters of desktop PCs.

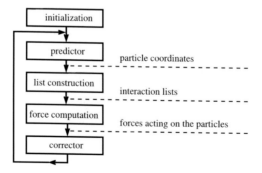

Fig. 2.46. Parallelization using the data replication model. The dashed horizontal lines indicate exchange of data amongst the processors

Data Decomposition Model

The data decomposition model requires significantly less communication; however, its implementation, and load balancing in particular, is much more complicated.

For this parallelization method the spatial simulation area is subdivided into sub-areas (boxes) which are assigned to the processors of the parallel machine. The trajectories of the particles in a certain box are then computed by the corresponding processor. Each processor keeps in its memory the radii, positions, velocities, etc. of all particles which belong to its box. As soon as a particle leaves its present box, the data of that particle are transferred to the processor which is responsible for the box into which the particle moves. For the computation of the forces which act on particles that are located close to the boundary of a box, data of close particles in the adjacent boxes are also needed. Therefore, on top of the data of all particles in the box, the corresponding processor also needs the data of particles in a certain surrounding of the box whose size depends on the particle size (see Fig. 2.47). Thus, the boxes may be considered to overlap each other. At each time step the information on the particles which are located in the overlapping region are exchanged between the processors concerned. Each processor has, therefore, the necessary information to compute the trajectories of the particles that are located in its box independent of the other processors.

The data of the particles in the overlapping areas are exchanged after the predictor step. This way for each box-changing particle, the data are transferred to the corresponding processor and the administration of this particle is taken over by this processor. Each particle entering the box of a processor is inserted into the Verlet lists of the particles located there. As soon as the communication phase is accomplished the next iteration step may be performed. The algorithm is sketched in Fig. 2.48.

The efficiency of the parallel algorithm depends sensitively on the spatial decomposition of the simulation area. If the boxes are chosen to be too small the ratio of the box area and the overlapping area decreases and, therefore,

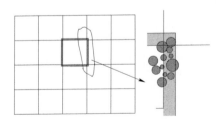

Fig. 2.47. Decomposition of the simulation area into boxes. Each box is assigned to a processor which computes the particles in the box. To this end the processor needs the data of all particles inside its box and in a certain surrounding of the box (in the gray shadowed area)

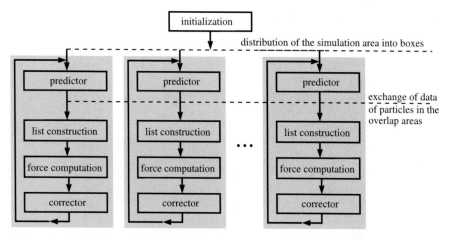

Fig. 2.48. Parallelization using the data decomposition model. Each processor computes the trajectories of particles in its box independently. The necessary data are exchanged after the predictor step

an increasing number of operations is performed twice. At the same time the number of particles leaving the box increases, i.e., the communication load increases further which diminishes the performance as well.

To keep the communication load small it is useful to check for regular flows of particles. In the example of granular pipe flow, the collective flow velocity may exceed the relative velocities of the particles largely. In such cases it is useful to consider a set of co-moving boxes.

For the case that the simulation area changes in the coarse of time (e.g., in the case of hopper flow), the sizes and shapes of the boxes have to be adapted dynamically to assure load balancing between the processors, otherwise a few processors may compute the majority of the trajectories while other processors run idle. Figure 2.49 shows the automatic load balancing for the case of a sedimentation process. An animation can be found on the book's web site.

The redistribution of the simulation area which is necessary to achieve optimal load balancing for systems of spatially-inhomogeneous distributed particles is always accompanied by serious algorithmic complications. There is a simple trick to avoid this redistribution of the simulation area, called data scattering: The simulation area is subdivided into many small boxes such that each processor is assigned n boxes instead of one. These n boxes should be distributed such that underpopulation of certain regions affect all processors equally. Figure 2.50 shows the subdivision of the simulation area for hopper flow. Assume there are four processors. The space is subdivided into 4×7 equal boxes. Initially the boxes are equally populated by particles. In the coarse of time the upper boxes are emptied. This, however, does not affect the equal load distribution of the processors. The disadvantage of this method is the increase of the total overlapping area which implies extra com-

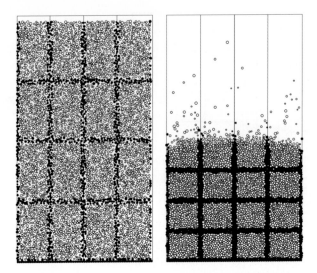

Fig. 2.49. Dynamic load balancing in the simulation of a sedimentation. Initially, the particles are homogeneously distributed in the simulation area (*left*), thus, the area is subdivided into 16 compartments of equal size. At the end of the sedimentation process (*right*), almost all particles reside in the lower half of the container. Therefore, the upper four compartments are significantly larger than the lower ones. The particles in the overlapping areas are drawn filled

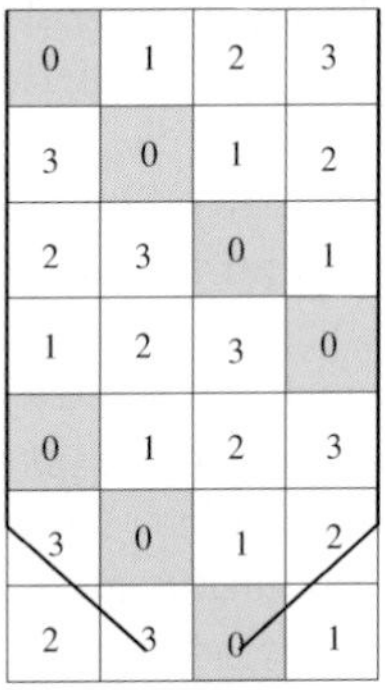

0	1	2	3
3	0	1	2
2	3	0	1
1	2	3	0
0	1	2	3
3	0	1	2
2	3	0	1

Fig. 2.50. Application of the data scattering method to hopper flow. There are 28 boxes distributed among four processors. The boxes belonging to processor 0 are gray-shadowed. Optimal load balancing is given independent of the filling height

munication effort, leading to decreasing efficiency. Hence, the described data scattering method is limited to relatively large systems where the total number of particles inside each box is large compared to the number of particles in the overlapping area.

2.11 Vector Class

The mathematical description of Molecular Dynamics algorithms can be almost completely written in terms of vector operations. Therefore, it is convenient to use vector operations also in the program code. To this end we introduce a class which allows for applying the basic operations +, -, +=, and -= to vectors as well and the operations * and *= for the element-wise multiplication of a vector by a scalar and functions for the computation of the inner and outer product and the norm of a vector. Moreover, the class defines the vector $null \equiv (0, 0, 0)$. By means of this class, the algorithms can be written in a compact and concise way. Examples for the application of the vector class are given in Sect. 2.1.7.

The vector class presented here is specific for two-dimensional systems. Its generalization to three dimensions is straightforward. Although the vector notation of the program code is convenient, an element-wise notation is preferable for performance optimized simulations.

The class Vector contains three data elements: the center of mass coordinates x and y, and one angle φ. The state of a particle, i.e., its position and its velocity, can be described by two variables of type Vector.

The components of the vector are declared as *private* data elements _x, _y, and _phi of type double, i.e., they are accessible only for methods and friends of the class. To read and modify their values, access functions are provided. Consider as an example the x-component (data element _x): The access function x() occurs in two versions:

1. double & x()
2. double x() const

The first version provides reading and writing access to the x-component. It returns a *reference* to the data element. A reference to a variable is in fact its address. Contrary to pointers, references are used in the same way as the variables themselves (i.e., no de-referencing operator needed). Any change of the reference changes the value of the referenced variable. The x-component of a Vector v is read, e.g., by xx=v.x() and written, e.g., by v.x()=xx.

Since the access function double & x() may modify the value of a data element, it cannot be applied to constant vectors, defined, e.g., by const Vector v. For this purpose, there is a second version of the access function, "double x() const" which cannot change the value of the vector element. The decision which variant of the access function is used, is made by the C++ compiler depending on the type of the data element concerned.

```
                              Vector.h
1  #ifndef _Vector_h
2  #define _Vector_h
3
4  #include <math.h>
5  #include <iostream>
6
7  using namespace std;
8
```

```
9  class Vector {
10   friend istream & operator >> (istream & is, Vector & v) {
11     is >> v._x >> v._y >> v._phi;
12     return is;
13   }
14   friend ostream & operator << (ostream & os, const Vector & v) {
15     os << v._x << " " << v._y << " " << v._phi;
16     return os;
17   }
18   friend Vector operator + (const Vector & v1, const Vector & v2) {
19     Vector res(v1);
20     res+=v2;
21     return res;
22   }
23   friend Vector operator - (const Vector & v1, const Vector & v2) {
24     Vector res(v1);
25     res-=v2;
26     return res;
27   }
28   friend Vector operator * (double c, const Vector & p) {
29     Vector res=p;
30     res*=c;
31     return res;
32   }
33   friend Vector operator * (const Vector & p, double c) {
34     return c*p;
35   }
36   friend double norm2d(const Vector & v) {
37     return sqrt(v._x*v._x+v._y*v._y);
38   }
39   friend double scalprod2d(const Vector & v1, const Vector & v2) {
40     return v1._x*v2._x + v1._y*v2._y;
41   }
42   friend double vecprod2d(const Vector & v1, const Vector & v2) {
43     return v1._x*v2._y-v1._y*v2._x;
44   }
45
46  public:
47   explicit Vector(double x=0,double y=0,double phi=0): _x(x), _y(y), _phi(phi){};
48
49   double & x() {return _x;}
50   double x() const {return _x;}
51   double & y() {return _y;}
52   double y() const {return _y;}
53   double & phi() {return _phi;}
54   double phi() const {return _phi;}
55
56   const Vector & operator += (const Vector & p){
57     _x+=p._x; _y+=p._y; _phi+=p._phi;
58     return *this;
59   }
60   const Vector & operator -= (const Vector & p){
61     _x-=p._x; _y-=p._y; _phi-=p._phi;
62     return *this;
63   }
64   const Vector & operator *= (double c){
65     _x*=c; _y*=c; _phi*=c;
66     return *this;
67   }
68  private:
69   double _x,_y,_phi;
70  };
71
72  const Vector null(0,0,0);
73  #endif
```

To apply vector operations in the same intuitive way as number operations, some operators are overloaded. Let us briefly describe the idea: whenever an operator is used in a C++ program it is replaced by the compiler with a corresponding function. When using, e.g., the operator "+" in "a+b", it is replaced by the function call "operator +(a,b)". In the same way all operators such as -, +=, -=, <<, >>, ==, etc. are replaced by the corresponding function calls. Overloading an operator means adding another operator of the same name, but with different (not yet implemented) types of its arguments. For each data type to which a certain operator is applied, a version of this operator with the operands of the type in question has to be implemented. Which versions of the operator is meant, is decided by the compiler according to the type of the arguments. For example, in the program

```
                        ────── v_example.cc ──────
1 │ double x,y,z;
2 │ Vector u,v,w;
3 │
4 │ z = x + y;
5 │ w = u + v;
```

the function operator + (double, double) is called in line 4, but in line 5 operator + (const Vector &, const Vector &) is used.

Hence, to be able to use the operator + for variables of type Vector, the function operator + (const Vector &, const Vector &) is defined, which has been done in the definition of the class Vector in lines 18–22. The new function specifies the action of the operator + for arguments of type Vector. In our case, + means element-wise addition.

Program Index

3

Event-Driven Molecular Dynamics

3.1 Idea and Motivation

The general idea of Molecular Dynamics, as it was discussed in detail in
Chap. 2, is to solve numerically Newton's equation of motion simultaneously
for all particles i:

$$\ddot{\vec{r}}_i = \frac{1}{m_i} \vec{F}_i \left(\vec{r}_1, \vec{v}_1, \dots, \vec{r}_N, \vec{v}_N \right), \tag{3.1}$$

This type of Molecular Dynamics is called *force-based*. This general scheme
has been successfully applied to the simulation of granular systems in many
situations; however, there are cases when force-based Molecular Dynamics is
less appropriate:

- In systems where the typical duration of a collision is much shorter than
  the mean time between successive collisions of a particle, particles are very
  rarely in contact with more than one other particle. Hence, most of the time
  each of the particles propagates along a ballistic trajectory, interrupted by
  collisions with other particles of very short duration. Therefore, the pairwise collisions of particles may be considered as instantaneous events and
  each of these events may be treated separately. Examples of such systems
  are granular gases [42], i.e., very dilute granular systems such as cosmic
  dust clouds (see Sect. 3.10). Although force-based Molecular Dynamics is
  applicable, in principle, to such systems, the computation is very inefficient.
  If Newton's equation of motion is integrated for an isolated pair of colliding particles, the post-collision velocities $\vec{v}'_{1/2}$ are obtained as functions of
  the pre-collision velocities $\vec{v}_{1/2}$. These functions can be used to set up an
  event-driven Molecular Dynamics simulation.
- In some cases the detailed interaction force of granular particles is not
  known as a function of the relative position $|\vec{r}_1 - \vec{r}_2|$, velocity $|\vec{v}_1 - \vec{v}_2|$, and
  orientation because to derive this function, a microscopic description of the
  particle material is required, i.e., a continuum mechanics description, which

may be *very* complicated, such as those for nonlinear materials or complicated particle shapes. Nevertheless, it is frequently possible to investigate the pairwise particle collision experimentally and, hence, to determine approximations for the post-collision velocities as functions of the pre-collision velocities.

The main assumption for applying event-driven Molecular Dynamics is that at any time instant in the entire system, there occurs at most one collision of infinitesimal duration. This collision alters the velocities of the involved particles according to a collision law which is characterized by coefficients of restitution (see Sect. 3.2). With the idealization of infinitesimal duration, these coefficients describe the mechanics of pairwise particle collisions exhaustively. Hence, the time-consuming simultaneous computation of the trajectories of all particles according to (3.1) is unnecessary. During the time intervals between collisions, the particles move along known ballistic trajectories. Therefore, the positions of the particles at the time of the next collision can be computed in one step. Obviously, this algorithm is much more efficient than force-based algorithms since the numerical integration of the equations of motion is avoided. Instead the dynamics of the system is determined by a sequence of discrete events.

The enormous increase of the simulation speed is the main motivation for replacing force-based algorithms by event-driven algorithms. Therefore it has been applied not only to dilute granular gases but also to more dense granular systems. The application of event-driven simulations is, however, not always justified. A heap of steel spheres, simulated by event-driven Molecular Dynamics could not conduct electrical current since at any time at most two of the spheres can be in touch. Obviously, this is not an adequate description. Under certain conditions, however, such algorithms may also be applied to dense systems (see Fig. 3.1); for each case it has to be checked whether the event-driven description of the system dynamics is appropriate. In cases where the idealization of isolated instantaneous collisions is not justified, force-based Molecular Dynamics is necessary, as described in Chap. 2, or in the section on Rigid-Body Dynamics (see Chap. 5).

In the present chapter event-driven Molecular Dynamics is applied to systems of identical spherical particles. The generalization to a polydisperse medium is straightforward. Applications of the algorithm to more complicatedly shaped particles can be found, e.g., in [10, 191].

First we introduce the coefficients of restitution which describe pairwise particle collisions and represent the central characteristics of event-driven algorithms. Section 3.6 describes a simple algorithm which may be used for the simulation of small systems. Finally, in Sect. 3.7 a more complex event-driven algorithm is discussed, which is suited for the simulation of large systems of millions of particles.

Fig. 3.1. For the example of a high-energy ball mill filled with steel spheres, the mean free flight time is certainly much larger than the duration of the collisions. Hence, event-driven Molecular Dynamics is appropriate. This system has been simulated by Reichardt et al. [232] to partially replace experimental investigations. See the book's web site for animated sequences. The picture is taken from [232]

3.2 Collision of Particles

The relative velocity of colliding particles i and j at the point of contact, $\vec{g}_{ij}$, is determined by the translational and rotational particle velocities:

$$\vec{g}_{ij} = \left(\vec{v}_i - \vec{\omega}_i \times R\vec{e}_{ij}^{\,n}\right) - \left(\vec{v}_j + \vec{\omega}_j \times R\vec{e}_{ij}^{\,n}\right) = \vec{v}_{ij} - R\left(\vec{\omega}_i + \vec{\omega}_j\right) \times \vec{e}_{ij}^{\,n} \ , \ (3.2)$$

with $\vec{v}_{ij} \equiv \vec{v}_i - \vec{v}_j$, R as the particle radius, and $\vec{e}_{ij}^{\,n}$ as the unit vector pointing from particle j to particle i. The normal and tangential collision velocities are given by the projections

$$\vec{g}_{ij}^{\,n} = \left(\vec{g}_{ij} \cdot \vec{e}_{ij}^{\,n}\right)\vec{e}_{ij}^{\,n} \qquad \text{and} \qquad \vec{g}_{ij}^{\,t} = -\vec{e}_{ij}^{\,n} \times \left(\vec{e}_{ij}^{\,n} \times \vec{g}_{ij}\right) \ . \qquad (3.3)$$

The coefficients of restitution in normal and tangential direction, ε^n and ε^t, are defined by

$$\begin{aligned} \left(\vec{g}_{ij}^{\,n}\right)' &= -\varepsilon^n \vec{g}_{ij}^{\,n} \ , & \text{with} & \quad 0 \leq \varepsilon^n \leq 1 \\ \left(\vec{g}_{ij}^{\,t}\right)' &= \varepsilon^t \vec{g}_{ij}^{\,t} \ , & \text{with} & \quad -1 \leq \varepsilon^t \leq 1 \ , \end{aligned} \qquad (3.4)$$

where (as everywhere else in this section) the primed symbols denote the post-collision values. (See the second item in Sect. 3.5.3 for the limits of ε^t.) By means of the coefficients of restitution, the post-collision velocities can be written as functions of the pre-collision velocities. We call this set of functions *collision rule*. For its derivation consider the point of contact as the center of rotation. Since the contact area is assumed to be infinitesimally small, none of the forces acting on the particles causes a torque. Consequently, there is no angular momentum transfer between the particles. Thus, the angular momentum relative to the contact point as the center of rotation is conserved for each particle separately. Therefore, there are three vectorial conservation laws, one for the linear momentum and two for the angular momomenta:

$$\vec{v}_i' + \vec{v}_j' = \vec{v}_i + \vec{v}_j \tag{3.5}$$

$$mR\vec{e}_{ij}^{\,\mathrm{n}} \times \vec{v}_i' + J\vec{\omega}_i' = mR\vec{e}_{ij}^{\,\mathrm{n}} \times \vec{v}_i + J\vec{\omega}_i \tag{3.6}$$

$$mR\vec{e}_{ij}^{\,\mathrm{n}} \times \vec{v}_j' - J\vec{\omega}_j' = mR\vec{e}_{ij}^{\,\mathrm{n}} \times \vec{v}_j - J\vec{\omega}_j \;. \tag{3.7}$$

Solving (3.6) and (3.7) for $\vec{\omega}_i'$ and $\vec{\omega}_j'$, we obtain

$$\vec{\omega}_i' = \vec{\omega}_i - \frac{mR}{J}\vec{e}_{ij}^{\,\mathrm{n}} \times (\vec{v}_i' - \vec{v}_i) \;, \quad \vec{\omega}_j' = \vec{\omega}_j + \frac{mR}{J}\vec{e}_{ij}^{\,\mathrm{n}} \times (\vec{v}_j' - \vec{v}_j) \;, \tag{3.8}$$

and using the definition of $\vec{g}_{ij}'$ (analogue to (3.2) but all quantities primed) we find

$$\begin{aligned}
\vec{g}_{ij}' &= \vec{v}_{ij}' - R\left(\vec{\omega}_i' + \vec{\omega}_j'\right) \times \vec{e}_{ij}^{\,\mathrm{n}} \\
&= \vec{v}_{ij}' - R\left(\vec{\omega}_i + \vec{\omega}_j\right) \times \vec{e}_{ij}^{\,\mathrm{n}} + \frac{mR^2}{J}\vec{e}_{ij}^{\,\mathrm{n}} \times \left[\vec{e}_{ij}^{\,\mathrm{n}} \times (\vec{v}_{ij} - \vec{v}_{ij}')\right] \;.
\end{aligned} \tag{3.9}$$

Subtracting (3.9) from (3.2) yields

$$\vec{g}_{ij}' - \vec{g}_{ij} = \vec{v}_{ij}' - \vec{v}_{ij} - \frac{1}{\tilde{J}}\vec{e}_{ij}^{\,\mathrm{n}} \times \left[\vec{e}_{ij}^{\,\mathrm{n}} \times (\vec{v}_{ij}' - \vec{v}_{ij})\right] \;, \tag{3.10}$$

with the reduced moment of inertia $\tilde{J} \equiv J/mR^2$. This result implies

$$\vec{e}_{ij}^{\,\mathrm{n}} \cdot \left(\vec{g}_{ij}' - \vec{g}_{ij}\right) = \vec{e}_{ij}^{\,\mathrm{n}} \cdot \left(\vec{v}_{ij}' - \vec{v}_{ij}\right) \tag{3.11}$$

since the second term of the right hand side of (3.10) is perpendicular to $\vec{e}_{ij}^{\,\mathrm{n}}$. The corresponding vector product reads

$$\begin{aligned}
\vec{e}_{ij}^{\,\mathrm{n}} \times \left(\vec{g}_{ij}' - \vec{g}_{ij}\right) &= \vec{e}_{ij}^{\,\mathrm{n}} \times \left[\vec{v}_{ij}' - \vec{v}_{ij} - \frac{1}{\tilde{J}}\left\{\vec{e}_{ij}^{\,\mathrm{n}}\left[\vec{e}_{ij}^{\,\mathrm{n}}\left(\vec{v}_{ij}' - \vec{v}_{ij}\right)\right] - \left(\vec{v}_{ij}' - \vec{v}_{ij}\right)\right\}\right] \\
&= \frac{\tilde{J}+1}{\tilde{J}}\vec{e}_{ij}^{\,\mathrm{n}} \times \left(\vec{v}_{ij}' - \vec{v}_{ij}\right) - \frac{1}{\tilde{J}}\vec{e}_{ij}^{\,\mathrm{n}} \times \left\{\vec{e}_{ij}^{\,\mathrm{n}}\left[\vec{e}_{ij}^{\,\mathrm{n}} \cdot (\vec{v}_{ij}' - \vec{v}_{ij})\right]\right\} \\
&= \frac{\tilde{J}+1}{\tilde{J}}\vec{e}_{ij}^{\,\mathrm{n}} \times \left(\vec{v}_{ij}' - \vec{v}_{ij}\right) \;,
\end{aligned} \tag{3.12}$$

where the general representation of an arbitrary vector $\vec{x}$,

$$\vec{x} = \vec{e}\,(\vec{e} \cdot \vec{x}) - \vec{e} \times (\vec{e} \times \vec{x}) \;, \tag{3.13}$$

with $\vec{e}$ being a unit vector, is used. Since $\vec{e}_{ij}^n \times \vec{e}_{ij}^n = 0$, the second term in the second line of (3.12) does not contribute to the right hand side. By means of (3.11), (3.12), and the definitions (3.3) and (3.4), $\vec{v}_{ij}' - \vec{v}_{ij}$ can be represented in the form (3.13):

$$
\begin{aligned}
\vec{v}_{ij}' - \vec{v}_{ij} &= \vec{e}_{ij}^n \left[\vec{e}_{ij}^n \cdot \left(\vec{g}_{ij}' - \vec{g}_{ij} \right) \right] + \frac{\tilde{J}}{1 + \tilde{J}} \vec{e}_{ij}^n \times \left[\vec{e}_{ij}^n \times \left(\vec{g}_{ij}' - \vec{g}_{ij} \right) \right] \\
&= \left(\vec{g}_{ij}^n \right)' - \vec{g}_{ij}^n + \frac{\tilde{J}}{1 + \tilde{J}} \left[\left(\vec{g}_{ij}^t \right)' - \vec{g}_{ij}^t \right] \\
&= -(1 + \varepsilon^n) \vec{g}_{ij}^n + \frac{\tilde{J}}{1 + \tilde{J}} \left(\varepsilon^t - 1 \right) \vec{g}_{ij}^t .
\end{aligned}
\tag{3.14}
$$

Using the definition $\vec{v}_{ij}' \equiv \vec{v}_i' - \vec{v}_j'$ and conservation of momentum, $\vec{v}_i' + \vec{v}_j' = \vec{v}_i + \vec{v}_j$, the post-collision velocities can be expressed in terms of the pre-collision velocities and $\vec{v}_{ij}' - \vec{v}_{ij}$:

$$
\begin{aligned}
2\vec{v}_i' &= 2\vec{v}_i + \vec{v}_{ij}' - \vec{v}_{ij} \\
2\vec{v}_j' &= 2\vec{v}_j + \vec{v}_{ij} - \vec{v}_{ij}' .
\end{aligned}
\tag{3.15}
$$

The post-collision angular velocities follow from the conservation of angular momentum, according to (3.8).

Combining (3.14) with (3.15) for the linear motion, with (3.8) for the rotation, and using $\vec{e}_{ij}^n \times \vec{g}_{ij}^n = 0$ yields the collision rule:

$$
\begin{aligned}
\vec{v}_i' &= \vec{v}_i - \frac{1 + \varepsilon^n}{2} \vec{g}_{ij}^n + \frac{\tilde{J} \left(\varepsilon^t - 1 \right)}{2 \left(\tilde{J} + 1 \right)} \vec{g}_{ij}^t \\
\vec{v}_j' &= \vec{v}_j + \frac{1 + \varepsilon^n}{2} \vec{g}_{ij}^n - \frac{\tilde{J} \left(\varepsilon^t - 1 \right)}{2 \left(\tilde{J} + 1 \right)} \vec{g}_{ij}^t \\
\vec{\omega}_i' &= \vec{\omega}_i - \frac{\varepsilon^t - 1}{2R \left(\tilde{J} + 1 \right)} \left(\vec{e}_{ij}^n \times \vec{g}_{ij}^t \right) \\
\vec{\omega}_j' &= \vec{\omega}_j - \frac{\varepsilon^t - 1}{2R \left(\tilde{J} + 1 \right)} \left(\vec{e}_{ij}^n \times \vec{g}_{ij}^t \right)
\end{aligned}
\tag{3.16}
$$

3.3 Uniqueness of the Collision Rule

Let us reconsider the problem solved in this section: the equations (3.16) provide a solution for 12 unknown variables, namely 2×3 velocity components and 2×3 components of the angular velocity. To provide a unique solution, 12 scalar equations are, thus, necessary. So let us count the number of equations.

For a general collision, in the absence of external forces and moments, there are 6 equations corresponding to the conservation of total linear momentum (3.5) and total angular momentum. For conservative systems there is an additional equation for the total energy. But granular systems are not conservative. The latter equation is, thus, replaced by equations involving the coefficients of restitution (see below).

Consequently, at first glance the system looks underdetermined, as 6 equations are missing. For the derivation of (3.16), however, further assumptions have been made: The hard-sphere assumption implies point-like contact of the colliding particles which implies in its turn that the angular momentum is conserved for both particles separately. Thus, instead of three equations for the total angular momentum, there are six equations (3.6) and (3.7). Still, three equations are missing.

These three equations correspond to energy. As a particularity, here it is distinguished between energy dissipation of the normal motion, characterized by ε^n and energy dissipation of the tangential motion, characterized by ε^t. How can these two coefficients represent three equations? Let us define a coordinate system whose z-axis is aligned with the direction $\vec{r}_i - \vec{r}_j$, i.e., with the direction normal to the particle contact. The x-axis is parallel with the tangential velocity, g^t, and the y-axis is perpendicular to g^t and $\vec{r}_i - \vec{r}_j$. Therefore, the pre- and post-collision velocities in the z-direction are related by ε^n, the velocities in the x-direction are governed by ε^t, and the velocity in the y-direction is zero before and after the collision. Thus, indeed, ε^n and ε^t stand for three equations.

Hence, there is no magic behind the derivation of (3.16): from 12 equations, 12 post-collision velocities are determined.

There are two idealizing assumptions which have been exploited for this derivation. First, it was assumed that the energy dissipation is governed by three separate equations, whereas in real mechanical systems the energy consumption is governed by one equation only. This statement implies that the coefficients of restitution in the normal and tangential direction, ε^n and ε^t, are not independent from each other (see also Sect. 3.5.3). The second assumption concerns the point-like contact: realistic materials interact by deforming each other, thus the contact area is not point-like. Therefore, the particles exert torques on each other and the angular momentum is not conserved separately for each particle. To abstain from the assumption of point-like contact requires the full solution of the contact of rotating soft spheres (including surface interaction), which is, as of now, an unsolved problem. Insofar, (3.16) is an approximation.

3.4 Sketch of the Algorithm

Using the collision rule (3.16), an algorithm for the simulation of a many-particle system is obtained:

1. Initialize the positions $\vec{r}_i$, velocities $\vec{v}_i$, and angular velocities $\vec{\omega}_i$ of all N particles at time $t = 0$.
2. Determine the time $t^* > 0$ when the next collision in the system occurs, i.e., the time

$$t^* = \min\left(t_{ij} > 0 : |\vec{r}_i\,(t_{ij}) - \vec{r}_j\,(t_{ij})| = 2R\,,\ i, j = 1, \ldots, N\right)\,. \quad (3.17)$$

3. Determine the positions of all particles at time t^*. In the absence of external fields these positions are given by

$$\vec{r}_i := \vec{r}_i + (t^* - t)\,\vec{v}_i\,, \qquad \text{where} \qquad i = 1, \ldots, N\,, \quad (3.18)$$

or for a constant external field, e.g., gravity $\vec{G}$, by

$$\vec{r}_i := \vec{r}_i + (t^* - t)\,\vec{v}_i + \frac{1}{2}\vec{G}\,(t^* - t)^2\,, \qquad \text{where} \qquad i = 1, \ldots, N\,. \quad (3.19)$$

4. Compute the new velocities and angular velocities of the colliding particles I and J by means of (3.16):

$$\begin{aligned}
\vec{v}_I &:= \vec{v}_I\,(\vec{v}_I, \vec{v}_J, \vec{\omega}_I, \vec{\omega}_J) \\
\vec{v}_J &:= \vec{v}_J\,(\vec{v}_I, \vec{v}_J, \vec{\omega}_I, \vec{\omega}_J) \\
\omega_I &:= \omega_I\,(\vec{v}_I, \vec{v}_J, \vec{\omega}_I, \vec{\omega}_J) \\
\omega_J &:= \omega_J\,(\vec{v}_I, \vec{v}_J, \vec{\omega}_I, \vec{\omega}_J)\,.
\end{aligned} \qquad (3.20)$$

5. Update the system time

$$t := t^*\,. \quad (3.21)$$

6. Proceed with step 2 of the algorithm.

Compared with force-based algorithms (see Chap. 2) this algorithm is favorable in several respects:

1. The interaction force as a function of the particle positions and velocities is not needed. The coefficients of restitution ε^n and ε^t may be determined experimentally as functions of the relative velocity $\vec{g}_{ij}$ and can be used in the form of fit-formulae.
2. The computational effort is determined by the number of collisions. No computer time is needed to compute the particle positions and velocities between collisions. In contrast to force-based Molecular Dynamics, the system time does not propagate by a fixed (or adaptive) time step but according to the sequence of collision events. Therefore, such algorithms are called event-driven algorithms. The lower the particle number density of the system, the more efficiently event-driven Molecular Dynamics performs as compared with force-based Molecular Dynamics.

As the main condition for the application of event-driven Molecular Dynamics it is required that at each point in time no more than one collision occurs. This condition is identical with the premise that collisions occur instantaneously, i.e., the duration of collisions is negligible. This assumption is an idealization since elastic collisions of spheres ($\varepsilon^n = 1$) last for [119]

$$t_{\text{coll}}^{\text{el}} \propto g^{-1/5} , \tag{3.22}$$

and (at the same elastic constant) dissipative collisions last even longer [261, 262]. A method which partially considers the duration of collisions was proposed in [167].

3.5 Coefficients of Restitution

3.5.1 Coefficient of Normal Restitution ε^n

Frequently it is assumed that the coefficients of restitution ε^n and ε^t are material constants. We wish to stress that this assumption is in contradiction with empirical data. Experiments showed that both coefficients vary with the impact velocity. Figure 3.2 shows the normal coefficient of restitution of colliding ice spheres at very low temperature [36]. Clearly this dependency should not be approximated by a constant. Moreover, the assumption of a constant coefficient of restitution is not in agreement with the mechanics of materials [165, 166, 229, 274].

The coefficients of restitution can be analytically derived, provided the interaction forces between colliding spheres are known. The most simple approximation for the deformation of a material, *viscoelastic deformation*, assumes

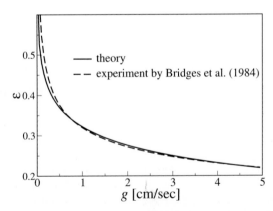

Fig. 3.2. The normal coefficient of restitution as a function of the normal impact velocity ($g_{ij}^t = 0$). *Dashed line*: experimental data for ice spheres at very low temperature [36]; *solid line*: theoretical result for viscoelastic particles in Padé approximation of (3.25) [229]

the conservative part of the stress depending linearly on the deformation and the dissipative part of the stress depending linearly on the deformation rate. The assumption of viscoelastic deformation is justified for dry and electrically neutral granular material, provided the collision velocity is small enough to avoid plastic deformation, fragmentation, and a nonlinear stress-strain relation. On the other hand it must be large enough to render surface forces such as van der Waals forces negligible.

For colliding viscoelastic spheres with interaction force (2.12), the equation of motion reads [44]

$$\ddot{\xi} + \frac{2\,Y\,\sqrt{R^{\mathrm{eff}}}}{3\,m^{\mathrm{eff}}\,(1-\nu^2)}\left(\xi^{3/2} + A\sqrt{\xi}\dot{\xi}\right) = 0\,, \quad \dot{\xi}(0) = g^{\mathrm{n}}\,, \quad \xi(0) = 0\,, \quad (3.23)$$

where $\xi(t) \equiv R_i + R_j - |\vec{r}_i(t) - \vec{r}_j(t)|$ is the deformation of the particles during the collision (see Sect. 2.1.4 and [44] for further details).

Solving the differential equation (3.23) for $\xi(t)$, with the definition (3.4) yields

$$\varepsilon^{\mathrm{n}} = \frac{\dot{\xi}(t_c)}{\dot{\xi}(0)}\,, \tag{3.24}$$

where t_c is the duration of the collision. The integration of (3.23) yields the result [262]

$$\begin{aligned}
\varepsilon^{\mathrm{n}}\,(g^{\mathrm{n}}) &= 1 + C_1\,A\rho^{2/5}g^{1/5} + C_2\,A^2\rho^{4/5}g^{2/5} + \cdots \\
&\equiv 1 + C_1^*\,g^{1/5} + C_2^*\,g^{2/5} + \cdots\,,
\end{aligned} \tag{3.25}$$

with

$$\rho \equiv \frac{2Y}{3(1-\nu^2)}\frac{\sqrt{R^{\mathrm{eff}}}}{m^{\mathrm{eff}}}\,. \tag{3.26}$$

The coefficients C_i are known numerical constants:

$$C_1 \approx -1.15344\,, \quad C_2 \approx 0.79826\,, \quad C_3 \approx -0.48358\,, \quad C_4 \approx 0.28528\,. \tag{3.27}$$

From the very good agreement of the experimental and theoretical data it may be concluded that the assumption of viscoelastic material properties is justified in a wide interval of impact velocities. For practical purposes it might be favorable to use a Padé approximation of (3.25) as given in [229] to assure the correct limit $\lim_{g^{\mathrm{n}}\to\infty}\varepsilon^{\mathrm{n}}\,(g^{\mathrm{n}}) = 0$. Figure 3.2 shows the function (3.25) in Padé approximation together with the experimental result.

Using (3.25) the normal coefficients of restitution can be computed directly from the particle radii and from material properties, namely, the Young modulus Y, the Poisson ratio ν, and the dissipative constant A which is a function of the material viscosity [44]. The latter quantity, however, is difficult to find in tables of material properties. The measurement of the coefficient of restitution is a complicated experimental problem, in particular for small impact velocity (e.g., [35, 36]).

It has been shown recently that the material deformation due to a collision of spheres is intimately related to the deformation due to the rolling motion of a sphere on a plane [37]. Correspondingly, the coefficient of restitution describing the loss of energy in collisions is directly related to the rolling friction coefficient μ_{roll} for a viscous sphere on a hard plane. This coefficient characterizes the torque acting opposite to the rolling motion due to rolling friction, $M = \mu_{roll} F^n$. (Here F^n is the normal force exerted by the plane onto the sphere caused by the sphere's own weight m.) Both coefficients are related via [38]

$$k = \frac{1 - \varepsilon^n}{(\rho/m)^{2/5} (g^n)^{1/5}} = \frac{\mu_{roll}}{V} ,$$

(3.28)

where

$$k = \frac{2^{4/5} \, 5^{2/5}}{3\sqrt{\pi}} \frac{\Gamma(21/10)}{\Gamma(3/5)} \approx 0.438$$

(3.29)

and V is the sphere's linear velocity in the direction of rolling.

The relation (3.28) allows us to determine the coefficient of restitution by measuring the resistance of a sphere against rolling on a hard plane which might be experimentally less complicated.

3.5.2 Tangential Coefficient of Restitution ε^t

The interaction force in the tangential direction of colliding viscoelastic spheres whose surfaces are covered by small asperities of random size has been derived in [44]. The derivation is rather technical and we do not want to discuss the details here. Instead we restrict ourselves to a more qualitative discussion of the corresponding tangential coefficient of restitution. Contrary to the normal coefficient, the tangential coefficient is a function of both components of the impact velocity.

Figure 3.3 shows ε^t as a function of the components of the impact velocity according to the analytical expression taken from [44]. The g^n–g^t plane reveals two domains that characterize qualitatively different types of collisions:

(a) Sliding motion ($\varepsilon^t > 0$) for small g^n and large g^t
(b) Reversion of the rotation ($\varepsilon^t < 0$) for small g^t and large g^n

For small normal and large tangential velocity (case a), the particles slide upon each other, i.e., $\varepsilon^t > 0$. This behavior is plausible since a small normal component g^n implies small normal force and, thus, with regard to Coulomb's friction law, a small maximal tangential force is the result. Therefore, $\varepsilon^t \to 1$ for $g^n \to 0$ due to the vanishing tangential force. The size of the area in

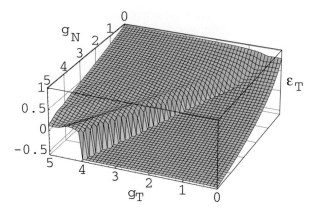

Fig. 3.3. Tangential coefficient of restitution as a function of the normal and tangential component of the impact velocity for certain material and surface properties [44]

the g^n–g^t plane which corresponds to the sliding particle contact depends on the surface roughness (characterized by the amplitude and frequency of asperities, see the model used in [44]). In the opposite case (b) a large normal force allows for the transfer of a large torque between the surfaces of the particles. Consequently the particles do not slide on each other but just roll as gear wheels. Depending on the tangential velocity the resulting torque may be large enough to revert the orientation of the rotation, provided the normal force caused by g^n is large enough. This mechanism explains why ε^t may be negative. The reversion of the rotation direction can actually be observed when throwing a rubber ball (superball) at a certain angle against the ground. The behavior of the ball may be so surprising that we are not able to catch it. In the case of the superball the negative tangential coefficient of restitution is not caused by surface roughness but by strong adhesive forces between the rubber and a smooth surface; however, the result is very similar. Perfect reversion of the rotation, $g^t(t^*) = -g^t(0)$, cannot be achieved since the necessary pressure in the normal direction (to avoid sliding and allow for a large torque) also causes viscous deformation of the particle in the normal direction, i.e., loss of kinetic energy.

The domains of different behavior in the g^n–g^t plane are separated by a rather sharp border which becomes more obvious when plotting the data from Fig. 3.3 in a different representation (Fig. 3.4). The larger g^n, the larger the critical tangential component of the impact velocity g^t_{cr} for which the surfaces start to slide. A larger normal component g^n causes a larger normal force, hence, a larger tangential velocity is necessary to overcome the surface asperities and to start the sliding motion.

The model includes a continuous transition from negative ε^t, corresponding to very rough spheres (like gear wheels), to the limit of perfectly sliding

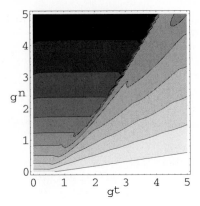

Fig. 3.4. Same data as in Fig. 3.3 drawn as contour plot. The two domains of qualitatively distinct behavior are separated by a rather sharp border

spheres ($\varepsilon^t \to 1$). In (force-based) Molecular Dynamics of granular materials the tangential motion is frequently modeled by the force law (2.18), which has been proposed in a similar form in [107]:

$$F^t = \min\left\{ -\gamma^t \left| g^t \right|, \ \mu F^n \right\} . \tag{3.30}$$

This force law implies a transition between two different regimes, which agrees with the discussed domains of ε^t in the g^n–g^t plane. A detailed quantitative discussion of the presented model for the tangential coefficient of restitution can be found in [44].

3.5.3 Relation between ε^n and ε^t

Let us add two comments on the relation between ε^n and ε^t:

1. Frequently in literature the tangential coefficient of restitution is neglected, i.e., $\varepsilon^t = 1$. It is said the particles are assumed to be "ideally smooth." However, even for the case of ideally smooth spheres, ε^t is smaller than 1 except for perfectly elastic particles ($\varepsilon^n = 1$) for the following reason: the normal coefficient of restitution ε^n as discussed in Sect. 3.5.1 is a direct consequence of the deformation rate $\dot{\xi}$ of the particle material. Hence if the particle surfaces have a tangential relative velocity, i.e., if they rotate in a compressed state during the collision, an additional deformation rate arises due to the rotation. This causes a viscous stress which counteracts the rotation [37, 38], i.e., there is a torque which decelerates the rotation and, thus, slows down the relative tangential motion. According to the definition of ε^t, (3.4), the slowing of the tangential motion at the collision implies $\varepsilon^t < 1$. Therefore, two spheres which meet at $g^t \neq 0$ always exert a torque on each other, even if they are perfectly smooth,

i.e., the collision always changes their angular velocities. The simultaneous assumptions $\varepsilon^n \neq 1$ and $\varepsilon^t = 1$ are, thus, inconsistent.

Virtually the entire kinetic theory of granular gases (see, e.g., [42]) is based on the assumption $\varepsilon^n < 1$ while $\varepsilon^t = 1$. This simplification is owed exclusively to the enormous mathematical complication which comes along with the incorporation of the correct (velocity dependent) tangential coefficient of restitution. At the present time it is not feasible to develop a full kinetic theory of granular gases including a realistic $\varepsilon^t (g^n, g^t)$.

2. The limits of the coefficients ε^n and ε^t as given in (3.4) are deduced from the natural assumption that both components of the relative velocity diminish after a dissipative impact. It has been recently shown, however, that the normal coefficient of restitution can be larger than one under conditions of strong coupling between the rotational and translational motion [164].

3.6 Simple Algorithm for Event-Driven Molecular Dynamics

3.6.1 Overview

The compilation of an efficient algorithm for event-driven Molecular Dynamics is an ambitious project. Before such an algorithm is discussed in Sect. 3.7, a very simple algorithm is first described, which serves to explain the essential ingredients and also the problems which appear in event-driven Molecular Dynamics algorithms. To this end we strictly adhere to the algorithm as outlined in Sect. 3.4.

For simplicity of the notation and brevity of the program code, we restrict ourselves to identical spherical particles and assume that the rotational degree of freedom may be neglected (see Sect. 3.5.3 for a discussion of this idealization). This idealization is frequently made in the Kinetic Theory of granular gases [42]. Consequently only the normal coefficient of restitution $\varepsilon \equiv \varepsilon^n$ appears. Moreover it is assumed that the particles move in a quadratic box of size 2×Lbox. The corners of the box are located at (-Lbox,-Lbox), (-Lbox,Lbox), (Lbox,Lbox), and (Lbox,-Lbox). For the generalization to rotating spheres and to individual particle radii R_i, the collision law (3.16) has to be generalized accordingly. The generalization to different container geometry and even to three-dimensional systems is straightforward as well; the algorithm follows the same idea as outlined in the present chapter.

The following Sects. 3.6.2–3.6.6 first describe some useful functions which are assembled in Sect. 3.6.7 to obtain the complete event-driven Molecular Dynamics program.

3.6.2 Force-Free Motion of Particles

In the absence of external fields the particles move between particle–particle or particle–wall collisions with constant velocity along straight lines. During the time tprop they cover the distance $\vec{v}_i \, t_{\text{prop}}$.

```
eventsimple.cc
1   void propagation(double tprop)
2   {
3     for(int i=0; i<N; i++){
4       x[i]+=tprop*vx[i];
5       y[i]+=tprop*vy[i];
6     }
7   }
to be continued
```

The function propagation() shown above is only applicable if there are no external forces. For the case of a constant force with acceleration $\vec{a}$, the propagation rule

$$\vec{r}_i := \vec{r}_i + t_{\text{prop}} \, \vec{v}_i + \frac{1}{2} t_{\text{prop}}^2 \, \vec{a} \tag{3.31}$$

has to be implemented.

3.6.3 Pairwise Particle Collisions

The collision law (3.16) allows us to compute the post-collision velocities of a pair of rough spherical particles as functions of their pre-collision velocities. For ideally hard spheres which are assumed to not transfer torques, the post-collision velocities of colliding spheres are determined from the pre-collision velocities by the simpler rule:

$$\vec{v}_i = \vec{v}_i - \frac{1+\varepsilon}{2} \left[(\vec{v}_i - \vec{v}_j) \cdot \vec{e}_{ij}^{\,n} \right] \vec{e}_{ij}^{\,n}$$

$$\vec{v}_j = \vec{v}_j + \frac{1+\varepsilon}{2} \left[(\vec{v}_i - \vec{v}_j) \cdot \vec{e}_{ij}^{\,n} \right] \vec{e}_{ij}^{\,n} \,. \tag{3.32}$$

This transformation is computed by collision():

```
eventsimple.cc
8    void collision(const pair<int,int> & ij)
9    {
10     int i=ij.first;
11     int j=ij.second;
12     double dx = x[i]-x[j];
13     double dy = y[i]-y[j];
14     double dist=sqrt(dx*dx+dy*dy);
15     double ndx=dx/dist;
16     double ndy=dy/dist;
17     double h=(1+eps)*((vx[i]-vx[j])*ndx+(vy[i]-vy[j])*ndy)/2;
18
19     vx[i]-=h*ndx;
20     vy[i]-=h*ndy;
21     vx[j]+=h*ndx;
22     vy[j]+=h*ndy;
23   }
to be continued
```

The argument `ij` of this function is the `pair<int,int>` of indices i and j of the collision partners. The program modifies the velocities of the colliding particles, `vx[i]`, `vy[i]`, `vx[j]`, and `vy[j]` according to (3.32).

To determine the following collision, a function is needed to assess whether two particles i and j collide if they continue their motion at their present velocities. If they collide, the equation

$$\left| [\vec{r}_i + (t^* - t)\,\vec{v}_i] - [\vec{r}_j + (t^* - t)\,\vec{v}_j] \right| = R_i + R_j \tag{3.33}$$

has a solution t^* with $t^* > t$, where t is the present time. With $\vec{r}_{ij} \equiv \vec{r}_i - \vec{r}_j$, $\vec{v}_{ij} \equiv \vec{v}_i - \vec{v}_j$, and $R_{ij} \equiv R_i + R_j$ we obtain

$$(t^* - t)^2 + 2\,(t^* - t)\,\frac{\vec{r}_{ij} \cdot \vec{v}_{ij}}{\vec{v}_{ij}^2} + \frac{\vec{r}_{ij}^2 - R_{ij}^2}{\vec{v}_{ij}^2} = 0\,. \tag{3.34}$$

The necessary condition for a collision is

$$\vec{r}_{ij} \cdot \vec{v}_{ij} < 0\,, \tag{3.35}$$

i.e., the particles must approach each other. The particles meet if the quadratic equation (3.34) has a solution, i.e., if

$$\left[\frac{\vec{r}_{ij} \cdot \vec{v}_{ij}}{\vec{v}_{ij}^2} \right]^2 + \frac{R_{ij}^2 - \vec{r}_{ij}^2}{\vec{v}_{ij}^2} > 0\,. \tag{3.36}$$

If both conditions (3.35) and (3.36) are fulfilled, the particles collide at time t^*.

The function `ppcoll()` computes the collision time t^* in case the particles collide, otherwise the function yields ∞ (constant `infty` in the program), i.e., a *very* large number which is considered as infinity. Section 3.6.6 describes how to setup a list of all possibly occurring collisions. To keep this list short, we aim to exclude redundant entries. Therefore, the function `ppcoll()` yields `infty` too if the location of the collision is outside the simulation area.

```
                             ──────── eventsimple.cc ────────
24  double ppcoll(int i, int j, double tol)
25  {
26      double dx,dy,dvx,dvy,xci,xcj,yci,ycj,scalar,h,p,q,w,ct;
27      double dist2,RR,vv2;
28
29      dx=x[i]-x[j];
30      dy=y[i]-y[j];
31      dvx=vx[i]-vx[j];
32      dvy=vy[i]-vy[j];
33      scalar=dx*dvx+dy*dvy;
34      if(scalar>0){
35          return infty;
36      } else{
37          dist2 = dx*dx+dy*dy;
38          RR = (dist2>=R*R ? R : R-tol);
39          vv2 = dvx*dvx+dvy*dvy;
40          q = dist2-4*RR*RR;
41          w = scalar*scalar - q*vv;
```

```
42    if(w<0){
43       return infty;
44    } else {
45       ct=Time+q/(-scalar + sqrt(w));
46       xci=x[i]+(ct-Time)*vx[i]; xcj=x[j]+(ct-Time)*vx[j];
47       if((xci>Lbox-R)||(xci<-Lbox+R)||(xcj>Lbox-R)||(xcj<-Lbox+R)){
48          return infty;
49       } else {
50          yci=y[i]+(ct-Time)*vy[i]; ycj=y[j]+(ct-Time)*vy[j];
51          if((yci>Lbox-R)||(yci<-Lbox+R)||(ycj>Lbox-R)||(ycj<-Lbox+R)){
52             return infty;
53          } else {
54             return ct;
55          }
56       }
57    }
58 }
59 }
```

———————— to be continued ————————

The third argument `tol` of `ppcoll()` is a very small number of the order 10^{-10} (as compared with the particle radius), which is used for the prevention of numerical errors. Its application will be explained below. For `tol`=0 the program computes the collision time t^* as given by (3.34). The latter equation has two real solutions, provided the condition (3.36) is fulfilled; for our purpose only the solution with the smaller value is relevant. The quadratic equation (3.34) can be solved by the standard formula

$$t^* - t = -\frac{\vec{r}_{ij} \cdot \vec{v}_{ij}}{\vec{v}_{ij}^2} - \sqrt{\left(\frac{\vec{r}_{ij} \cdot \vec{v}_{ij}}{\vec{v}_{ij}^2}\right)^2 + \frac{R_{ij}^2 - r_{ij}^2}{\vec{v}_{ij}^2}} \, . \tag{3.37}$$

The numerical evaluation of (3.37) in a computer algorithm is, however, problematic for very small distances on the particle surfaces ($r_{ij}^2 - R_{ij}^2 \ll R_{ij}^2$). In this case the difference of two almost equal numbers has to be computed, which may lead to large numerical errors. The expression

$$t^* - t = \frac{r_{ij}^2 - R_{ij}^2}{-\vec{r}_{ij} \cdot \vec{v}_{ij} + \sqrt{(\vec{r}_{ij} \cdot \vec{v}_{ij})^2 + \vec{v}_{ij}^2 (R_{ij}^2 - r_{ij}^2)}} \tag{3.38}$$

is mathematically equivalent to (3.37), but avoids numerical problems even for small distances between the particles [91].

When employing this trick the numerical error is only *reduced*, it can never be eliminated completely. Due to such errors it may happen that the numerical value of (3.38) is a tiny bit larger than the exact value. Therefore, if the system is propagated to the new positions at time t^* (which is a tiny bit too large as well) both particles overlap by a small fraction of their radii (which can be as small as 10^{-12}). This situation is not allowed as per the assumptions of event-driven Molecular Dynamics. These overlaps are a serious problem. In Sect. 3.6.9 the effect of numerical errors are discussed in detail.

One method to avoid such illegal overlaps is to assign each particle two radii. One is the normal radius R, the other radius is smaller by the small num-

ber `tol`. If two particles are closer than $2R$ to each other at the time of calculation of their potential collision, the radii of both are temporarily reduced. This
allows us to process the collision of both particles in the regular way. An error
can only occur if the numerical problem persists and the particles collide with
another particle (with overlap) before separating from each other, i.e., before
acquiring a distance of $2R$. This situation is extremely unlikely if the energy
loss per collision is small. Since `tol` is a very small number, the diminution of
a few particles does not affect the system properties. In the main program the
function `ppcoll()` is called with `null` as the tolerance. The parameter `null`
is defined as a constant there, e.g., `const double null=10e-10;`.

For the case of a constant external field such as gravity, (3.33) remains
unaffected since the acceleration $\vec{a}$ drops out of the equation:

$$\left\| \left[\vec{r}_i + (t^* - t)\, \vec{v}_i + (t^* - t)\, \frac{\vec{a}}{2} \right] - \left[\vec{r}_j + (t^* - t)\, \vec{v}_j + (t^* - t)\, \frac{\vec{a}}{2} \right] \right\|$$

$$= \| [\vec{r}_i + (t^* - t)\, \vec{v}_i] - [\vec{r}_j + (t^* - t)\, \vec{v}_j] \| = R_i + R_j . \quad (3.39)$$

Hence, the first part of `ppcoll()` which determines the collision time t^* is
also correct if there are constant and uniform external fields. The locations,
where the collisions take place, change if external fields are taken into account, however. Therefore, the last part of the function has to be modified
accordingly.

If the external field is not uniform, e.g., for central forces, the collision
time can be computed only approximately.

3.6.4 Wall Collisions

It is assumed here that particles are reflected elastically when they collide
with the wall:

```
                          ──────── eventsimple.cc ────────
60  void wcollision(int i)
61  {
62    if( y[i]<0){
63       if((y[i] > x[i]) || (y[i] > -x[i])){
64          vx[i]=-vx[i];
65       } else{
66          vy[i]=-vy[i];
67       }
68    } else {
69       if((y[i] < -x[i]) || (y[i] < x[i])){
70          vx[i]=-vx[i];
71       } else{
72          vy[i]=-vy[i];
73       }
74    }
75  }
                          ──────── to be continued ────────
```

In the case of an elastic collision only the concerned velocity component
changes its sign. To simulate inelastic reflection or to apply a certain granular

temperature at the wall (see Sect. 3.8.3), wcollision() has to be modified. Periodic boundary conditions can also be simulated (see Sect. 3.8.2), the implementation is discussed in Sect. 3.10.

Similar to particle–particle collisions, wall collisions are *events* too since velocities of particles are modified. Hence, for each particle of the system the time has to be determined when it collides with one of the walls, provided it continues its motion at its present velocity. The collision time is determined by

$$
t_x = \begin{cases} \text{collision time right,} & \text{if } v_x(i) > 0 \\ \text{collision time left,} & \text{if } v_x(i) < 0 \end{cases}
$$

$$
t_y = \begin{cases} \text{collision time top,} & \text{if } v_y(i) > 0 \\ \text{collision time bottom,} & \text{if } v_y(i) < 0 \end{cases}
$$

$$
t^* = \min(t_x, t_y) .
$$

(3.40)

This time is computed by pwcoll():

```
                            eventsimple.cc
76  double pwcoll(int i,double tol)
77  {
78     double tx, ty;
79     double RR = R;
80
81     if((Lbox+x[i]>R) || (Lbox-x[i]<R) || (Lbox+y[i]>R) || (Lbox-y[i]<R)) {
82        RR=R-tol;
83     }
84     if(vx[i]==0){
85        tx=infty;
86     } else {
87        if(vx[i] >0) {
88           tx=(Lbox-x[i]-RR)/vx[i];
89        } else {
90           tx=(-Lbox-x[i]+RR)/vx[i];
91        }
92     }
93     if(vy[i]==0){
94        ty=infty;
95     } else {
96        if(vy[i] >0){
97           ty=(Lbox-y[i]-RR)/vy[i];
98        } else {
99           ty=(-Lbox-y[i]+RR)/vy[i];
100       }
101    }
102    return Time+min(tx,ty);
103 }
                            to be continued
```

Again, the parameter tol is used to prevent numerical errors (see Sect. 3.6.3). For the case of external fields, the acceleration has to be taken into account for the computation of the collision times of the particles with the wall.

3.6.5 Initialization

As a simple random initialization of the particle positions and velocities, in `init()` the particles are assigned random positions inside the simulation area with the only constraint that mutual overlaps are not permitted. Finally `update()` is called (discussed later) for all particles to initialize the collision list.

```
                              eventsimple.cc
104  void init(double tol)
105  {
106    bool overlap;
107    int j;
108
109    x[0]=ranf(Lbox-R-tol);
110    y[0]=ranf(Lbox-R-tol);
111    vx[0]=ranf(1);
112    vy[0]=ranf(1);
113    for(int i=1; i<N; i++){
114      if(!(i % 100 )) cout << "Init "<< i << endl;
115      do{
116        overlap=false;
117        x[i]=ranf(Lbox-R-tol); y[i]=ranf(Lbox-R-tol);
118        j=0;
119        do{
120          overlap=((x[i]-x[j])*(x[i]-x[j])+(y[i]-y[j])*(y[i]-y[j])<
121                        4*(R+tol)*(R+tol));
122        } while((++j<i) && !overlap);
123      } while(overlap);
124      vx[i]=ranf(1);
125      vy[i]=ranf(1);
126    }
127    for(int i=0; i<N; i++) update(i);
128  }
                          to be continued
```

This simple initialization function is of numerical complexity $\mathcal{O}(N^2)$, therefore, it is applicable only for dilute systems with a rather small particle number of up to about $N = 10^4$. For larger number of particles or for systems of larger density, more sophisticated initialization schemes are necessary.

3.6.6 Schedule of Collision Times

As outlined in Sect. 3.2 at each moment one needs to know the next occurring particle–particle or particle–wall collision. In principle, after each collision the collision times for all $N(N-1)/2$ pairs of particles could be recomputed using `ppcoll()` and all N times of wall collisions using `wcollision()`. The succeeding event is then the particle–particle or the particle–wall collision that corresponds to the minimum of these times.

Although formally correct, this algorithm is very inefficient. Here we assume we have a list of all possible collisions, i.e., their times and the involved collision partners. When a collision of particles i and j occurs, the collision list stays almost correct. Only entries where either i or j is a collision partner become invalid since their velocity vectors have just changed. Instead of reconstructing the entire event list it is sufficient to remove all invalid entries from the list and insert all new events involving i and j.

To remove the invalid entries of particle i, the entire collision list could be searched for entries where i is involved. These entries can be found more efficiently if each particle has a vector `clist` which contains the times when this particle is involved in collisions.

In the main program (see p. 156) the collision list `cseq` is defined as `map<double,pair<int,int>>`. The first template parameter (the key) is the argument (the index), the second parameter is the value. The argument of the collision list is the collision time and the according value is the pair of colliding particles. The function `update()` keeps the list `cseq` and the vectors `vector<double> clist[N]` updated.

```
──────────────── eventsimple.cc ────────────────
129  void update(int i)
130  {
131    double ct;
132
133    for(unsigned int ii=0; ii!=clist[i].size(); ii++) cseq.erase(clist[i][ii]);
134    clist[i].clear();
135    for(int j=0; j<N; j++){
136      if(i!=j){
137        ct=ppcoll(i,j,null);
138        if(ct < infty){
139          cseq[ct]=pair <int, int> (i,j);
140          clist[i].push_back(ct);
141          clist[j].push_back(ct);
142        }
143      }
144    }
145    ct=pwcoll(i,null);
146    if(ct < infty){
147      cseq[ct]=pair<int, int>(i,-1);
148      clist[i].push_back(ct);
149    }
150  }
──────────────── to be continued ────────────────
```

The argument of the function is the index i of a particle whose velocity has changed. First, `cseq.erase(clist[i][ii])` removes all entries from the collision list `cseq` whose collision times are recorded in the vector `clist[i]`. Since at each time there occurs at most one collision, i.e., the keys are unique, this procedure is an elegant and efficient way to erase all invalid entries from the collision list. Finally in line 134 the vector `clist[i]` is emptied. Then the collisions of particle i with all other particles $j = 0 \ldots N - 1$ are predicted. If the particles collide (`ct!=infty`), the pair `pa` of i and j is stored in `cseq`. The corresponding collision time `ct` is added to the vectors `clist[i]` and `clist[j]`, which allows us to remove this entry at a later time as described above, in the same way the time of the wall collision is recorded. To distinguish between a particle–particle collision and a collision with the wall, the second collision partner `pa.second` is assigned the value -1 for the case of a wall collision.

Whenever the velocity of a particle i is modified it is sufficient to call the function `update(i)` to update the collision list. Consequently for each particle–particle collision, `update()` has to be called twice, i.e., once for each collision partner.

3.6.7 Main Program

In the preceding sections, all important parts of the event-driven Molecular Dynamics have been described and can now be assembled to a complete program. First we define the function prototypes and the constants and variables:

N   Number of particles.
nstep   Total number of collisions to be computed.
nps   Interval for the output of snapshots of the simulation in Postscript format.
nprint   Interval for the output of the total energy.
noverlap   Interval for the test if particles overlap.
pssize   Scaling factor for the snapshots.
R   Particle radius.
eps   Coefficient of normal restitution.
Lbox   Size of the simulation area. The area ranges from -Lbox to Lbox in both spatial directions.
infty   Very large number which is interpreted by the program as ∞.
null   Very small number which is used to prevent numerical errors.

```
                              ──── eventsimple.cc ────
151  #include <iostream>
152  #include <fstream>
153  #include <vector>
154  #include <map>
155
156  const int N=3000, nstep=2000000000, nps=50000, nprint=1000, noverlap=10000;
157  const double R=1, eps=0.95, Lbox=150, pssize=500;
158  const double infty=1e20, null=1e-10;
159
160  vector<double> x(N),y(N),vx(N),vy(N);
161  vector<double> clist[N];
162  double Time=0;
163  map<double,pair<int,int> > cseq;
164
165  ofstream fps("event.ps"), fenergy("kinenergy");
166
167  void init(double);
168  void propagation(double);
169  void collision(const pair<int,int> &);
170  void wcollision(int);
171  void update(int);
172  void psplot(int, ofstream &);
173  bool checkoverlap();
174  double ppcoll(int,int,double);
175  double pwcoll(int,double);
176  double ranf(double x){ return (2*double(rand())/(1+double(RAND_MAX))-1)*x;}
177  double kinenergy();
                              ──── to be continued ────
```

In the main program in each iteration step (variable it) the time tnext of the next occurring collision and the pair ijnext of the corresponding collision partners are read from the first entry of the collision list cseq. If the second collision partner has the index -1, the next collision is a wall collision of the first partner. In this case we determine the time tn of the collision by pwcoll(). Then the new particle positions are computed by propagation(), and in wcollision() the velocity of the colliding particle is changed according to the collision rule. Finally update() is called to update the collision list.

The collision of two particles is handled analogously in lines 198–203, except for calling collision() and ppcoll() instead of wcollision() and pwcoll(), and executing update() for both collision partners.

```
                           ──────── eventsimple.cc ────────
178  int main()
179  {
180    double tnext, tn=0;
181    pair<int,int> ijnext;
182
183    init(null);
184
185    for(int it=0; it<nstep; it++){
186      if(it % noverlap==0) checkoverlap();
187      if(it % nprint ==0) fenergy << it<<" "<<kinenergy() << endl;
188      tnext=cseq.begin()->first;
189      if(!(it%nprint)) cout << "IT: " << it << "  time= " << Time << endl;
190      ijnext=cseq.begin()->second;
191      if(ijnext.second==-1){
192        tn=pwcoll(ijnext.first,null);
193        propagation(tn-Time);
194        Time=tnext;
195        wcollision(ijnext.first);
196        update(ijnext.first);
197      } else{
198        tn=ppcoll(ijnext.first,ijnext.second,null);
199        propagation(tn-Time);
200        Time=tnext;
201        collision(ijnext);
202        update(ijnext.first);
203        update(ijnext.second);
204      }
205      if(it%nps==0) psplot(it/nps,fps);
206    }
207  }
                           ──────── to be continued ────────
```

Why is the collision time computed again in the main program although it is already contained in the collision list? There may be a large time lag between the recording of a forthcoming collision of particles i and j in the collision list and the execution of that collision. If thousands of collisions occur in between for which propagation() is called, tiny numerical errors may accumulate. Hence, using the collision times which have been computed earlier may result in an overlap of the particles, i.e., the distance of the particles may be a tiny bit smaller than the sum of their radii $2R$. Recomputing the collision time may prevent this error.

3.6.8 Output

We add two simple functions for data output. The function psplot() generates a sequence of Postscript snapshots:

```
                           ──────── eventsimple.cc ────────
208  void psplot(int page, ofstream & psout)
209  {
210    if(!page){
211      psout << "%!PS-Adobe-2.0" << endl;
```

```
212   psout << "%%BoundingBox: 0 0 "<<pssize+20<<" "<< pssize+20<< endl;
213   psout << "%%EndComments" << endl;
214   psout << "/frame {10 10 translate " <<Lbox/pssize<<" setlinewidth ";
215   psout << pssize/(2*Lbox) <<" "<< pssize/(2*Lbox) <<" scale ";
216   psout << Lbox<<" "<<Lbox<<" translate newpath "
217          << -Lbox<<" "<<-Lbox<<" moveto "
218          << 2*Lbox;
219   psout <<" 0 rlineto 0 "<<2*Lbox<<" rlineto "<<-2*Lbox
220          <<" 0 rlineto closepath stroke}def" << endl;
221   psout << "/c { 1 0 360 arc stroke} def" << endl;
222   }
223   psout << "%%Page: " << page << " " << page << endl;
224   psout << "frame " <<endl;
225   for(int i=0; i<N; i++) psout << x[i]<<" "<<y[i]<<" c" << endl;
226   psout << "stroke showpage " << endl;
227   }
```
———— to be continued ————

These snapshots can be used to generate animated sequences of simulations. The function `kinenergy()`

———— eventsimple.cc ————
```
228   double kinenergy() {
229     double ekin=0;
230
231     for(int i=0; i<N; i++) ekin+=vx[i]*vx[i]+vy[i]*vy[i];
232     return ekin;
233   }
```
———— to be continued ————

is an example for a data output function. The simple program for event-driven Molecular Dynamics is now complete.

3.6.9 A Note on Numerical Errors

When describing the algorithm for event-driven Molecular Dynamics, numerical errors have been mentioned at several places. These errors may have serious consequences as soon as they cause particles to penetrate each other, i.e., when $r_{ij} \equiv |\vec{r}_i - \vec{r}_j| < R_i + R_j$. As a consequence, the collision time for the pair (i, j) may be incorrectly determined. Since there are always rounding errors in numerical computations, we must take care to avoid the described situation.

In the preceeding sections some methods have already been described to prevent overlapping of particles. Let us discuss this problem in a more systematic way: assume at time t two particles at position $\vec{r}_i$ and $\vec{r}_j$ and assume that the equation $|\vec{r}_i(t^*) - \vec{r}_j(t^*)| = R_i + R_j$ has two real solutions, t_1^* and t_2^*, with $t_1^* < t_2^*$ (see (3.34)) so that

$$(t^* - t)^2 + 2(t^* - t)\frac{\vec{r}_{ij} \cdot \vec{v}_{ij}}{\vec{v}_{ij}^2} + \frac{\vec{r}_{ij}^2 - R_{ij}^2}{\vec{v}_{ij}^2} = 0 . \tag{3.41}$$

There are three qualitatively different cases:

(a) $t_1^* > t$ and $t_2^* > t$. This case corresponds to a regular collision. From (3.41) we conclude $\vec{r}_{ij} \cdot \vec{v}_{ij} < 0$ as a necessary condition for a collision. Obviously, the collision takes place at time t_1^*.

(b) $t_1^* < t$ and $t_2^* < t$, which implies $\vec{r}_{ij} \cdot \vec{v}_{ij} > 0$. There is no collision.
(c) $t_1^* < t$ and $t_2^* > t$. This case requires a negative last term in (3.41), i.e., $|\vec{r}_i - \vec{r}_j| < R_i + R_j$ (overlap). This solution may occur for both $\vec{r}_{ij} \cdot \vec{v}_{ij} < 0$ and $\vec{r}_{ij} \cdot \vec{v}_{ij} > 0$.

There is no simple way to detect overlapping pairs of particles corresponding to case (c). Assume the current (regular) collision in a simulation takes place between particles i and j. After propagating to their new positions their new velocities are computed according to the collision rule (3.32). Assume further that these particles overlap now due to a numerical error. By subsequently solving (3.41) this overlap can be easily detected, since then $t_1^* < t$ and $t_2^* > t$. However, the propagation step corresponding to the collision (i, j) may also lead to an overlap of particles k and l somewhere else in the system. This may happen, e.g., if k and l undergo a grazing collision due to a small numerical error. If their positions were computed at infinite precision k and l would not collide. Such cases cannot be detected easily. Fortunately they are very rare.

Let us elucidate the consequences of such overlaps. First, we look at the case where the just collided pair (i, j) overlaps. After the collision the velocities change such that $\vec{r}_{ij} \cdot \vec{v}_{ij} > 0$, i.e., the particles depart from each other. As explained in Sect. 3.6.3 the collision times are only determined if $\vec{r}_{ij} \cdot \vec{v}_{ij} < 0$, therefore, the (erroneous) collision (i, j) is not recorded in the collision list. The extremely small overlap resulting from numerical errors vanishes automatically, except if either i or j collide with a third particle in the small time span when i and j overlap. In this case the particles i and j may be sent to an approaching trajectory before they are separated. For each pair (i, j) this case is very unlikely, but as the number of particles becomes large there is a non-vanishing probability for this scenario to occur in the system. Thus, the second type of illegal situations has to be discussed when the overlapping particles approach each other (which also applies for the case when the pair (k, l) is concerned, as discussed in the previous paragraph).

As stated above, for overlapping particles we obtain $t_1^* < t$ and $t_2^* > t$. If t_2^* is taken as the time of collision the particles do not separate again since in a collision the normal component of the relative velocity changes its sign. Thus, these particles stay together just as chained rings (see Fig. 3.5).

Fig. 3.5. If two particles overlap once due to a numerical rounding error (*left* figures, overlap strongly exaggerated) and $\vec{r}_{ij} \cdot \vec{v}_{ij} < 0$ (see text for explanation), the next collision (*middle*) occurs as for chained rings if t_2^* is used erroneously. Once overlapping, the particles stay entangled and the *right* figure shows the next collision

Alternatively, when investigating the collision time of (i, j) with $\vec{r}_{ij} \cdot \vec{v}_{ij} < 0$, $t_1^* < t$, and $t_2^* > t$, we could assume t_1^* (which is negative!) as the time of the collision, i.e., the event-driven algorithm would progress backwards in time. This is possible and may resolve the situation, but it may lead to illegal overlaps of other pairs. In particular in situations of clustering when the particles are closely packed, going back in time may lead to a cascade of overlap situations. Thus, if applying that method, we recommend checking the subsequent configuration (see below).

In our own scientific work we prefer a different procedure: many overlaps may be prevented by performing the propagation step not in full length, but a tiny bit shorter, such that after the propagation the particles keep a small distance. For this reason, before calling `propagation()` the collision time `pwcoll()` or `ppcoll()` is determined, respectively, using `tol` as the third argument.

Unfortunately, there is no perfect protection against the mutual overlapping of particles due to numerical rounding errors. As described, once the particles are overlapping, in general there is no simple way to resolve this situation. The best method to cope with such situations is to backup the particle coordinates and velocities in certain time intervals and to check periodically to see whether particles overlap. This can be done by `checkoverlap()`:

```
                            ── eventsimple.cc ──
234  bool checkoverlap()
235  {
236    double dist;
237
238    for(int i=1; i<N; i++){
239      for(int j=0; j<i; j++){
240        dist=(x[i]-x[j])*(x[i]-x[j])+(y[i]-y[j])*(y[i]-y[j])-4*R*R;
241        if(dist<0){
242          cout <<" OVERLAP "<<i<<" "<<j<< " " << dist << endl;
243          return true;
244        }
245      }
246    }
247    for(int i=0; i<N; i++){
248      if((x[i]-R<-Lbox)||(y[i]-R<-Lbox)||(x[i]+R>Lbox)||(y[i]+R>Lbox)){
249        cout<<"OVERLAP WALL"<<i<< " "<<y[i]+R-Lbox << endl;
250        return true;
251      }
252    }
253    return false;
254  }
```

If particles overlap, the function returns `true`. A previously saved data set of coordinates and velocities is restored and the simulation proceeds with the initialization of the collision list. Restarting the simulation eliminates accumulated rounding errors and may avoid particles overlapping. This restoring procedure is time-consuming, here we do not give an implementation but it can be found on the book's web site. Since overlapping particles often stay overlapping forever it is not necessary to call the function `checkoverlap()` frequently. There is only a small risk to miss such a situation if it occurs just once.

3.6.10 Critical Discussion of the Algorithm

The described program implements exactly the algorithm outlined on p. 141. It should be understood as a sketch of an event-driven Molecular Dynamics algorithm rather than a practically-applicable tool. The program is inefficient and can be improved in several respects:

1. The initialization function distributes the particles randomly in the simulation area. To find the initial position of the ith particle we have to check for overlap with all $i-1$ particles which have been initialized before. This $\mathcal{O}\left(N^2\right)$ complexity can be avoided by using more efficient algorithms.
2. The list of collision times contains $\mathcal{O}\left(N^2\right)$ entries, among them are also collisions of pairs of particles which are distant from each other. It is very unlikely that such particles collide at all. Hence, keeping such improbable collisions in the list is a waste of time and memory.
3. For each collision all particles in the system are propagated to their new positions, even if they are distant from the colliding particles. If potential collisions of distant particles are not recorded in the collision list (see item 2) only particles close to the location of the last collision have to be propagated.
4. In the function `checkoverlap` all possible pairs of particles are checked for overlap, although most of them are distant pairs. Employing methods similar to force-based Molecular Dynamics, like the link cell method (see Sect. 2.4.3), the numerical effort could be drastically reduced.

The mentioned problems can be solved by a more sophisticated algorithm as explained in the next section.

3.7 Improved Algorithm for Event-Driven Molecular Dynamics

3.7.1 Reduction of the Collision List

The simple event-driven algorithm presented in the preceding section can be improved considerably in many respects by relatively simple means. Most of these improvements are based on the fact that distant particles have only a small chance of colliding in the near future. Consequently, the entry in the collision list for a very distant pair of particles (i, j) will very likely be erased due to collisions of either i or j with other particles before the scheduled collision of (i, j) occurs (see Fig. 3.6). With very small probability, however, the scheduled collision (i, j) may take place, i.e., neither i nor j collide with any other particle before. Hence, a correct event-driven Molecular Dynamics algorithm must take this very unlikely situation into account.

To eliminate distant pairs of particles from the collision list, the simulation area is subdivided into N_b quadratic boxes; only particles j which are in

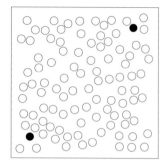

Fig. 3.6. Assume the collision of the black particles is recorded in the collision list. With large probability this collision will not occur since it may be expected that at least one of them will collide with other particles before they actually meet

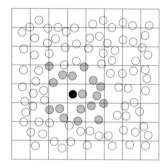

Fig. 3.7. The simulation area is subdivided into boxes. The neighbors of the black particle are drawn in gray. Only neighbors are considered as potential collision partners

the same box as particle i or in one of the adjacent boxes are called neighbors of i and are considered as potential collision partners (Fig. 3.7). The computational effort to assign all N particles to their boxes grows as $\mathcal{O}(N)$.

By subdividing the simulation area into N_b boxes the number of entries in the collision list is reduced drastically. Since each box contains on average N/N_b particles, each particle has approximately $9N/N_b$ neighbors in two dimensions, or $27N/N_b$ neighbors in three dimensions, which have to be considered when constructing the collision list. In two dimensions, therefore, the collision list contains about $9N^2/2N_b$ collisions.[1] Even for medium-sized particle systems the benefit is considerable: In the simple algorithm, for a simulation of 100,000 particles about 5×10^9 pairs of particles have to be con-

---

[1] For this crude estimate it is assumed that *all* pairs of neighboring particles have an entry in the collision list. The actual number is smaller since only such pairs of particles are recorded which are neighbors *and* collide when continuing their motion with the present velocities. This excludes particles which move away from each other.

sidered. If the area is subdivided into 100×100 boxes this number reduces to about 4.5×10^6 for spatially homogeneous particle distribution. The number of pairs is, therefore, 1000 times smaller, which reduces the size of the collision list drastically.

As soon as one of the particles leaves its present box for one of the adjacent boxes the set of its neighbors changes too. Some of the previous neighbors are now too far to be considered as neighbors. Moreover, there will be new neighbors which have not been considered as such yet. Hence, if a particle changes its box the collision list has to be modified, similar to the case after a collision. Consequently, besides the information about expected collisions, the collision list must also contain the information on box changes for each particle. Therefore, this list is better called *event list* instead of collision list.

 In the following sections the algorithm will be explained in detail. The program is not provided here since it is rather long and technical. It can be found on the book's web site.

3.7.2 Data Organization

The central problem of event-driven Molecular Dynamics is to keep the event list up to date. After each event, i.e., whenever a particle changes its velocity due to a collision or when a particle leaves its box, the event list must be updated. This comprises addition of new potential events and deletion of events that have become irrelevant.

In the simple program all events have been recorded in a single list. Adding a new event is non-critical, the pair of particle indices is simply inserted into the map cseq (complexity $\mathcal{O}(\log N_{\text{ev}})$, with N_{ev} being the number of events) and the collision time is added to the lists clist of the concerned particles (complexity $\mathcal{O}(1)$). Removing obsolete events is also simple, since the event times are stored in the lists clist and are, therefore, easily accessible.

However, this data organization can be further improved by splitting the global event list into smaller lists. Thus each particle owns a list of events in which it is involved. Since particle–particle collisions involve two particles, such events are assigned to the particle with the smaller index to avoid double storage. If the boxes are small enough, the list of each particle contains only a few potential collision partners. The event list must allow for fast access to the event with smallest time consumption (i.e., to the next occurring event in which the owner particle is involved), therefore, a priority queue as provided by the Standard Template Library (type priority_queue) is employed. Adding entries to a priority queue is a fast operation. This operation belongs to the same complexity class ($\mathcal{O}(\log n)$ (with n being the number of elements of the queue) as the corresponding operation in the simple algorithm. Nevertheless, the search is faster since each of the event lists is much smaller than the global list. Deleting entries from the lists is a non-trivial task which is discussed in the next section.

The next event occurring in the system corresponds to the minimum time of all entries in the event lists. The time of the next event in which a particle is involved can be retrieved efficiently since priority queues are optimized for this task. These times can be collected in any search tree, as, e.g., the STL-container set. There is, however, a more efficient alternative called *tournament tree* which is described in Sect. 3.7.6.

3.7.3 Removal of Invalid Entries from the Event Lists

Each event invalidates entries in the event lists, either because particles change their velocities or because pairs of particles become distant due to particles changing boxes. Consequently, after each event the event lists have to be adjusted.

When particles i and j collide, all entries in the event lists of i and j become invalid. Moreover, entries in the lists of other particles k also become invalid where either i or j is one of the potential collision partners. In the simple algorithm each particle is assigned a vector clist which allows removal of these entries from the global event list.

There is a simple trick by Marín et al. [177] to bypass the removal of invalid entries: After a collision the lists of events owned by the collision partners is erased, as in the simple program. The events not owned by the partners are more problematic. On average this concerns one half of all events where the partners are involved. Each particle counts the total number of collisions it was involved in so far. When adding an entry for a particle–particle collision to the event list, the number of collisions in which the *other* partner (not the one that administers the event) was involved is now also stored. The entries in the collision list of particle i contain now the index of the partner j, the collision time, and the value of the collision counter of j at the time when the entry was recorded. When reading an entry in the list of particle i its validity can be easily checked: If the present counter value of the partner j is larger than the number in the event list of i, the entry is not valid since j has meanwhile suffered collisions with other particles or with the wall.

Hence, when particles i and j collide, there is no need to remove the entries from other particles k which refer to i or j. Whenever an event is read which turns out to be invalid after comparing the recorded counter value with the actual one, this entry is disregarded. Note that after a collision of i and j, still all events in the event lists of i and j are deleted. By evaluating the collision counters we only avoid searching in the lists of uninvolved particles for possibly invalidated events.

3.7.4 Collision-Free Motion of Particles

When a particle i transits into an adjacent box, the program computes the times of collisions with all new neighbors and the time when the particle will leave the just entered box. What to do with entries in the collision list of

i referring to such particles j which are not neighbors of i anymore? Those entries can be deleted since whenever j joins the neighborhood of i again, it is reinserted into the collision list of i anyway. Deleting such entries needs computer time, so let us discuss what happens if the collision with j is not deleted when i leaves its. If the entry of j in the list of i is valid (i.e., neither suffered a collision since recording this event), the particles are approaching each other. Hence, it may happen that the loss of neighborhood of i and j is only temporary. After the next box change of j, both particles may become neighbors again. Since i and j approach each other, the following sequence may occur:

1. i changes its box and is not neighbor of j anymore
2. j changes its box, i and j are neighbors again
3. i changes its box and is not neighbor of j anymore
4. j changes its box, i and j are neighbors again
5. ...

This sequence proceeds until i and j collide with each other or either i or j collides with another particle. According to the described algorithm, in steps 2 and 4 of the sequence above, new entries in the collision list of either i or j are created. This way multiple entries in collision lists may appear, referring to the same collision. Due to tiny numerical errors these entries may contain slightly different collision times. When the collision of i and j actually takes place the entry with the earliest collision time is automatically chosen. If the entry with the earliest collision time is not the most recent entry in the collision list, the collision time is underestimated by a tiny amount, i.e., the particles collide too early. This situation can be avoided by checking whether there already exists an entry in the list of i concerning a collision with j before recording the collision into the list. If there is no entry so far, the collision is added to the list, otherwise only the collision time is updated.

On the other hand, this check is not really necessary since if the particles collide too early, they have a *very small* distance on the order of the double-precision number resolution, which can be tolerated. The collision is never scheduled too late, which would be a more critical situation as discussed in Sect. 3.6.9.

Therefore, for the benefit of high performance multiple entries of collisions in the collision lists are tolerated. We have to admit that the arguments are not completely conclusive here since leaving invalid entries in the collision list increases the computer time for inserting and retrieving information from the lists. However, the computer time for these operations increases only very slowly (logarithmically) with the lengths of the lists.

3.7.5 Optimal Box Size

There is a significant advantage to not deleting irrelevant events: it allows for very small box size. Deleting all irrelevant events would require larger boxes in order to minimize the number of box change events, which in this case would require significant computation time. By choosing smaller box sizes the event lists are shorter since only the immediately forthcoming potential events are recorded in the lists. The frequency of box-change events increases with decreasing box size, which implies frequent updating of the event lists, including the check for new collision partners. Nevertheless, the number of computations of possible event times is minimal, since only events which have a good chance to actually take place (due to close proximity of the partners) are considered. For larger boxes the event lists need to be updated less frequently, but the number of potential events grows. Most of these events are irrelevant because the particles collide with other partners before they come even close. Thus the additional effort due to the large number of box changes is offset by a smaller number of potential partners. If invalid events are not deleted as described above, the optimal box size for a homogeneous gas is such that each box contains one particle on average, i.e., there are as many boxes as particles. For inhomogeneous gases, i.e., gases which contain areas of high density, a number of boxes contain many particles. In this case it is preferable to choose smaller box sizes to keep the event lists short, even as small as one particle diameter. The box size cannot be smaller, otherwise collisions may be overlooked (see Fig. 3.8). For simulations of particles of different sizes the largest particle diameter determines the box size. Retaining information on irrelevant events incurs larger memory requirement. If memory is the limiting resource, irrelevant events must be deleted, resulting in increased computer time.

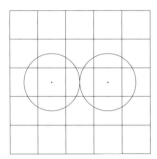

Fig. 3.8. If the box size is chosen smaller than the particle diameter, particles may collide although their centers are not located in adjacent boxes

3.7.6 Scheduling Events

In each iteration loop the next event in the system is determined, i.e., the entry in the collision list with smallest execution time must be found. Rapaport [230] proposed organizing the events in a binary tree. This idea is implemented in the simple program (Sect. 3.6) by using the STL-container map. The organization of the events in particle-associated small lists offers further opportunities to optimize the event bookkeeping [177].

The particle-associated event lists are very short. For optimal box size, they typically contain fewer than 5 entries since each box contains only a few particles. Therefore, using a priority list, one can easily determine the next occurring event for each particle i, scheduled for time t_i, which this particle administers. To determine the next event in the entire system requires finding among all these events the one with the smallest time, $t^* = \min_{i=1...N} t_i$. Hence, the smallest number t^* from the set $t_1, \ldots, t_N$ of invariant size N is determined. Moreover, after each event no more than two elements of the set can change. In [11, 176, 177] a *tournament tree* was proposed as the most efficient method for organizing the next local (i.e., particle associated) events t_i. A binary tree is constructed whose terminal leaves contain the times of the earliest particle-associated events t_i. In the nodes the minimum of their descending nodes' values (see Fig. 3.9, left) is recorded. Therefore each node contains the next event time of all particles in its branch of the tree. Consequently the root node contains the next event occurring in the system. This data structure can be implemented very efficiently.

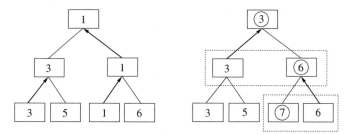

Fig. 3.9. *Left*: each node of the tournament tree contains the minimum of its descendant nodes' values. The arrows indicate the upward propagation of the minimum values. *Right*: update procedure after execution of an event at time 1. The new next event in which that particle is involved, is scheduled for time 7. The dotted boxes mark the pairs of nodes which have to be compared, the circles mark the updated values

Assume at time $t_i = 1$ particle i is involved in an event. After updating its private collision list the next event concerning i takes place at time $t_i = 7$. Hence, the corresponding entry in the tournament tree is updated. The parent node of the respective leaf is outdated as well and must be updated by comparing the two values of the particles which belong to this node. This comparison may be regarded as a competition between the descendant nodes. The node with the smallest event time wins the competition, therefore this data structure is called a "tournament tree." This way all levels of the tree are considered. The modification of any t_i, hence, triggers a sequence of updates up to the root node. The maximum number of comparison operations is determined by the depth of the tree, i.e., $\log_2 N$. The idea of the update procedure is shown in the right hand part of Fig. 3.9.

3.7.7 Update of Particle Positions

In the simple algorithm (Sect. 3.6) after each collision all particles are propagated to their new positions since these positions are needed for the update of the global collision list.

This global propagation step is obsolete now since the search for possible collision partners is restricted to a certain neighborhood [230]. Therefore, it is sufficient to propagate the colliding particles and the particles in their neighborhood to the new positions. Only the coordinates of these particles are needed for the update of the particle-associated collision lists. Whereas in the simple program the new positions of all N particles are computed, now only the positions of the particles which are located in 9–14 boxes have to be computed. (In general the collision partners need not be located in the same box.) For optimal box size the positions of about 10 particles (for the case of a monodisperse system) are calculated. The numerical effort for this step is therefore constant, i.e., independent of N. Aside from an enhanced performance, as an additional benefit of the delayed propagation, the numerical errors accumulate at a slower rate since particle positions are computed more rarely.

Each box is assigned the time of its last update t_{upd}. If at time t the computation of the positions of the particles in this box becomes necessary, the particles are propagated with their current velocity according to $\vec{r}_{ij} := \vec{r}_{ij} + \vec{v}_i\,(t - t_{\mathrm{upd}})$, where t is the current (global) system time. Alternatively the time of the last update can be also assigned to the particles themselves. This allows for propagating a particle without access to any of the boxes.

3.7.8 Efficiency of the Algorithm

Using the described data structures and algorithms the computational effort for one event rises as $\mathcal{O}(\log N)$. In most applications a certain number of events per particle has to be computed. Thus, the total effort rises as $\mathcal{O}\,(N \log N)$. For data output all particles are propagated to their correct positions. For such steps the effort is at least $\mathcal{O}(N)$. In principle, for a large number of boxes, the performance of these steps is determined by the number of boxes too. However, for a given volume fraction η (total volume of the particles divided by the system's volume) the number of boxes cannot be larger than $\pi N/4\eta$, therefore, the effort for this operation is $\mathcal{O}(N)$.

To estimate the efficiency of the algorithm, the memory requirement and computer time consumption for an event-driven Molecular Dynamics simulations were determined using a computer with Intel P4 processor (3.0 GHz) and 1 Gb memory.

We simulated 10^7 collisions in a granular gas in the absence of gravity with periodic boundary conditions for a fixed particle number density. Figure 3.10 (right) shows the memory requirement growing approximately linearly with the number of particles. The computer time (left) depends weaker than linearly on the number of particles. For comparison the solid line shows $T \sim \log(N)$.

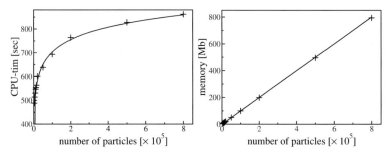

Fig. 3.10. Memory (*right*) and computer time (*left*) consumption of an event-driven Molecular Dynamics simulation using the improved algorithm. For comparison, a logarithmic function (*left*) and a linear function (*right*) is shown. Input/Output operations have been disregarded

For the measurement the initialization procedure was disregarded as well as the costs for data output. The output of the particle coordinates (snapshot) or any function which depends on the positions of the particles needs a full update of the particle positions which requires computer time $\mathcal{O}(N)$. Therefore, a large part of the simulation time is frequently spent on data output.

For our research we always emphasize the optimization of the algorithm with respect to CPU time but not with respect to memory. If the code is optimized with respect to low memory consumption, e.g., by choosing a smaller number of boxes or by removing obsolete entries from the collision lists, the number of particles which fit into a given memory capacity can be increased significantly.

3.8 Boundary Conditions

The behavior of particles close to the system borders is described by boundary conditions. As in hydrodynamics, boundary conditions can have a profound effect on the properties of a particle system. For many applications either reflecting or periodic boundary conditions or heated walls are appropriate. In this section the methods for implementing boundary conditions and their physical implications are discussed.

3.8.1 Reflecting Boundaries

Applying reflecting boundary conditions means that each particle which collides with the system walls reverts its velocity component perpendicular to the wall, whereas the velocity component parallel to the wall stays invariant. This way the kinetic energy of the particle is conserved. This type of boundaries is used for the simple event-driven algorithm (Sect. 3.6.4).

Simulations show that particles accumulate close to reflecting walls. This effect occurs even in the absence of external forces such as gravity and is caused by an enhanced collision frequency close to the walls. Moreover, at the walls the typical relative velocity is larger than in the bulk of the system. Here we cannot strictly prove this claim, however, we provide two instructive and typical examples: Consider two particles i and j as sketched in Fig. 3.11 (left). The particles collide due to the presence of the reflecting wall. In another region more distant from the wall the collision would not occur. In contrast, there are no situations where the presence of a reflecting wall prevents a collision of otherwise colliding particles. Thus, close to a reflecting wall the collision frequency is enhanced. The right part of Fig. 3.11 shows a pair of particles which in the absence of a reflecting wall would collide with relative velocity $v_i - v_j$. Due to the presence of the wall this collision occurs with relative velocity $v_i + v_j$ which corresponds to a larger amount of dissipated energy and, thus, to a faster decay of the granular temperature.

On average, two particles i and j moving toward the wall suffer more collisions and more intensive collisions (larger relative velocity) than an equivalent pair i' and j' moving with the same velocities and the same mutual spatial relation somewhere else, i.e., not close to the wall.

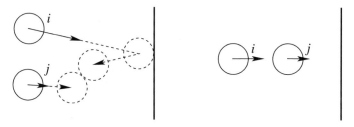

Fig. 3.11. Reflecting walls lead to increased collision frequency and to more intensive collisions. *Left*: the particles meet due to the presence of the reflecting wall. In the absence of the wall these particles would not collide. *Right*: without the wall the particles would collide with relative velocity $v_i - v_j$. Because of the wall this collision takes place with relative velocity $v_i + v_j$, dissipating more energy

Both effects, the enhanced collision frequency and the higher intensity of the collisions lead to a more rapid decay of temperature in the vicinity of reflecting walls and consequently to a reduced pressure [93] (see Sect. 3.10.2), resulting in enhanced particle density. After a long period of time most of the particles are located close to the system wall. The pressure instability also occurs in a freely cooling gas, but with reflecting walls the instability is more pronounced close to the walls.

Figure 3.12 shows a simulation of a dilute force-free granular system ($N = 20,000$, $\varepsilon = 0.999$) in a circular container with reflecting boundary conditions. Under certain conditions the particles do not accumulate uniformly close to the wall but form stationary or irregularly moving clusters close to the reflecting wall [185].

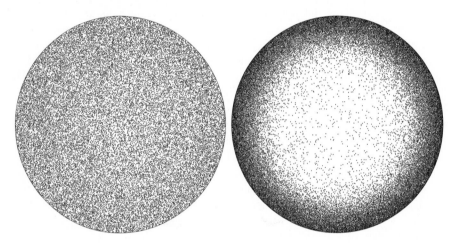

Fig. 3.12. A dilute granular system evolves in a circular container in the absence of external forces. Initialized homogeneously (*left*), the particles accumulate close to the reflecting wall after a long time (*right*)

3.8.2 Periodic Boundary Conditions

Periodic boundary conditions are intended to mimic very large systems, i.e., the investigated system is thought to be much larger than the simulated number of particles. The particles are contained within a primary simulation volume. When a particle leaves one side of this volume, it re-enters from the opposite side. Thus, periodic boundary conditions allow the simulation to proceed as if the primary volume was surrounded by identical copies of itself. The primary simulation volume is thus periodically replicated in all directions to form a quasi-infinite volume. Any particle i at position $\vec{r}_i$ in the primary volume thus represents an infinite set of particles located at

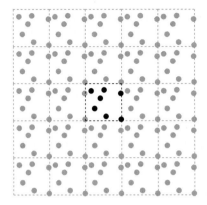

Fig. 3.13. The basic simulation volume (drawn in *black*) is repeated in all dimensions an infinite number of times. Here the neighbor volumes are drawn in *gray*

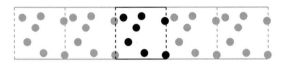

Fig. 3.14. The simulation area may be periodically continued also in one dimension only. Here the basic simulation volume is bound by reflecting horizontal walls

$$\vec{r}_i^{\,k,l,m} = \vec{r}_i + k\vec{X} + l\vec{Y} + m\vec{Z} \ , \tag{3.42}$$

where $\vec{X}$, $\vec{Y}$, and $\vec{Z}$ are the edge vectors of the primary volume and $k, l, m = -\infty \ldots \infty$ are integers (Fig. 3.13). Each particle i in the primary volume is thought as interacting not only with other particles j from the same volume, but also with their periodic images.

The simulation area of a two-dimensional system with periodic boundaries in both directions is topologically equivalent to the surface of a torus. If only one direction is periodic, it is equivalent to the surface of a cylinder (Fig. 3.14).

Using periodic boundary conditions, surface effects can be eliminated from the system, but a number of complications arise:

1. Since each particle has an infinite number of images, there is an infinite number of interactions with each other particle (having an infinite number of images too).
2. If the primary volume is chosen too small, there appear correlations between opposite edges of the primary volume, i.e., the spatial structures are of the same characteristic size as the system itself. This way a particle may interact directly or indirectly with itself across the primary volume.

The first item causes problems when long-range forces, such as electrostatic interactions, are involved. Sophisticated techniques have been developed, such as Ewald-summation to simulate systems of charged or gravitating particles.

For granular systems these problems are not relevant since particles interact only through mechanical collisions. For such systems, the *minimum image criterion* is frequently applied which states that a particle i can interact at most with the closest image of particle j. This convention is safe if the primary volume size exceeds the interaction distance, e.g., the double cutoff length of a potential or in our case the particle diameter. The allowed range of the coefficients in (3.42) is thus reduced to $k, l, m \in \{-1, 0, 1\}$.

For the prediction of collisions in an event-driven simulation, however, the minimum image criterion is not applicable since it may well happen that the next collision in the system occurs between particle i (from the basic simulation volume) and an image of particle j which is not the closest to particle i. This image may even correspond to $k, l, m \notin \{-1, 0, 1\}$, i.e., this image is outside the basic volume and its direct neighboring volumes.

Therefore, the simple algorithm presented in Sect. 3.6 is not directly suited for the simulation of periodic boundary conditions. Its generalization is, however, simple: For the collision list we consider for each particle i from the basic volume all collision partners j from the same volume and their images from the neighboring volumes. Whenever a particle leaves the simulation volume and is reinserted at the opposite edge, its entries in the collision list become invalid and have to be recomputed. Hence, when a particle leaves the volume, the same values are computed as if it had collided with a wall.

The efficient algorithm sketched in Sect. 3.7 is directly suited for periodic boundary conditions since the simulation area here (the basic volume) is subdivided into smaller boxes. When a particle leaves one of these boxes (also close to the system borders) the according collision lists are updated. Since particles and their images are located in very distant boxes, the problem of ambiguous particle position is avoided.

The second complication listed above is more substantial since it is system-inherent and cannot be resolved by improving the algorithm. Let us give a simple example: Fig. 3.15 shows systems of 10,000 and 100,000 particles in a quadratic container with periodic boundary conditions. The systems are identical with respect to the number density and particle properties. Therefore, equivalent behavior may be expected naively. But in contrast, the systems develop quite differently. At time T which corresponds to $\#C_{\text{small}} = 10^6$ collisions of the small systems and $\#C_{\text{large}} = 10^7$ for the large system (hence 100 collisions per particle), both systems look identical qualitatively, keeping in mind that the larger system (ten times the area of the small system) is scaled down in Fig. 3.15. At time $1.5\,T$, corresponding to $\#C_{\text{small}} = 1.5 \times 10^6$ and $\#C_{\text{large}} = 1.5 \times 10^7$ the systems behave differently. Whereas there are medium sized clusters in the large system which can continue to grow, the small system reveals a system-spanning cluster that will not grow further. This cluster interacts with itself, which contradicts the precondition that the system size must exceed the correlation length significantly. Therefore, while the larger system still shows realistic behavior, the state of the small system is highly artificial. The results of simulating the small system become incorrect

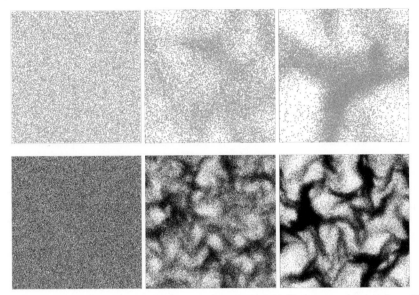

Fig. 3.15. Two systems of 10,000 (*top* row) and 100,000 (*bottom* row) particles with periodic boundary conditions at the same number density. Both systems are intended to simulate an infinitely extended system and ideally should behave identically. The *left* snapshots show the systems at initialization time, the *middle* at time T, and the *right* figure at time $1.5\,T$

as soon as system-spanning clusters emerge. The important effect of clustering in granular gases is discussed in more detail in Sect. 3.10.2.

Further artifacts emerge if the system is too dilute. In this case between successive collisions, the particles travel on average many times through the system. Such systems are said to be in the Knudsen-regime [156]. The Knudsen number is defined as the ratio of the characteristic system size and the mean free path of the particles between collisions. To apply periodic boundary conditions, the Knudsen number must be significantly larger than one.

Consequently, the basic volume size has to be chosen large enough to avoid undesired artificial correlations. As soon as such correlations are observed, the simulation results are dubious. In Sect. 3.10.3 we show that even very large systems may be too small to simulate granular systems in certain situations. For granular gases with $\varepsilon = $ const, *any* clustering system will reveal system-spanning clusters after some time. For other systems (granular gases of viscoelastic particles) it is unclear whether such artifacts can be avoided by choosing a larger system [43, 214] (see Sect. 3.10.3).

3.8.3 Heated Walls

In many applications the system is supplied with energy through the boundaries. An important example is a granular system driven by a vibrating wall.

If the frequency of the oscillating wall is large enough, the energy supply may be modeled by a *heated wall*.

The idea of heated walls is simple: whenever a particle touches the wall, its velocity is chosen from a Maxwell distribution according to a certain temperature T.

There is, however, a complication which is frequently disregarded in the literature, therefore, we devote some discussion to this problem. Let us consider a system of *elastically* colliding particles of unit mass in a quadratic container with one heated side. If at wall collisions both components of the particle velocity are chosen from a Gaussian distribution, after a short time the temperature of the gas approaches an incorrect final temperature $T' < T$, although the gas is supposed to be in thermal equilibrium with the wall of temperature T (see Fig. 3.16).

For the explanation of this unexpected behavior, assume for a moment that the particles do not obey a Maxwell distribution, but a simplified distribution where only two values of the velocity normal to the heated wall are admitted, $v_1^n = 1$ and $v_2^n = 2$. Assume that half of the particles move with velocity v_1^n and the other half with v_2^n. Particle–particle collisions are ignored. At wall collisions the normal velocities are set randomly to v_1^n or v_2^n in such a way that the average number of particles with v_1^n and v_2^n stays constant. The particles which collide in the time interval $(t, t + \Delta t)$ come from an area of width $v^n \Delta t$. Obviously, this area is twice as large for particles traveling with $v_2^n = 2$ than for particles moving at $v_1^n = 1$. Consequently, in the time span Δt two times

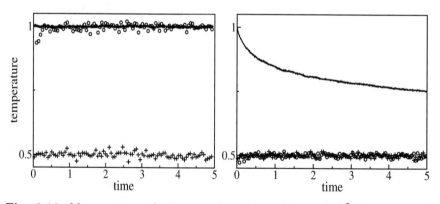

Fig. 3.16. Mean square velocity over time for a system of 10^5 elastic particles ($\varepsilon = 1$) of radius $R = 0.0005$ in a quadratic container $L = 1$ with one heated side (temperature $T = 1$). The system is initialized at temperature $T_0 = 1$. The *solid* lines show the total energy per particle (averaged over all particles), the *circles* and *crosses* display the energy of the motion in normal and tangential direction with respect to the heated wall (recorded after wall contact, each data point shows on average over 1,000 wall collisions). *Left*: correct heating according to (3.43) and (3.44). *Right*: incorrect heating; at wall collisions both components of the velocity are chosen from a Maxwell distribution at temperature T

more particles of velocity v_2^n collide with the wall than those with velocity v_1^n. In other words, faster particles collide more frequently with the wall.

To keep the velocity distribution constant, on average the same number of particles which collide with the wall has to be given the velocity v_1 as there are incoming particles of velocity v_1, and respectively the same for particles of velocity v_2. Therefore, among all particles colliding with the wall, 2/3 must adopt v_2 and 1/3 adopt v_1 to keep the distribution constant. Erroneously initializing half of the particles with v_1^n and v_2^n (misled by the fact that half of the particles in the gas moves with v_1^n and half with v_2^n) would distort the velocity distribution and *cool* the system, since the number of particles moving with v_2^n after wall contact is too small.

Now, we release the non-physical assumption of a bimodal velocity distribution and generalize: Assume a system contains on average $N(v^n)$ particles of velocity v^n normal to the heated wall. Then $\propto v^n N(v^n)$ particles collide with the wall per time unit. If the particles close to the wall are in thermal equilibrium with the wall, the random normal velocities have to be chosen such that the statistical properties are conserved, i.e., $\propto v^n N(v^n)$ particles must be assigned the normal velocity v^n. Consequently, at wall collisions the normal velocities have to be determined according to the probability distribution

$$p(v^n) = \frac{m}{T} v^n \exp\left[-\frac{m\,(v^n)^2}{2T}\right], \tag{3.43}$$

where the prefactor comes from the normalization condition $\int_0^\infty p(v^n)\mathrm{d}p = 1$.

The above arguments do not apply to the tangential velocities since the rate of wall collisions depends only on the normal velocity component, but is independent of the tangential component. Hence, the tangential velocities have to be chosen from a Maxwell distribution

$$p(v^t) = \sqrt{\frac{m}{2\pi T}} \exp\left[-\frac{m\,(v^t)^2}{2T}\right], \tag{3.44}$$

which is normalized as $\int_{-\infty}^\infty p(v^t)\mathrm{d}p = 1$. Checking again the energy evolution of the heated container as described above, the average energy fluctuates around the expected value T for the normal and $T/2$ for the tangential component (see Fig. 3.16, right).

Random numbers of the distribution (3.43) can be generated from equidistributed random numbers $z \in [0, 1)$ by means of the transformation

$$v^n = \sqrt{-\frac{2T}{m} \ln(1 - z)}. \tag{3.45}$$

The general method to compute such transformations is shown in detail in Sect. 7.2 on p. 275. Random numbers according to a Gaussian distribution for the tangential velocities (3.44) can be efficiently generated using the Box–Muller algorithm (see p. 299). Again we stress that in order to model a heated

wall with a prescribed temperature T, one *must not* initialize the normal velocity component according to a Maxwell distribution. Using this wrong initialization would (1) not lead to thermal equilibrium of a gas of elastically
colliding particles (since the velocity distribution close to the wall is always
different from the distribution far away); and (2) yield a smaller gas temperature than T. The question of which temperature the gas adopts asymptoti-
cally is a difficult problem. It has been solved by Goldhirsch [92]. For a general
mathematical description of the particle–boundary interaction see [62].

As an example for heated-wall boundary conditions in Fig. 3.17 a system
of $N = 20{,}000$ particles is shown. The container has periodic boundary conditions in the vertical directions, reflecting boundary conditions at the right
wall, and is heated at the left wall. Characteristic profiles of the particle number density and the granular temperature can be observed. This and similar
systems have been numerically and theoretically investigated, e.g., in [105].

For systems of particles that collide via a constant (velocity-independent)
coefficient of restitution in the absence of external forces, any multiplication
of the particle velocities by a constant factor leads to precisely the same
particle trajectories. Therefore, the absolute value of the heating temperature
is irrelevant for such systems, only *relative* changes are meaningful.

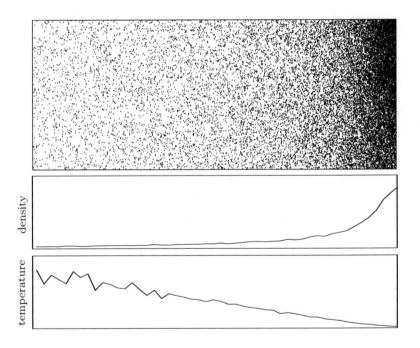

Fig. 3.17. Snapshot of a simulation of $N = 20{,}000$ particles in a container with a
heated left wall. The box is periodic in vertical direction and the right wall is elastic.
The *lower* figures show the corresponding density and temperature profiles

From the kinetic theory of granular gases (e.g., [42]) it is known that the particle velocities in a force-free granular gas do not perfectly obey a Maxwell distribution due to inelastic collisions. The deviations from the Maxwell distribution concern the high energy tail [75] and also the main part of the distribution [39, 41, 95, 283]. Thus if we talk about "heating a granular gas" we have to specify the meaning of this term. There are two possible interpretations:

1. The granular gas is coupled to an external thermodynamic reservoir at equilibrium temperature T; one may think of a molecular gas. In this case the granular particles hitting the wall have to be reinjected according to the velocity distribution of the particles in the reservoir, just as described in this section. Here the velocity components have to be determined according to to (3.43) and (3.44).
2. The thermal wall is intended to mimic the coupling to a reservoir of granular particles. In this case, the velocities have to be chosen from the steady state velocity distribution of a granular gas at temperature T with the mentioned deviations from the Maxwell distribution. Let us give an example: Assume a system of size L in the x-direction. The system is heated as described in the previous item at $x = 0$ with reflecting boundary conditions at $x = L$. From theory, e.g., [105], we know the profile of the (granular) temperature distribution and in particular $T(x = 0.8L)$. If we are interested in a simulation in the region $0.8L \leq x \leq L$, we can replace the gas in the area $0 \leq x < 0.8L$ by a heated wall at position $x = 0.8L$. In this case, the velocities should not be determined according to (3.43) and (3.44), which generate a Maxwell distribution, but in the same way as described above, where different formulae have to be derived to generate a distribution that agrees with the gas' inherent velocity distribution at $T(x = 0.8L)$.

The mentioned deviations from the Maxwell distribution are, however, small such that for most purposes they can be neglected for the simulation of heated-wall boundary discussions. A detailed discussion on the velocity distribution of dilute granular systems can be found in [42].

3.9 Inelastic Collapse

For event-driven Molecular Dynamics simulations instantaneous particle collisions are assumed, i.e., the duration of a collision is $t_c = 0$. If, moreover, we assume a constant coefficient of restitution $\varepsilon = $ const (see Sect. 3.5 for a detailed discussion) there may occur a rather artificial effect which is called *inelastic collapse* [180]: Consider three identical particles which move along a line, see Fig. 3.18. Provided the coefficient of restitution is small enough, there exist initial conditions $v_{1,2,3}$ for which the particles suffer an infinite number of collisions in finite time and dissipate the kinetic energy of their relative motion completely. After undergoing an inelastic collapse the particles move

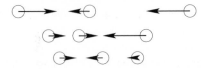

Fig. 3.18. Sketch of an inelastic collapse. There exist initial conditions such that the *left* and the *right* particle collide in alternating sequence with the particle in the *middle* until the energy of the relative motion is exhausted completely

with common velocity according to the total initial momentum. This effect was described by Shida and Kawai [266]. A necessary condition for a collapse of three particles is $\varepsilon \leq 7 - 4\sqrt{3} \approx 0.07$. For more than three particles a collapse can occur in a much wider range of the coefficient of restitution [180].

The assumption $\varepsilon = $ const is, however, neither in agreement with experiments nor with mechanics of materials. For more realistic particle models, such as viscoelastic particles, the coefficient of restitution is not constant but increases with decreasing impact velocity, $\varepsilon_{g \to 0} = 1$ (see Sect. 3.5.1). Consequently the necessary condition for an inelastic collapse cannot persist for an infinite number of collisions. As soon as the relative velocity drops below a certain value, ε exceeds the critical value and the collision sequence is interrupted. Consequently the inelastic collapse cannot exist [94, 260].

Moreover, even with the assumption $\varepsilon = $ const for two- and three-dimensional systems, it is still unclear whether the inelastic collapse exists since the phase space fraction of initial conditions which lead to a collapse is possibly of smaller dimension than the total phase space. In this case the probability to find a perfect collapse in a simulation is zero.

On the other hand, in two- and three-dimensional numerical simulations the inelastic collapse has been reported, e.g., [72, 181, 308]. With certain assumptions on the tangential coefficient of restitution, it has been shown that the cluster probability, i.e., the size of the corresponding phase space volume increases largely if the particles have rotational degrees of freedom [258]. Furthermore it has been found that the critical interval for ε which allows for inelastic collapse is significantly larger than for particles without rotational degrees of freedom. Unfortunately the method used in [258] is not conclusive because of the finite precision of number representation in the computer. In any case, if the inelastic collapse exists in two- and three-dimensional systems, it proceeds much less spectacularly than in one dimension since the total dissipation of the energy of the relative motion concerns only the normal component of the relative velocities; the tangential component stays unaffected. Since a collapse occurs in finite time we could not even identify such an event in a simple experiment, even if we were able to manufacture particles with $\varepsilon = $ const.

Nevertheless, even an *incomplete* inelastic collapse (as it can be found frequently in numerical simulations) causes problems since the collisions take place with increasing frequency. The particles undergo a very large number of collisions in a short time, i.e., the simulation practically cannot proceed beyond that time.

There are several numerical tricks to avoid an inelastic collapse: The "TC model" [167] takes into account that each collisions lasts for a finite time t_c. If a particle collides again within the time t_c, the second collision occurs elastically, i.e., with $\varepsilon = 1$. This procedure interrupts the infinite sequence of collisions since for the second collision the necessary condition $\varepsilon \leq 7 - 4\sqrt{3}$ is violated.

The TC model is intended to resolve two more problems related to hard sphere systems: it incorporates many-particle interactions and offers a possibility to simulate static systems by introducing an elastic energy of the contact. We do not discuss this model in detail but refer to [167].

Another way to avoid the inelastic collapse was proposed in [69, 104]: As soon as an approaching collapse is detected the velocity vectors of the involved particles are rotated by a small angle (e.g., $5°$). This rotation transforms part of the normal relative velocity into tangential relative velocity and interrupts the collision sequence.

In our own research we have been mainly interested in simulations of almost elastic particles, i.e., $\varepsilon \in (0.99 \ldots 0.9999)$. For such small damping, coherent motion of very many particles is necessary to allow for an inelastic collapse. Hence, the effect was not observed.

3.10 Granular Gases

3.10.1 What are Granular Gases?

The simulation of granular gases is the main application of event-driven Molecular Dynamics. The term *granular gas* is not uniquely defined in the literature. Frequently, granular gases are defined as rarefied granular systems for which the mean free path between collisions is much larger than the particle diameter. With this precondition the assumption of *molecular chaos* is justified, i.e., the assumption that particle collisions occur temporally and spatially uncorrelated. At the present time the statistical mechanics and kinetics of dilute granular gases belong to the most intensively studied subjects in the field of granular matter. A review on the rapidly developing state of the art can be found in [42, 215, 218].

In a less strict definition the spatial configuration of the particles is disregarded: Granular gases are systems of dissipatively colliding particles for which the typical time lag between successive collisions of a particle is much larger than the duration of a collision. In this case the duration of the collisions can be neglected, i.e., collisions are considered to occur instantaneously. The latter definition, hence, also includes dense systems, provided the particles are stiff enough to assure that the duration of the collisions is negligible for the description of the dynamics.

Although both definitions are very different with respect to the physical properties, any system which is described by either of these definitions may be simulated using event-driven Molecular Dynamics.

Consider a uniformly distributed granular gas in the absence of external influences, such as gravity or container walls. Due to dissipative particle collisions, the particles slow down eventually, i.e., they lose part of their kinetic energy. During the first stage of this process the gas stays homogeneous. This part of the gas evolution is called the *homogeneous cooling state*. The term cooling refers to the *granular temperature*, which is defined as the mean kinetic energy of the random motion of the particles, in analogy to the temperature of common molecular gases (see (4.2) in Sect. 4). Contrary to molecular gases, the temperature of undriven granular gases decays persistently. Dissipative collisions are also the reason for the time-dependence of the kinetic coefficients, such as the coefficients of diffusion, viscosity, and thermal conductivity. Even the velocity distribution of the particles is time-dependent for granular gases. Some of these characteristics evolve in a rather complicated way.

Using methods of statistical physics, the exciting and uncommon physical effects in granular gases have been intensively studied, such as deviations from the Maxwell velocity distribution [39, 41, 75, 95, 283], anomalous diffusion [40], shock waves [96, 97], self-organized formation of vortices (large scale velocity correlations) [45, 201, 285], and others.

The most exciting effect, however, is the spontaneous formation of large scale clusters in freely cooling granular gases, which has been investigated first by McNamara [179] and by Goldhirsch and Zanetti [93].

Most of the theoretical results have been obtained for infinite systems and with certain simplifying assumptions which have to be checked numerically. The larger the simulated systems the more precise numerical measurements are possible. A detailed introduction into the kinetic theory of granular gases can be found in [42].

3.10.2 Cluster Instability

An initially homogeneous granular gas at a certain granular temperature cools down due to inelastic particle collisions. At early times of its evolution the gas stays homogeneous and isotropic, i.e., there are no macroscopic inhomogeneities nor flows (homogeneous cooling state).

At later stages of its evolution, the gas becomes spatially inhomogeneous and pronounced density inhomogeneities develop. For small systems the formation of such clusters can be observed after few minutes of computer time using the simple algorithm as described in Sect. 3.6. Figure 3.19 shows a granular gas of $N = 3,000$ particles which interact via a constant coefficient of restitution $\varepsilon = 0.95$.

The formation of clusters can be explained qualitatively by simple arguments: Due to statistical properties of the gas the density field undergoes fluctuations. In denser regions the particles collide more frequently than in

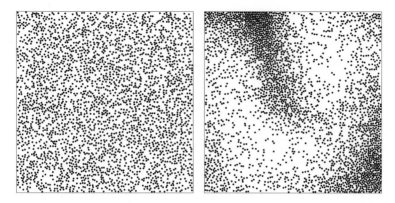

Fig. 3.19. Formation of clusters in a granular gas of $N = 3{,}000$ particles with reflecting boundary conditions. *Left*: initial state. *Right*: snapshot after 550,000 collisions

more dilute regions. Therefore, dense regions of the system cool down faster than dilute regions, and the local pressure in these regions decays as compared with the pressure in dilute regions. The resulting pressure gradient causes a flux of particles into denser region, i.e., the flux further increases the density. From these argument follows that an initially small fluctuation of the density is enhanced which results in the formation of clusters. The homogeneous cooling state of a force-free granular gas is, therefore, unstable. The described scenario is called a pressure instability [93].

While the pressure instability was numerically investigated in 1993 using a system of 40,000 particles, where up to a total of 10^7 collisions (500 collisions per particle) were observed [93], the rapid evolution of computer technology together with efficient algorithms allows now the investigation of much larger systems over longer time. Figure 3.20 shows the evolution of a homogeneously initialized granular gas with periodic boundary conditions of $N = 10^6$ particles over a period of 10^8 collisions, i.e., on average each particle was involved in 200 collisions. The simulation was performed on a PC with a 3-GHz Intel P4 CPU and lasted for about one hour.

3.10.3 Some Open Problems

While the kinetic theory of molecular gases is complete (except for some very complex mathematical issues), there is a wide variety of open problems regarding the physics of granular gases. It may be expected that the complex behavior of dissipative (granular) gases will still be of interest to theoretical physicists for a long time. Explaining most of these questions appears to be difficult without giving a profound introduction into the kinetic theory which would go far beyond the scope of this book. Here we name only three rather different problems which are still waiting for solutions.

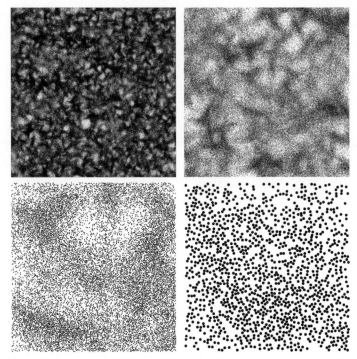

Fig. 3.20. A force-free granular gas of $N = 10^6$ granular particles after 10^8 collisions (*top left*). The other frames show magnifications. *Top right*: zoom factor 3; *bottom left*: zoom factor 9; *bottom right*: zoom factor 27

Long-Term Behavior of a Granular Gas

While there is a number of theoretical results for the homogeneous cooling state, not much is known about the kinetics of clustering. For a cooling gas with $\varepsilon =$ const, it is obvious that after the transition to the cluster state, the clusters will persist forever. This is due to the fact that the evolution of a hot gas differs from the evolution of a cool gas only by the time unit. The sequence of collisions, however, and thus the evolution of all macroscopic properties is independent of temperature. Therefore, for the case $\varepsilon =$ const, the absolute value of the temperature is unimportant since scaling of the particle velocities does not change the sequence of particle collisions.

The situation is different for more realistic gases of viscoelastic particles: Since the coefficient of restitution ε is a decaying function of the impact velocity (see Sect. 3.2), for decaying temperature the particles collide more and more elastically. For $t \to \infty$ the particles collide perfectly elastically ($\varepsilon = 1$), just as in a molecular gas. A molecular gas, however, stays homogeneous in the absence of external forces.

Hence, the described scenario of cluster formation refers to a granular gas whose particles collide with a constant coefficient of restitution, but not necessarily to a gas of viscoelastic particles. Since the particles in dense (cool) regions dissipate on average a smaller fraction of their kinetic energy per collision than particles in dilute regions. In analogy to molecular gases, they tend to dissolve inhomogeneities. It is not obvious à priori whether the cluster regime is stable after a long time or whether the system returns to the homogeneous cooling state.

To solve this problem theoretically, the time scales of cluster formation and spreading of particles, i.e., cluster dissolution, have to be estimated. This problem is difficult and so far has not been solved.

Figure 3.21 shows the evolution of granular gases of particles which collide

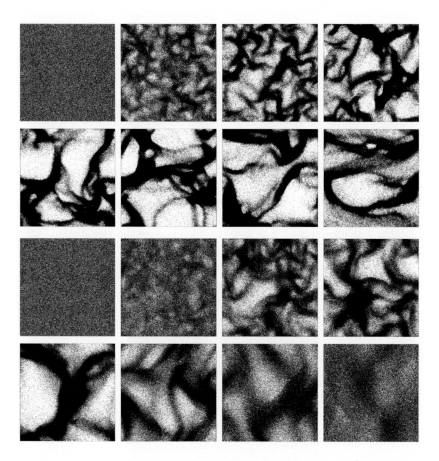

Fig. 3.21. Snapshots of the evolution of granular gases of $N = 10^5$ particles. *Top series* (8 figures): $\varepsilon = 0.95 = \text{const}$ is assumed. The snapshots are taken after 0, 10^7, 1.5×10^7, 2.5×10^7, 1.5×10^7, 10^8, 2×10^8, and 3×10^8 collisions. *Bottom series*: viscoelastic particles with $\varepsilon = \varepsilon(g)$. The snapshots are taken after 0, 10^7, 2×10^7, 3×10^7, 10^8, 5×10^8, 10^9, and 2×10^9 collisions

with $\varepsilon = 0.95 = $ const (top) and of viscoelastic particles with $\varepsilon = \varepsilon(g)$. Both gases consist of $N = 100{,}000$ particles. The parameters of the gas of viscoelastic particles are chosen such that initially $\varepsilon\left(v_{T(t=0)}\right) = 0.95$, where $v_{T(t=0)}$ is the thermal velocity at the initial temperature $T(t = 0) = 1$. Therefore, initially both gases behave identically. As expected, for $\varepsilon = $ const clusters occur earlier since after some cooling the viscoelastic particles collide less dissipatively.

The last snapshots of the sequence may suggest the conclusion that the clusters dissolve for the case of viscoelastic particles, i.e., the simulation supports the hypothesis that cluster formation is a transient phenomenon (see also [43, 214]). Note, that in the sequence shown in Fig. 3.21 system-spanning clusters appear, which contradicts the precondition for applying periodic boundary conditions (see Sect. 3.8.2). Therefore, the impression of Fig. 3.21 may be misleading.

Heated Granular Gas with Gravity

Consider a granular gas which is heated from below in a gravity field. In a certain range of system parameters we observe a dense cluster of almost perfectly hexagonally-packed particles, which floats on top of a dilute granular gas (see [184] for details). A snapshot of this system is shown in Fig. 3.22. The upper part of the cluster is organized in a crystalline structure, including defects as domain boundaries and voids. For a given container geometry, (constant) coefficient of restitution, and driving temperature, there exists a minimum number of particles for the emergence of the floating-cluster configuration. For smaller particle number a density profile close to the barometric formula is found [74].

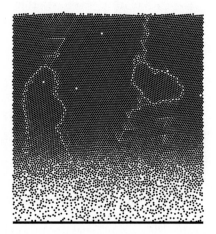

Fig. 3.22. A granular gas heated from below ($N = 10{,}000$; $\varepsilon = 0.98815$). In a certain range of temperature a dense cluster floats on top of a dilute granular gas [184]. Periodic boundary conditions have been assumed in the horizontal direction

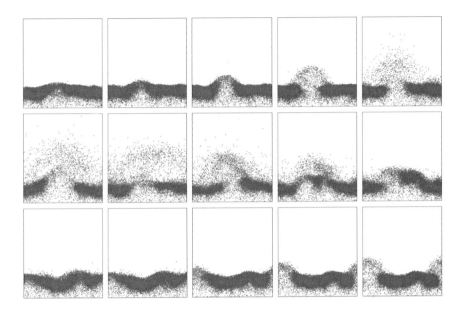

Fig. 3.23. Snapshots of the system's evolution. Time increases from the *top left* to the *bottom right*. Large eruptions occur in irregular intervals

For moderate driving temperature the cluster floats at a stable height. For somewhat higher temperatures, a relatively regular oscillation of the height is found. For yet higher temperatures, recurrent eruptions occur in irregular intervals. Figure 3.23 shows snapshots of the system.

While the time-independent behavior of the system has been described by a hydrodynamic model up to very good accuracy [184], so far neither the height oscillations nor the nature of the eruptions could be explained theoretically in quantitative terms.

Collision Chains

In dense systems it is frequently observed that a subset of almost linearly aligned particles is involved in many more collisions than the particles in the direct neighborhood of these structures. Figure 3.24

shows a granular gas of 5,000 particles in a circular container with a heated wall. The gray scale encodes the number of collisions the particle has suffered. A more detailed analysis [75] reveals that 5% of the most frequently colliding particles take part in about 96% of all the collisions in the system. Figure 3.24 shows that these particles are almost linearly aligned, therefore, we call them collision chains. Such collision chains appear in a self-organized way and are reorganized persistently. Figure 3.25 shows a collection of collision-chain skeletons at different instances of the simulation (see [213] for details).

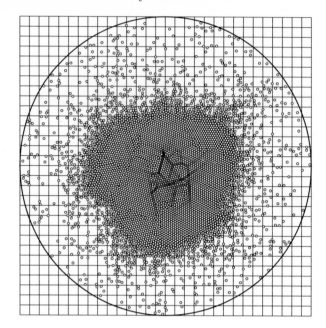

Fig. 3.24. A force-free granular gas of $N = 5{,}000$ particles in a circular container with a heated wall. The gray scale of the particles encodes the number of collisions in which it was involved during the past 10^5 collisions of the total system. Dark particles have suffered many collisions. The lattice used for the event-driven simulation is also shown

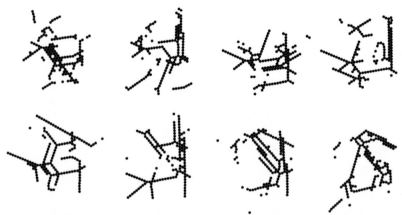

Fig. 3.25. Collision chain skeletons from the system shown in Fig. 3.24 at different times. While the typical sizes of such chains remain invariant, the concrete realization undergoes large fluctuations

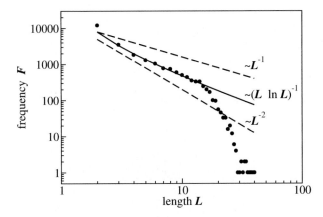

Fig. 3.26. The distribution of collision chain lengths. The *dashed lines* show $F \propto$ L^{-1} and $F \propto L^{-2}$

The collision chains from the simulation shown in Fig. 3.24 have been investigated with respect to their length distribution. From the analysis of about 28,000 collision chains, the length distribution shown in Fig. 3.26 was obtained. Obviously the distribution does not obey a power law, however, it can be fitted by $F \propto (L \log L)^{-1}$. So far neither the conditions for the appearance of collision chains nor their length distribution has been clarified sufficiently. Whether these collision chains are related to the force chains, which are observed in force-based Molecular Dynamics simulations (see p. 5), is not known either.

Program Index

4

Direct Simulation Monte Carlo

4.1 Idea of Direct Simulation Monte Carlo

Chapter 2 deals with Molecular Dynamics for the simulation of granular systems. This method allows us, in principle, to simulate systems of complicatedly shaped particles in dynamic and static situations, and also in the case of multi-particle contacts. Molecular Dynamics requires only one precondition: one needs to know the forces and torques acting between contacting particles as functions of the particles' positions, their velocities, their angular orientation, and their angular velocities, i.e., $\vec{F}_{ij}(\vec{r}_i, \vec{r}_j, \vec{v}_i, \vec{v}_j, \vec{\varphi}_i, \vec{\varphi}_j, \vec{\omega}_i, \vec{\omega}_j)$ and $\vec{M}(\dots)$ (depending on the same arguments). Although (force-based) Molecular Dynamics is a universal simulation method, its application is restricted to rather small system size due to the time-intensive numerical solution of Newton's equation of motion (2.1). Molecular Dynamics provides an exact description of granular systems as the trajectory of each grain is computed.

The second type of Molecular Dynamics simulations, event-driven Molecular Dynamics as described in Chap. 3, is numerically much more efficient than force-based Molecular Dynamics. It is applicable if the dynamics of a system is mainly determined by pairwise particle collisions of vanishing duration, i.e., if multi-particle contacts as well as long-lasting contacts are much less important for the dynamics as compared with pairwise collisions (see Sect. 3.1 for a more detailed discussion). Although event-driven Molecular Dynamics is less universal than force-based simulations, it is also an exact description since it computes the particle trajectories. Event-driven Molecular Dynamics is very efficient; it allows for the simulation of up to millions of particles over moderate real times. However, the simulation may become very time consuming for large three-dimensional systems.

The present chapter describes a simulation method which is yet more efficient than event-driven Molecular Dynamics. This non-deterministic method was proposed by Bird [30] for the simulation of rarefied molecular gases and is termed "Direct Simulation Monte Carlo" (DSMC). Although the particle trajectories are not computed, it has been shown that under certain preconditions, which are discussed below, DSMC describes many experimental observations for equilibrium and non-equilibrium systems (e.g., [193]).

Just like Molecular Dynamics algorithms, DSMC also deals with the motion of particles, although these particles are of different nature [34]. To explain this, we briefly introduce the very fundamentals of kinetic gas theory.

The state of a dilute molecular or granular gas is described by the *one-particle distribution function* $f(\vec{r}, \vec{v}, t)$. It quantifies the (infinitesimal) number of particles, $f(\vec{r}, \vec{v}, t)\, d\vec{r}\, d\vec{v}$, which are located in an infinitesimal volume $d\vec{r}$ at position $\vec{r}$ and whose velocities are in the infinitesimal interval $d\vec{v}$ around $\vec{v}$. (For brevity of the notation we do not consider the rotational degree of freedom, otherwise the distribution function would have two more arguments $f(\vec{r}, \vec{v}, \vec{\varphi}, \vec{\omega}, t)$ and the mathematical expressions would be lengthier.) Thus, integration over the full phase space yields the total number of particles:

$$\int f(\vec{r}, \vec{v}, t)\, d\vec{r}\, d\vec{v} = N \ . \tag{4.1}$$

The distribution function describes the system completely on a coarse grained level. The hydrodynamic fields of particle number density n, flow velocity $\vec{u}(\vec{r}, t)$, and granular temperature $T(\vec{r}, t)$ follow simply as moments of the distribution function:

$$n(\vec{r}, t) = \int f(\vec{v}, \vec{r}, t)\, d\vec{v}$$

$$n(\vec{r}, t)\, \vec{u}(\vec{r}, t) = \int \vec{v} f(\vec{v}, \vec{r}, t)\, d\vec{v} \tag{4.2}$$

$$\frac{3}{2} n(\vec{r}, t)\, k_B T(\vec{r}, t) = \frac{m}{2} \int [\vec{v} - \vec{u}(\vec{r}, t)]^2 f(\vec{v}, \vec{r}, t)\, d\vec{v} \ .$$

This is the starting point of the hydrodynamic theory of (granular) gases. A detailed introduction to this exciting field is given in [42]. All hydrodynamics is based on the distribution function $f(\vec{r}, \vec{v}, t)$, which in its turn depends on the microscopic properties of the particles and their interaction.

The idea of DSMC is to determine the time-dependent distribution function $f(\vec{r}, \vec{v}, t)$ by means of a quasi-particle simulation. The dynamics of the distribution function is governed by the Boltzmann equation which is discussed in the next section. Naturally, this equation does not describe the motion of particles (as Newton's equation), but the time-dependent flow of probability. Thus, the quasi-particles in DSMC (or direct simulation Monte Carlo) are probability quanta rather than real particles.

Obviously, each real particle of a system contributes to the probability field $f(\vec{r}, \vec{v}, t)$, according to its coordinates and its velocity. Thus, the Boltzmann equation may, on one hand, be derived from the dynamics of particles. On the other hand, the Boltzmann equation may be transformed into a dynamics for quasi-particles (probability quanta). This is the idea behind DSMC.

4.2 Boltzmann Equation

In the absence of external forces there are three different mechanisms by which the number of particles in a certain phase space interval, $f(\vec{v}, \vec{r}, t)\, \mathrm{d}\vec{r}\, \mathrm{d}\vec{v}$, may change:

1. Particles from the interval $\mathrm{d}\vec{v}$ around $\vec{v}$ collide with other particles. Their post-collision velocity is $\vec{v}' \neq \vec{v}$, i.e., the particles leave the velocity interval $\mathrm{d}\vec{v}$ around $\vec{v}$. The number of particles in this interval is, thus, reduced by this mechanism.
2. Particles of velocity $\vec{v}''$ collide in such a way that their post-collision velocity is in the interval $\mathrm{d}\vec{v}$ around $\vec{v}$, i.e., the particles enter the interval. The number of particles at $\vec{v}$ is, therefore, increased by this mechanism.
3. Particles may enter or leave the interval $\mathrm{d}\vec{r}$ around $\vec{r}$ by free streaming, i.e., without interaction with other particles. Their velocities remain, therefore, unchanged.

Let us first discuss the evolution of the velocity distribution function due to mechanisms 1 and 2. For inelastic particles of hard-sphere gases, the post-collision velocities $\vec{v}_1'$ and $\vec{v}_2'$ of a pair of colliding particles relate to the present velocities $\vec{v}_1$ and $\vec{v}_2$ via the collision rule (3.32):

$$\vec{v}_1' = \vec{v}_1 - \frac{1+\varepsilon}{2}\,(\vec{v}_{12} \cdot \vec{e})\,\vec{e}\,, \qquad \vec{v}_2' = \vec{v}_2 + \frac{1+\varepsilon}{2}\,(\vec{v}_{12} \cdot \vec{e})\,\vec{e}\,, \tag{4.3}$$

which is a simplification of the full collision rule (3.16) for non-rotating spheres. In (4.3) the definitions of the relative particle velocity, $\vec{v}_{12} \equiv \vec{v}_1 - \vec{v}_2$, and the unit vector at contact, $\vec{e} \equiv (\vec{r}_1 - \vec{r}_2)/|\vec{r}_1 - \vec{r}_2|$, have been used.

For the inverse collision (mechanism 2), the *pre-collision* velocities $\vec{v}_1''$ and $\vec{v}_2''$ are related to the present velocities via the inverse collision rule:

$$\vec{v}_1'' = \vec{v}_1 - \frac{1+\varepsilon}{2\,\varepsilon}\,(\vec{v}_{12} \cdot \vec{e})\,\vec{e}\,, \qquad \vec{v}_2'' = \vec{v}_2 + \frac{1+\varepsilon}{2\,\varepsilon}\,(\vec{v}_{12} \cdot \vec{e})\,\vec{e}\,, \tag{4.4}$$

as it follows from inverting the collision rule (4.3).

Let us change the notation slightly by looking at the change of the distribution function during the small time span $\mathrm{d}t$ from the point of view of particle 1, i.e., $f(\vec{v}_1, t)$ is considered. The number of particles $f(\vec{v}_1, t)\, \mathrm{d}\vec{v}$ in the phase space volume $\mathrm{d}\vec{v}$ around $\vec{v}_1$ changes due to a collision if either

1. $(\vec{v}_1, \vec{v}_2) \to (\vec{v}_1', \vec{v}_2')$, i.e., particle 1 leaves this interval after collision with any particle of velocity $\vec{v}_2$, or
2. $(\vec{v}_1'', \vec{v}_2'') \to (\vec{v}_1, \vec{v}_2)$, i.e., particle 1 enters the interval.

The total change of f during Δt, therefore, reads

$$\Delta f(\vec{v}_1, t)\, \mathrm{d}\vec{v}_1 = -\,\mathrm{d}\vec{v}_1 \int \mathrm{d}\vec{v}_2\, \mathrm{d}\vec{e}\, \nu^-\,(\vec{v}_1, \vec{v}_2, \vec{e})\, \Delta t$$

$$+ \int \mathrm{d}\vec{v}_1''\, \mathrm{d}\vec{v}_2''\, \mathrm{d}\vec{e}\, \nu^+\,(\vec{v}_1'', \vec{v}_2'', \vec{e})\, \Delta t\,. \tag{4.5}$$

The first integral is the contribution of collisions suffered by particles moving at velocity $\vec{v}_1$, thus reducing the number of particles in the velocity interval $d\vec{v}$ around $\vec{v}_1$. The expression $\nu^- (\vec{v}_1, \vec{v}_2, \vec{e})$ is the number of collisions per time (collision frequency) between particles of velocities $\vec{v}_1$ and $\vec{v}_2$ which occur at the normal vector $\vec{e}$. We integrate over the velocity of the collision partner $\vec{v}_2$ and the normal vector to account for all possible particle orientations due to mechanism 1.

The second integral in (4.5) describes the opposite mechanism 2, i.e., particles moving at velocities $\vec{v}_1''$ and $\vec{v}_2''$ collide and end up at velocities $\vec{v}_1$ and $\vec{v}_2$. The corresponding collision rate is $\nu^+ (\vec{v}_1'', \vec{v}_2'', \vec{e})$. The integration is performed over those values of $\vec{v}_1''$, $\vec{v}_2''$, and $\vec{e}$ which yield $\vec{v}_1$ and an arbitrary $\vec{v}_2$ by the inverse collision rule (4.4).

The collision frequencies[1] may be computed by considering the geometric relation between particles 1 and 2. This derivation can be found in detail in [42]; only a brief motivation and the final result are given here.

To estimate the number of collisions $\nu^- (\vec{v}_1, \vec{v}_2, \vec{e})\, d\vec{v}_1\, d\vec{v}_2\, d\vec{e}$ at orientation $\vec{e}$ between two particles of radius R and velocities $\vec{v}_1$ and $\vec{v}_2$, consider the equivalent problem of a resting particle of radius $2R$ in a bath of point particles moving at velocity $\vec{v}_1 - \vec{v}_2$. The required number of collisions with relative orientation $\vec{e}$ can be found by counting the number of point particles which collide in the infinitesimal interval dt. These point particles are contained in an inclined generalized cylinder of base $(2R)^2 \vec{e}\, d\vec{e}$ and (oriented) height $\vec{v}_{12}\, dt = (\vec{v}_1 - \vec{v}_2)\, dt$ (see Fig. 4.1). The volume of this generalized cylinder is

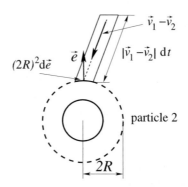

Fig. 4.1. The collision frequency is determined by counting the number of point particles moving at velocity $\vec{v}_{12} = \vec{v}_1 - \vec{v}_2$ which collide in the interval dt with a resting particle of radius $R_2 = 2R$. All point particles in the generalized cylinder of base $(2R)^2 d\vec{e}$ and height $|\vec{v}_{12}|\, dt$ collide with the large particle during the interval dt

---

[1] More precisely, the expressions ν^- and ν^+ in (4.9) are densities of collision frequencies with respect to $\vec{e}$, $\vec{v}_1$, and $\vec{v}_2$. For compatibility with the standard literature, however, we call them collision frequencies.

$$dV = \Theta\left[-\vec{v}_{12} \cdot \vec{e}\right](2R)^2 \left|\vec{v}_{12} \cdot \vec{e}\right| d\vec{e}\, dt \ . \tag{4.6}$$

The Heaviside function

$$\Theta(x) \equiv \begin{cases} 0 & \text{for} \quad x < 0 \\ 1 & \text{for} \quad x \geq 0 \end{cases} \tag{4.7}$$

discriminates approaching particles, where $\vec{v}_{12} \cdot \vec{e} < 0$, from departing particles which do not collide. Thus the desired collision frequency is

$$\nu^-\left(\vec{v}_1, \vec{v}_2, \vec{e}\right) d\vec{v}_1 d\vec{v}_2 d\vec{e} = \frac{dV}{dt} f\left(\vec{v}_1\right) f\left(\vec{v}_2\right) d\vec{v}_1 d\vec{v}_2 \ . \tag{4.8}$$

A similar derivation can be applied to compute the collision frequency ν^+ for collisions of particles traveling at $\vec{v}_1''$ and $\vec{v}_2''$. The final result reads

$$\nu^-(\vec{v}_1, \vec{v}_2, \vec{e})\, d\vec{v}_1 d\vec{v}_2 d\vec{e} = f(\vec{v}_1, t) f(\vec{v}_2, t)\, \Theta[-\vec{v}_{12} \cdot \vec{e}](2R)^2 \left|\vec{v}_{12} \cdot \vec{e}\right| d\vec{v}_1 d\vec{v}_2 d\vec{e}$$
$$\nu^+(\vec{v}_1'', \vec{v}_2'', \vec{e})\, d\vec{v}_1'' d\vec{v}_2'' d\vec{e} = f(\vec{v}_1'', t) f(\vec{v}_2'', t)\, \Theta[-\vec{v}_{12}'' \cdot \vec{e}](2R)^2 \left|\vec{v}_{12}'' \cdot \vec{e}\right| d\vec{v}_1'' d\vec{v}_2'' d\vec{e} \ . \tag{4.9}$$

With the collision rule (4.3) and the rule for the inverse collision (4.4) the differential volume $d\vec{v}_1'' d\vec{v}_2''$ can be expressed by the Jacobian,

$$d\vec{v}_1'' d\vec{v}_2'' = \frac{\mathcal{D}\left(\vec{v}_1'', \vec{v}_2''\right)}{\mathcal{D}\left(\vec{v}_1, \vec{v}_2\right)} d\vec{v}_1\, d\vec{v}_2 = \frac{1}{\varepsilon}\, d\vec{v}_1\, d\vec{v}_2 \ . \tag{4.10}$$

The detailed computation can be found in [42]. The Jacobian given in (4.10) applies to particles which collide with a coefficient of restitution that is independent of the impact velocity, $\varepsilon = \text{const}$ (see Sect. 3.5.1 for a discussion of this assumption). For different collision rules, this expression has to be modified.

Moreover, using the definition of the coefficient of restitution

$$\frac{-\vec{v}_{12}'' \cdot \vec{e}}{\vec{v}_{12} \cdot \vec{e}} = \frac{\left|\vec{v}_{12}'' \cdot \vec{e}\right|}{\left|\vec{v}_{12} \cdot \vec{e}\right|} = \frac{1}{\varepsilon} \ , \tag{4.11}$$

the property of the Theta-function

$$\Theta\left(-\vec{v}_{12}'' \cdot \vec{e}\right) = \Theta\left(\vec{v}_{12} \cdot \vec{e}\right) \ , \tag{4.12}$$

and changing the integration variable $\vec{e} \to -\vec{e}$, the second equation (4.9) turns into

$$\nu^+\left(\vec{v}_1'', \vec{v}_2'', \vec{e}\right) d\vec{v}_1'' d\vec{v}_2'' d\vec{e}$$
$$= -\frac{1}{\varepsilon^2} f\left(\vec{v}_1'', t\right) f\left(\vec{v}_2'', t\right) \Theta\left[-\vec{v}_{12} \cdot \vec{e}\right](2R)^2 \left|\vec{v}_{12} \cdot \vec{e}\right| d\vec{v}_1 d\vec{v}_2 d\vec{e} \ . \tag{4.13}$$

In the limit $\Delta t \to 0$, equation (4.5), together with the collision frequencies (4.9) and (4.13), turns into the Boltzmann equation for the evolution of the velocity distribution function due to particle collisions, disregarding the rotational degree of freedom

$$\frac{\partial}{\partial t} f\left(\vec{v}_1, t\right) = I , \tag{4.14}$$

with the collision integral

$$I \equiv (2R)^2 \int d\vec{v}_2 \int d\vec{e}\, \Theta\left(-\vec{e} \cdot \vec{v}_{12}\right) |\vec{e} \cdot \vec{v}_{12}| \left[\frac{1}{\varepsilon^2} f(\vec{v}_1'', t) f(\vec{v}_2'', t) - f(\vec{v}_1, t) f(\vec{v}_2, t)\right]. \tag{4.15}$$

So far the free streaming motion of particles has been disregarded (mechanism 3 on p. 193). This corresponds to the evolution of a homogeneous force-free granular gas, where $f\left(\vec{v}, \vec{r}, t\right) = f\left(\vec{v}, t\right)$. In general, however, collision-free (ballistic) particle transport also modifies f because the probability to find a particle in the phase space volume $f\left(\vec{v}, \vec{r}, t\right) d\vec{v}\, d\vec{r}$ changes when particles enter or leave the differential volume element $d\vec{r}$. This contribution leads to an additional term in the Boltzmann equation which becomes

$$\left(\frac{\partial}{\partial t} + \vec{v}_1 \cdot \nabla\right) f\left(\vec{v}_1, \vec{r}, t\right) = I . \tag{4.16}$$

The collision integral (4.15) stays unaffected, except for the arguments of the distribution function which also depends on the position $\vec{r}$ now.

For the discussion so far, it has been assumed that the particles move independently from each other, i.e., the two-particle distribution function is the product of two one-particle distribution functions:

$$f_2\left(\vec{v}_1, \vec{v}_2, \vec{r}_1, \vec{r}_2, t\right) = f\left(\vec{v}_1, t\right) f\left(\vec{v}_2, t\right) . \tag{4.17}$$

This assumption is called *molecular chaos hypothesis* or *Stoßzahlansatz*, which is the most important precondition for the Boltzmann equation. In real systems, the interaction of particles, however, cause correlations which may require consideration of the two-particle distribution function,

$$f_2\left(\vec{v}_1, \vec{v}_2, \vec{r}_1, \vec{r}_2, t\right) = f\left(\vec{v}_1, \vec{v}_2, |\vec{r}_1 - \vec{r}_2|, t\right) = f\left(\vec{v}_1, \vec{v}_2, r_{12}, t\right) \tag{4.18}$$

(in the homogeneous isotropic case). The most simple correction is the Enskog factor g_2 (the pair correlation function at contact) to account for excluded volume effects in gases of finite density. Equation (4.17) reads then

$$f_2\left(\vec{v}_1, \vec{v}_2, 2R, t\right) \approx g_2\, f\left(\vec{v}_1, t\right) f\left(\vec{v}_2, t\right) \tag{4.19}$$

and the Boltzmann equation (4.14) turns into the Boltzmann–Enskog equation [63]:

$$\left(\frac{\partial}{\partial t} + \vec{v}_1 \cdot \nabla\right) f\left(\vec{v}_1, t\right) = g_2 I . \tag{4.20}$$

The Enskog factor depends on the particle number density n via [61]

$$g_2 = \frac{2 - \eta}{2(1 - \eta)^2} , \qquad \text{with} \qquad \eta = \frac{4}{3}\pi n R^3 \tag{4.21}$$

for the three-dimensional case. In the two-dimensional case different expressions apply [109]. Other deviations from molecular chaos which can be taken into account by additional terms originate from ring collisions [284] and from velocity correlations which emerge from the dissipative character of the collisions [216]. They will not be discussed here.

4.3 Collision Frequency of a Uniform Hard Sphere Gas

For the DSMC algorithm the total number of collisions per unit time dN_c/dt of N uniformly distributed particles in a finite volume V is needed. To this end, the density of the collision frequency $\nu^-\left(\vec{v}_1, \vec{v}_2, \vec{e}\right)$ given by (4.9), is integrated over all velocities $\vec{v}_1$ and $\vec{v}_2$, all orientations $\vec{e}$, and over the volume of the system. For a homogeneous gas, the latter integration is trivial:

$$\begin{aligned}
\frac{dN_c}{dt} &= \frac{V}{2} \int \nu^-\left(\vec{v}_1, \vec{v}_2, \vec{e}\right) d\vec{v}_1 \, d\vec{v}_2 \, d\vec{e} \\
&= \frac{V}{2} \int \Theta\left[-\vec{v}_{12} \cdot \vec{e}\right] f\left(\vec{v}_1, t\right) f\left(\vec{v}_2, t\right) (2R)^2 \left|\vec{v}_{12} \cdot \vec{e}\right| d\vec{v}_1 \, d\vec{v}_2 \, d\vec{e} .
\end{aligned} \tag{4.22}$$

Since each collision involves two particles, the factor $1/2$ is needed for the total number of collisions to avoid double counting. Using the rule

$$\int \Theta\left(-\vec{a} \cdot \vec{e}\right) \left|\vec{a} \cdot \vec{e}\right| d\vec{e} = \pi \left|\vec{a}\right| \tag{4.23}$$

for an arbitrary constant vector $\vec{a}$, the collision frequency reads

$$\frac{dN_c}{dt} = \frac{V}{2}\pi(2R)^2 \int f\left(\vec{v}_1, t\right) f\left(\vec{v}_2, t\right) \left|\vec{v}_{12}\right| d\vec{v}_1 d\vec{v}_2 . \tag{4.24}$$

The integral is (up to a constant factor) the definition of the average relative velocity of particles, $n^2 \langle v_{12} \rangle$, where $n \equiv N/V$ is the particle number density. Thus, the final result for the collision frequency in the system is

$$\frac{dN_c}{dt} = \frac{2\pi R^2 N^2 \langle v_{12} \rangle}{V} . \tag{4.25}$$

To evaluate the average relative velocity $\langle v_{12} \rangle$, one needs the distribution function which is unknown à priori.

4.4 Integration of the Boltzmann Equation

The method of Direct Simulation Monte Carlo is intended to solve the Boltzmann or Boltzmann–Enskog equation, i.e., it determines the velocity distribution as a function of the spatial position $\vec{r}$ and time t. This integration is performed by subjecting imaginary probability units $\Delta f(\vec{r}, t)$ to the action of the collision operator. Since the velocity distribution function represents the probability to find a particle in a certain phase space interval, in a sense, these probability units can be understood as quasi-particles. For systems of identical particles, these quasi-particles may be considered as representatives of a certain fixed number of real particles (see [86] for a quantitative discussion). Therefore, the simulation of the Boltzmann equation resembles a particle algorithm. Nevertheless, one should be aware that the simulated particles are not identical with the physical particles. For example, the particle number density can be increased to arbitrary values without formally leaving the low-density limit, since the Boltzmann equation is still solved by the algorithm [34]. Therefore, DSMC should be used with some care as inappropriate application may easily lead to non-physical results; an example is given in Sect. 4.6.

For simplicity, let us first discuss a homogeneous system. To simulate the Boltzmann–Enskog equation, pairs of particles are selected according to the actual collision frequency given by (4.9),

$$\nu^-(\vec{v}_1, \vec{v}_2, \vec{e}) = f(\vec{v}_1, t)\, f(\vec{v}_2, t)\, \Theta\left[-\vec{v}_{12} \cdot \vec{e}\right](2R)^2 \left|\vec{v}_{12} \cdot \vec{e}\right|\,, \tag{4.26}$$

i.e., the probability of selecting a pair of particles with certain velocities will be proportional to the number of particles of these velocities $f(\vec{v}_1, t)$ and $f(\vec{v}_2, t)$, and also proportional to the normal relative velocity $\left|\vec{e} \cdot \vec{v}_{12}\right|$ of these particles. The constant factor $(2R)^2$ as well as $\Theta\left[-\vec{v}_{12} \cdot \vec{e}\right]$ are not critical. The latter factor means that, on average, half of the pairs selected by the other proportionalities collide since the other half has positive relative normal velocity, i.e., they depart from each other.

There is no simple, direct way to select pairs of particles such that (4.26) is assured. The distribution can, however, be generated à posteriori by a three step process:

1. A pair of particles is selected randomly, where each pair is equally probable.
2. A direction $\vec{e}$ is selected randomly according to an equidistribution.
3. The pair is accepted if

$$\left|\vec{e} \cdot \vec{v}_{12}\right| = \left|\vec{e} \cdot (\vec{v}_1 - \vec{v}_2)\right| > \mathtt{rand}[0, 1)\, v_{12}^{\max}\,, \tag{4.27}$$

where $\mathtt{rand}[0, 1)$ is a uniformly distributed random number and $v_{12}^{\max}$ is an upper bound of the relative particle velocity. The question of how to select $v_{12}^{\max}$ is addressed below.

The number of candidate pairs which have to be considered to simulate the correct number of collisions, can be found by means of kinetic theory. Let us assume a sufficiently small time interval Δt, such that the system can be considered as stationary. Equation (4.25) yields the number of collisions which take place during Δt on average:

$$N_{\mathrm{c}}^{\Delta t} = \frac{\mathrm{d}N_{\mathrm{c}}}{\mathrm{d}t} \Delta t = \frac{2\pi R^2 N^2 \langle |\vec{v}_{12}| \rangle \Delta t}{V} . \tag{4.28}$$

Thus, the number of candidate pairs $N_{\mathrm{p}}^{\Delta t}$ is determined such that the number of *accepted* candidate pairs equals $N_{\mathrm{c}}^{\Delta t}$. Their ratio follows from the selection procedure described above:

$$\frac{N_{\mathrm{c}}^{\Delta t}}{N_{\mathrm{p}}^{\Delta t}} = \frac{\langle |\vec{e} \cdot \vec{v}_{12}| \rangle}{v_{12}^{\max}} = \frac{1}{v_{12}^{\max}} \frac{1}{4\pi n^2} \int \mathrm{d}\vec{e}\, \mathrm{d}\vec{v}_1 \mathrm{d}\vec{v}_2 f(\vec{v}_1) f(\vec{v}_2) |\vec{e} \cdot (\vec{v}_1 - \vec{v}_2)|$$

$$= \frac{1}{v_{12}^{\max}} \frac{\pi}{2\pi n^2} \int \mathrm{d}\vec{v}_1 \mathrm{d}\vec{v}_2 f(\vec{v}_1) f(\vec{v}_2) |\vec{v}_1 - \vec{v}_2| = \frac{1}{2v_{12}^{\max}} \langle |\vec{v}_{12}| \rangle , \tag{4.29}$$

where the definitions of $\langle |\vec{e} \cdot \vec{g}| \rangle$ and $\langle |\vec{v}_{12}| \rangle$ have been used. Inserting (4.28) yields the required number of candidate pairs

$$N_{\mathrm{p}}^{\Delta t} = \frac{4\pi R^2 N^2 v_{12}^{\max}}{V} \Delta t \equiv C_1 v_{12}^{\max} N^2 \Delta t , \tag{4.30}$$

which defines the constant C_1.

Fortunately, the average relative velocity of the particles $\langle |\vec{v}_{12}| \rangle$ is not contained in the final result. The numerical evaluation of this quantity would require serious computational effort which grows as $\mathcal{O}(N^2)$. The value of $v_{12}^{\max}$ can be chosen arbitrarily with the only restriction that it needs to be an upper bound of the average, normal relative velocity. However, due to (4.30), the number of candidate pairs to be considered increases linearly with $v_{12}^{\max}$ since the number of rejected candidate pairs grows when the chosen $v_{12}^{\max}$ is too large. Therefore, it is desirable to choose the smallest possible value for the upper bound $v_{12}^{\max}$ to accelerate the simulation. There are various heuristics for the choice of this parameter, either based on the velocities found in a sample, i.e., on the velocity values found for candidate particles in previous time steps (e.g., [86]), or based on assumptions about the velocity distribution function. If there are reasons to expect a velocity distribution similar to a Maxwell distribution, a good choice is $v_{12}^{\max} = C\, v_{\mathrm{T}}$, where v_{T} is the thermal velocity and C is a constant, e.g., $C = 5$.

The described procedure is applicable to the simulation of a uniform system only. Non-uniform systems are subdivided into boxes by a rectangular lattice such that each box can be considered as uniform (local equilibrium). Collision partners are always chosen from the same cell. The flow of probability due to density gradients, as described by the flow term in the Boltzmann equation (4.16), can be simulated by quasi-particles as well. During the time

Δt, each probability quantum moves according to the velocity of the corresponding quasi-particle,

$$\vec{r}_i = \vec{r}_i + \vec{v}_i \, \Delta t \,, \qquad i = 0, \ldots, N - 1 \,. \tag{4.31}$$

Since local equilibrium is assumed for each box, the box size must be larger than the mean free path.

Finally, the value of the time step Δt has to be discussed. In an event-driven Molecular Dynamics simulation the particles move on ballistic trajectories except for events of infinitesimal duration when the velocities of colliding particles are modified according to a collision law. Thus, periods of undisturbed motion are determined by the time lag δt between subsequent collisions in the entire system. For force-based Molecular Dynamics simulations the characteristic progression time is the time step of the integration algorithm, which must be smaller than δt since the duration of each collision will be resolved in many integration time steps.

For the numerical solution of the Boltzmann equation, which explicitly *disregards* correlations between particles, it is enough to assure that the collision frequency of each particle is in agreement with physical reality, while the sequence of collisions is disregarded. Thus, the main idea of the Direct Simulation Monte Carlo method is to decouple collisions and particle motion. Each particle moves and undergoes characteristic collisions according to its own position and velocity and according to the statistical properties of the gas, but disregards the positions and velocities of the other particles. Consequently, the characteristic time scale for a DSMC is the average time lag between subsequent collision of *the same particle*, Δt, which is much larger than the time lag between subsequent collisions in the entire system, $\Delta t \gg \delta t$.

4.5 Implementation

For the DSMC simulation of a given system, two constants have to be defined, the time step Δt and the box size $L_{\text{box}} = L/N_{\text{box}}$, where L is the size of the cubic simulation area and N_{box} is the number of boxes in each spatial dimension. For a system in equilibrium, L_{box} and Δt can be chosen as constants of the simulation according to the arguments given above.

In Sect. 4.6 simulation results for force-free granular gases are shown, where the thermal velocity decays over many orders of magnitude. For such systems that are far from equilibrium and, moreover, that undergo structure formation, Δt and/or L_{box} should be time-dependent as well. Here it is assumed that the box size is fixed. Whether the choice of L_{box} is justified can be checked à posteriori by repeating the simulation with a slightly smaller box size. This argument is similar to that for the choice of the integration time step for solving a differential equation: if the result does not change noticeably when taking half the time step, the original time step was sufficiently small.

Each iteration cycle of the DSMC algorithm consists of six steps:

1. The current time step Δt is chosen such that $L_{\text{box}}/\Delta t$ is several times larger than the average particle velocity. This value is estimated by means of the current temperature of the granular gas.
2. The particles are propagated at their current rate, disregarding collisions:

$$\vec{r}_i = \vec{r}_i + \vec{v}_i \, \Delta t \, , \qquad i = 0, \ldots, N - 1 \, . \tag{4.32}$$

3. The particle positions and velocities are modified according to the boundary conditions, again disregarding particle–particle collisions.
4. The particles are assigned to their boxes according to their positions $\vec{r}_i$.
5. The velocities of the particles are modified due to collisions. The collisions are determined randomly, depending on the local thermodynamic characteristics as described above.
6. The system time is advanced, $t = t + \Delta t$.

These steps are described in detail below. The iteration is repeated until $t \geq t_{\text{end}}$. In the program, first the constants and global variables as well as the function prototypes are defined.

```
                        ──────── DSMC.cc ────────
 1  #include <iostream>
 2  #include <fstream>
 3  #include <vector>
 4  #include <cmath>
 5  using namespace std;
 6
 7  const int N = 100000, Nbox = 1;
 8  const double R = 1, L = 215.44347, eps = 0.8, tend=1000;
 9  const double C1 = 4*M_PI * R*R * Nbox*Nbox*Nbox / (L*L*L);
10  vector <double> x(N),y(N),z(N),vx(N),vy(N),vz(N);
11  vector <vector<vector<vector<int> > > > latt;
12  vector <vector<vector<double> > > dcollrest;
13  double vmax=0, energy=0, dt;
14
15  double vmax_estimate(){ return sqrt(energy/N)*20;}
16  double ranf(double x){ return double(rand())*x/(double(RAND_MAX));}
17  double Gaussian(double mu, double sigma);
18  void init();
19  void sort();
20  void propagate();
21  int paircollision(int i, int j);
22  int collision();
                        ──────── to be continued ────────
```

N	Number of (quasi-) particles
Nbox	Number of boxes in each dimension
R	Particle radius
L	Linear system size of the cubic simulation area with periodic boundary conditions
eps	Coefficient of restitution
tend	Desired simulation time
dt	Simulation time step Δt
x(N)...vz(N)	Coordinates and velocities of the particles
latt	Lists of the particle indices that are located at each lattice site

dcollrest (Fractional) number of collisions that could not be handled in the present simulation step but are carried over to the next step (see below for detailed explanation)

energy   Total kinetic energy of the system

vmax   Maximal particle velocity as estimated in vmax_estimate() by means of the current temperature

C1   Constant numerical coefficient $C_1 \equiv 4\pi R^2 N_{\text{box}}^3/L^3$ (equivalent to (4.30) with L^3/N_{box}^3 as volume)

ranf(x)   Generates equidistributed random numbers in the range $[0, x)$

In main(), the initialization function init() is first called and three more variables are defined: the number of performed time steps it, the cumulative number of collisions ncols, and the current time time. Then main() performs simulation steps until $t \geq t_{\text{end}}$. In the main loop (lines 28–37) the six steps mentioned above are performed for each iteration cycle.

```
                              ───── DSMC.cc ─────
23 │ int main()
24 │ {
25 │   init();
26 │   int it=0, ncols=0;
27 │   double time=0;
28 │   while( time < tend){
29 │     vmax = vmax_estimate();
30 │     dt = 0.2 * L / Nbox / vmax;
31 │     time += dt;
32 │     propagate();
33 │     sort();
34 │     ncols += collision();
35 │     it++;
36 │     cout << time << " " << ncols << " " << energy << endl;
37 │   }
38 │ }
                              ───── to be continued ─────
```

In init() the field variables adopt their required dimensions, namely 0..Nbox for all indices of latt and dcollrest. Each element of latt, e.g., latt[3][5][7], is a vector itself whose size, however, is not specified yet. Later, each of these vectors will contain the indices of the particles that are located in the corresponding box.

```
                              ───── DSMC.cc ─────
39 │ void init()
40 │ {
41 │   latt.resize(Nbox);
42 │   dcollrest.resize(Nbox);
43 │   for(int ix=0;ix<Nbox;ix++){
44 │     latt[ix].resize(Nbox);
45 │     dcollrest[ix].resize(Nbox);
46 │     for(int iy=0;iy<Nbox;iy++){
47 │       latt[ix][iy].resize(Nbox);
48 │       dcollrest[ix][iy].resize(Nbox);
49 │     }
50 │   }
51 │   double  vcomx=0, vcomy=0, vcomz=0;
52 │   for(int i=0; i<N; i++){
53 │     x[i]=ranf(L); y[i]=ranf(L); z[i]=ranf(L);
54 │     vx[i]=Gaussian(0,1); vy[i]=Gaussian(0,1); vz[i]=Gaussian(0,1);
```

```
55      vcomx += vx[i]; vcomy += vy[i]; vcomz += vz[i];
56    }
57    vcomx /= N; vcomy /= N; vcomz /= N;
58    for(int i=0; i<N; i++){
59      vx[i] -= vcomx; vy[i] -= vcomy; vz[i] -= vcomz;
60      energy += vx[i]*vx[i]+vy[i]*vy[i]+vz[i]*vz[i];
61    }
62    energy /= 2;
63  }
```
_____ to be continued _____

In lines 52–61 the particles are distributed uniformly in the system and the velocities are initialized according to a Gaussian distribution with the condition of vanishing center of mass velocity,

$$\sum_{i=0}^{N-1} \vec{v}_i = 0 \, . \tag{4.33}$$

Finally the total energy is initialized. Gaussian random numbers are generated using the Box–Muller algorithm which is discussed in detail on p. 299.

_____ DSMC.cc _____
```
64  double Gaussian(double mu, double sigma)
65  {
66    static bool first=true;
67    static double sqrt2logx1,twopix2;
68
69    if(first){
70      first=false;
71      sqrt2logx1=sqrt(-2*log(drand48()));
72      twopix2=2*M_PI*drand48();
73      return mu+sigma*sqrt2logx1*cos(twopix2);
74    } else {
75      first=true;
76      return mu+sigma*sqrt2logx1*sin(twopix2);
77    }
78  }
```
_____ to be continued _____

In **propagate()** the coordinates of the particles are advanced by their current velocities due to (4.32). At the same time, periodic boundary conditions are applied.

_____ DSMC.cc _____
```
79  void propagate()
80  {
81    for(int i=0; i<N; i++){
82      x[i] = fmod(x[i] + vx[i] * dt + L, L);
83      y[i] = fmod(y[i] + vy[i] * dt + L, L);
84      z[i] = fmod(z[i] + vz[i] * dt + L, L);
85    }
86  }
```
_____ to be continued _____

For the simulation of reflecting boundary conditions or heated walls, the same procedure is applied as explained in detail in Sect. 3.8 for event-driven Molecular Dynamics. Since there are no particle–particle collisions in the propagation step of DSMC, the modeling of boundary conditions is straightforward.

For application of the collision rule, the function **sort()** assigns each particle to its box:

```
      ┌─────────────────────────── DSMC.cc ───────────────────────────┐
  87  │ void sort()                                                     │
  88  │ {                                                               │
  89  │   for(int ix=0; ix<Nbox; ix++){                                 │
  90  │     for(int iy=0; iy<Nbox; iy++){                               │
  91  │       for(int iz=0; iz<Nbox; iz++){                             │
  92  │         latt[ix][iy][iz].clear();                               │
  93  │       }                                                         │
  94  │     }                                                           │
  95  │   }                                                             │
  96  │   for(int i=0; i<N; i++){                                       │
  97  │     int ix=(int)(Nbox * x[i]/L);                                │
  98  │     int iy=(int)(Nbox * y[i]/L);                                │
  99  │     int iz=(int)(Nbox * z[i]/L);                                │
 100  │     latt[ix][iy][iz].push_back(i);                              │
 101  │   }                                                             │
 102  │ }                                                               │
      └───────────────────────── to be continued ─────────────────────┘
```

First the lists of the preceding time step are emptied. Then, starting at line 96, the particles are assigned to their boxes.

The function `collision()` is the central part of the DSMC algorithm. First the number of particles, `np`, in each lattice site is determined (line 109). If $np \geq 2$, the number of candidate collision pairs `dcoll` is computed in line 111. The first term on the right-hand side of line 111 is given by (4.30), where `np*np` has been replaced by the more accurate expression `np*(np-1)`, taking into account that particles may not collide with themselves.

The second term is left over from the previous time step: the variable `dcoll` is a floating point number. The actual number of candidate collision pairs in this lattice site is its integer part `ncoll`. The leftover part is stored in the array `dcollrest` where it is carried over to the next iteration.

```
      ┌─────────────────────────── DSMC.cc ───────────────────────────┐
 103  │ int collision()                                                 │
 104  │ {                                                               │
 105  │   int ncols=0;                                                  │
 106  │   for(int ix=0; ix<Nbox; ix++){                                 │
 107  │     for(int iy=0; iy<Nbox; iy++){                               │
 108  │       for(int iz=0; iz<Nbox; iz++){                             │
 109  │         int np = latt[ix][iy][iz].size();                       │
 110  │         if(np > 1 ){                                            │
 111  │           double dcoll = C1*np*(np-1)*vmax * dt + dcollrest[ix][iy][iz]; │
 112  │           int ncoll = int(dcoll);                               │
 113  │                                                                 │
 114  │           dcollrest[ix][iy][iz] = dcoll - ncoll;                │
 115  │           for(int icol=0; icol<ncoll; icol++){                  │
 116  │             int index1=(int)(ranf(np));                         │
 117  │             int index2=((int)(index1+ ranf(np-1)) + 1) % np;    │
 118  │             int p1=latt[ix][iy][iz][index1];                    │
 119  │             int p2=latt[ix][iy][iz][index2];                    │
 120  │                                                                 │
 121  │             ncols += paircollision(p1, p2);                     │
 122  │           }                                                     │
 123  │         }                                                       │
 124  │       }                                                         │
 125  │     }                                                           │
 126  │   }                                                             │
 127  │   return ncols;                                                 │
 128  │ }                                                               │
      └───────────────────────── to be continued ─────────────────────┘
```

In the loop in lines 115–122, `ncoll` candidate pairs are drawn and subjected to `paircollision()`, where the actual collision is carried out.

Whether a candidate collision pair is accepted can be decided by applying (4.27). To this end, first a unit vector $\vec{e}$ of random direction has to be determined. Because of the condition $|\vec{e}| = 1$, this three-dimensional vector has only two independent parameters. We choose the spherical coordinates $\varphi \in [0 \ldots 2\pi)$ and $\vartheta \in [0..\pi]$ (see Fig. 4.2), thus giving

$$\vec{e} = \begin{pmatrix} \vec{e}_x \\ \vec{e}_y \\ \vec{e}_z \end{pmatrix} = \begin{pmatrix} \sin\vartheta\cos\varphi \\ \sin\vartheta\sin\varphi \\ \cos\vartheta \end{pmatrix} . \tag{4.34}$$

To obtain a unit vector of random direction, φ and ϑ have to be chosen randomly in such a way that each direction is equally probable. If φ is chosen according to an equidistribution, ϑ is not equally distributed as shown in Fig. 4.2. Instead, the probability to draw a value from an interval $d\vartheta$ around ϑ is proportional to the area of the small band of width $d\vartheta$ at latitude ϑ, i.e., $dA = 2\pi\sin\vartheta d\vartheta$. Therefore,

$$P(\vartheta) = \frac{2\pi\sin\vartheta}{4\pi} = \frac{1}{2}\sin\vartheta , \tag{4.35}$$

since the total surface area of the sphere is 4π. Introducing the new random variable $q = \cos\vartheta$, i.e., $dq = -\sin\vartheta d\vartheta$, and requiring

$$P(q)\,|dq| = P(\vartheta)d\vartheta , \tag{4.36}$$

the distribution transforms to

$$P(q) = \frac{1}{2} \tag{4.37}$$

(for a detailed description of the transformation of random variables see pp. 274–276). Hence, the random variable $q \equiv \cos\vartheta$ is uniformly distributed in the interval $q = [-1 \ldots 1]$.

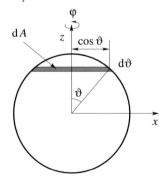

Fig. 4.2. The probability to draw a certain angle ϑ is given by the ratio of the area of the gray band to the total area of the unit sphere. The radius of the band is $\cos\vartheta$

In the program the components of the random unit vector, ndx, ndy, and ndz are determined in lines 135–140 according to (4.34), using two independent, equidistributed random numbers.

```
                                           DSMC.cc
129  int paircollision(int i, int j)
130  {
131    double dvx  = (vx[i]-vx[j]);
132    double dvy  = (vy[i]-vy[j]);
133    double dvz  = (vz[i]-vz[j]);
134
135    double phi = ranf(2*M_PI);
136    double costheta = ranf(2)-1;
137    double sintheta = sqrt(1-costheta*costheta);
138    double ndx = cos(phi)*sintheta;
139    double ndy = sin(phi)*sintheta;
140    double ndz = costheta;
141    double vnorm = dvx * ndx + dvy * ndy + dvz * ndz;
142
143    if(fabs(vnorm) < ranf(vmax) ) return 0;
144    double h=(1+eps) * vnorm / 2;
145    vx[i] -= h * ndx;   vy[i] -= h * ndy;   vz[i] -= h * ndz;
146    vx[j] += h * ndx;   vy[j] += h * ndy;   vz[j] += h * ndz;
147    energy -= (1-eps*eps)/4 * vnorm * vnorm;
148    return 1;
149  }
```

Then the rule (4.27) is applied to accept or reject the chosen pair, depending on the scalar product $\vec{e} \cdot \vec{v}_{12}$. Here, approaching and departing particles are not distinguished anymore since this distinction would only lead to rejection of half of all potential pairs, thus, requiring twice as many candidate pairs. It would not carry any benefit since the spatial relation of particles from the same box is, by assumption, of no significance. Therefore, treating all candidate pairs as approaching accelerates the simulation.

4.6 Application to a Force-Free Granular Gas

Virtually the entire kinetic theory of granular gases is based on the Boltzmann equation (or Boltzmann–Enskog equation), with the assumption of molecular chaos. Therefore, DSMC simulations are directly comparable with the results of kinetic theory. A comprehensive introduction into this exciting field can be found in [42]. To derive the kinetic theory of granular gases, many approximations have been made whose justification is not obvious à priori, thus, the main application of DSMC is checking the analytical results in the theory of granular gases. Other methods may yield ambiguous results since it may be difficult to decide whether possible discrepancies are a result of the underlying approximations or an inaccurate analysis of the kinetic theory.

As an application of DSMC, we describe here the simulation of a homogeneous granular gas in three dimensions with periodic boundary conditions in the absence of gravity, which is the simplest case of granular gas dynamics (see also Sect. 3.10). The coefficient of restitution is $\varepsilon = $ const. From kinetic theory three important results follow, which can be compared with DSMC:

1. The granular temperature, i.e., the mean kinetic energy of the random motion of the particles, decays according to Haff's law [106],

$$T = \frac{T_0}{1 + (t/\tau_H)^2} \, , \tag{4.38}$$

with T_0 as the initial temperature and the relaxation time is given by

$$\tau_H = \frac{8}{3} \left(1 - \varepsilon^2\right) R^2 n \sqrt{\pi T_0} \tag{4.39}$$

(e.g., [42]). Figure 4.3 shows the decay of temperature of a gas of $N = 10^5$ particles ($R = 1$, $\varepsilon = 0.8$). The value of τ_H as obtained from the simulation (by fitting this curve to Haff's law) is $\tau_H = 0.017$. Kinetic theory predicts $\tau_H = 0.017016$. Both values are, hence, in perfect agreement.

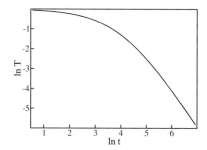

Fig. 4.3. Evolution of the temperature of a gas of 100,000 particles. The numerical data agree almost perfectly with the analytical prediction

2. The velocity distribution function deviates from the Maxwell velocity distribution, which was described in literature by a non-vanishing second (and higher) Sonine coefficient [39, 95, 283]. The Sonine-polynomials expansion for granular gases is explained, e.g., in [42].
   Figure 4.4 shows the velocity distribution function of a granular gas as derived from kinetic theory [95, 283] (solid line), together with a Maxwell distribution (dashed line) as it is valid for molecular gases, and DSMC results (dots). For this highly damped gas we find a systematic deviation from the Maxwell distribution; for smaller damping the deviation is less pronounced. DSMC and Kinetic Theory are again in perfect agreement.
3. For granular gases the high energy tail of the reduced distribution function $f(c) \equiv f(v/v_T)$ does not obey a Maxwell distribution ($\sim \exp(-c^2)$), as in the case of molecular gases, but decays as $\exp(-ac)$ [75]. This behavior can be understood if one considers that the number of particles of the reduced velocity $c = v/v_T$ is determined by a balance of *three* processes: (i) losses due to collisions of particles at velocity c (thus changing their velocity to some c'); (ii) gains due to particles resulting at c after

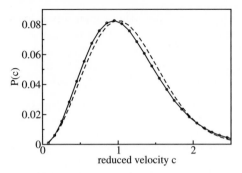

Fig. 4.4. Velocity distribution function of a granular gas ($N = 10^6$, $\varepsilon = 0.3$, $R = 1$) as derived from kinetic theory (*solid line*) together with DSMC results (*dots*). There are small but systematic deviations from the Maxwell distribution (*dashed line*)

a collision; and (iii) variation without collisions due to decaying thermal velocity v_T of a cooling gas, i.e., although the particle's velocity v stays the same its reduced velocity $c = v/v_T$ changes due to temperature decay. For gases of elastic particles, process (iii) is irrelevant and the balance of the processes (i) and (ii) yields the Maxwell distribution. For dissipative gases of particles which collide with $\varepsilon =$ const for the high velocity tail, process (ii) may be neglected as compared to (i). The process (iii) causes in this case an increase of the number of particles in the high velocity tail with $c \gg 1$. The resulting balance of (i) and (iii) yields the steady-state exponential overpopulation of the high energy tail [75].

Figure 4.5 shows the one-dimensional velocity distribution of the same gas as drawn in Fig. 4.4, but for larger velocities. The asymptotic behavior agrees with the theoretical result $\sim \exp(-ac)$ and can be clearly distin-

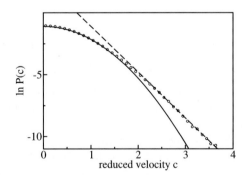

Fig. 4.5. The velocity distribution of the same gas as in Fig. 4.4 ($N = 10^6$, $R = 1$, $\varepsilon = 0.3$). As predicted by kinetic theory [75], the curve follows $\sim \exp(-ac)$ (*dashed line*) rather than a Maxwell distribution (*solid line*)

guished from $\sim \exp(-c^2)$, as expected for a molecular gas of elastically colliding particles.

The result presented in Fig. 4.5 was obtained using standard DSMC as described in this chapter. A variant of DSMC was designed by Rjasanov and Wagner [242, 243] to study high-energy tails of distributions with high accuracy.

So far, a homogeneous granular gas has been considered. Although we know from event-driven Molecular Dynamics that the homogeneous state of a granular gas is unstable (see Sect. 3.10, Figs. 3.19–3.21), the homogeneous state is sustained artificially by reducing the number of boxes to $N_{\text{box}} = 1$. Since inside each box the particles collide independent of their position, any inhomogeneities which may occur are irrelevant to the simulation. In the special case $N_{\text{box}} = 1$, the propagation step $\vec{r}_i = \vec{r}_i + \vec{v}_i \Delta t$ may be omitted, of course.

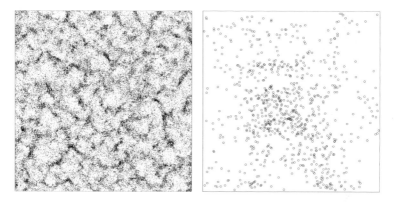

Fig. 4.6. *Left*: a snapshot of a granular gas after about 700 seconds. *Right*: a closeup of one of these clusters

If inhomogeneous systems are considered, the simulation area has to be subdivided into boxes. Figure 4.6 (left) shows a granular gas of $N = 10^5$ particles ($R = 1$, $\varepsilon = 0.8$) in a quadratic box ($L = 2,000$, corresponding to the particle number density $n = 0.025$). The system is subdivided into 100×100 boxes only.[2] As in event-driven Molecular Dynamics the gas is characterized by filament-like clusters. In the closeup, it becomes obvious that the DSMC algorithm allows for overlapping particles, i.e., its application is strictly limited to dilute systems. In this respect, DSMC simulations are very different from Molecular Dynamics simulations.

---

[2] The snapshots shown in Figs. 4.6 and 4.7 were generated by a two-dimensional variant of the program code given in this chapter. Since the transition from three dimensions to two dimensions is straightforward, it is not discussed here. The program for two-dimensional DSMC can be found on the web site.

DSMC for granular gases should be applied carefully as inappropriate application may easily lead to non-physical results. As an example, Fig. 4.7 shows the evolution of the same system as shown in Fig. 4.6, with the only difference being the subdivision of the simulation area into 10×10 boxes.

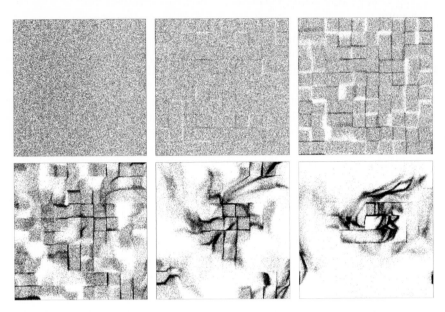

Fig. 4.7. Same system as shown in Fig. 4.6, but with 10×10 boxes. The snapshots show the system at initialization time and after 340, 3,000, 14,000, 59,000, and 290,000 seconds. After a short simulation time, the lattice structure becomes obvious. Thus, inappropriate application of DSMC may easily lead to non-physical results

After a short simulation time, the box structure becomes visible. This can be understood since due to the principle of DSMC, the particles inside a box are treated as uniformly distributed. Thus, spatial structures must be significantly larger than the box size to be correctly described by DSMC. From this point of view, a small box size should be chosen. On the other hand, the box size should not be too small since a certain ensemble size is needed to justify the assumption of local homogeneity. Finally, the box size is limited by the requirement that at least two particles are needed for a collision.

Consequently, DSMC is definitely suited to check the results of the kinetic theory of homogeneous systems—this has been its main application area in the context of granular systems. For non-homogeneous systems, the application of DSMC is not always justified.

5

Rigid-Body Dynamics

5.1 Rigid Bodies

In (force-based) Molecular Dynamics simulations, as described in Chap. 2 the particle trajectories are determined by numerically integrating Newton's equation of motion for all particles. Such simulations have been proven to be very useful and predictive in many applications, in particular when the *dynamical* behavior of the grains dominates the system properties. There are, however, situations where force-based Molecular Dynamics algorithms fail or, at least, become numerically inefficient. We are faced with three major problems:

1. *The interaction force of contacting particles must be known as a function of the particle positions and velocities.*
   The Hertzian contact force formula (2.10) for the normal component of the elastic interaction force is valid only for spheres. The formula, including the dissipative part (2.12), is further limited to smooth surfaces. For generally shaped colliding bodies, e.g., for sharp-edged grains, the interaction force as a function of the particle positions and velocities is unknown. Many of the well established formulae for interaction forces are based on sound reasoning but do not follow from a rigorous mathematical derivation, i.e., they may fail in some cases.
2. *Static friction is problematic.*
   Whereas the normal force for smooth bodies can be determined from bulk properties of the material, the tangential force is determined by bulk and surface properties. A natural (phenomenological) assumption is

$$F_{ij}^{t} \leq \mu F_{ij}^{n} , \tag{5.1}$$

where μ is the Coulomb friction parameter. Unfortunately, this model is not sufficient for static systems: assume a (non-spherical) particle resting on an inclined plane (angle α). Its weight force has a normal component of $F^{n} = mg \cos \alpha$ and a tangential component of $F^{t} = mg \sin \alpha$. As long as the tangential force is smaller than μF^{n} the particle remains at rest, i.e., its net tangential force is zero. Hence, the static frictional force acting in this system is $F^{\text{fric}} = -mg \sin \alpha = -F^n \tan \alpha$, not directly depending on any material parameter. Formulated more precisely, we can say there is no force law which relates the static frictional force with the normal force; the static frictional force takes any value necessary to keep the particle at rest. This property is subject to the condition that the static frictional force may not exceed the limit μF^n. It depends, thus, *indirectly* on material parameters. To model static situations correctly in Molecular Dynamics simulations, one has to construct special force laws which mimic real static friction, e.g., [253]. This force cannot be derived from material or surface properties, hence, it is to some extent arbitrary.

3. *The numerical integration of Newton's equation may become inefficient.*
   The particles of granular materials are stiff, which implies a steep gradient of the interaction force. Therefore, a very small integration time step is required. The more rigid the particle material the slower the simulation progresses. Simulating long-term behavior of systems of technologically relevant sizes can be very time consuming. Moreover, for certain combinations of material parameters the stability of common integration schemes may not be guaranteed. Integration schemes which do not suffer from this problem require much CPU time.

To apply force-based Molecular Dynamics, certain assumptions on the interaction forces of contacting particles have to be made. In some cases, as for the Hertzian contact, these forces may be mathematically derived from basic material properties. In other cases the forces (as functions of the particle positions and velocities) have to be determined heuristically such that the simulation results agree as closely as possible with the experiment. Whether this choice for the microscopic forces is appropriate, i.e., whether they reproduce the experimentally observed behavior, can be assessed only à posteriori, after performing the simulation.

While Molecular Dynamics simulations are always based on the evaluation of interaction forces, *Rigid-Body Dynamics* (sometimes also called *Contact Dynamics*) is based on the opposite idea: The interaction forces are determined from consistency requirements on the behavior of the particles. This method is, therefore, suited to simulate perfectly rigid particles without the necessity to specify a certain force-deformation law. For an illustration, consider the example of a rigid sphere that rests on a rigid flat surface (see Fig. 5.1). There are two forces acting on the sphere, gravity $(-mg)$ and the vertical contact force (F^n), both counted positive in the upward direction. For the choice $F^n = 0$, the sphere would move downward with acceleration g, i.e., it would penetrate the surface. This non-physical behavior has to be avoided by the proper choice of the force F^n. The correct behavior of the sphere is ensured by a simple set of conditions:

Condition 1: *Contact forces have to be chosen in such a way as to avoid mutual deformation of contacting particles.*

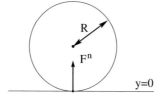

Fig. 5.1. A rigid sphere resting on a rigid flat surface

For our example, $F^n \geq mg$. On the other hand, a force $F^n > mg$ would cause spontaneous upward motion of the sphere, which is, of course, not physically possible. This is avoided by the second condition.

Condition 2: *A contact force vanishes when the contact breaks.*
A contact is said to break if the normal component of the relative acceleration of the concerned particles or their normal velocity is larger than zero. (The relative velocity is considered positive if the particles separate from each other.) In our example, $F^n > mg$ causes the contact to break, which implies that the contact force vanishes. The choice $F^n > mg$ is, hence, inconsistent with condition 2.

Condition 3: *There are no attractive normal forces.*
This excludes any negative contact forces. Since $F^n \geq mg$, was already concluded, this condition in a strict sense is redundant for the example problem, but it has to be taken into account in more general situations.

For the described example, thus, neither $F^n < mg$ (condition 1) nor $F^n > mg$ (condition 2) is admitted, leaving us with

$$F^n = mg .$$

For this choice the total force is zero and the sphere rests on the plane. The conditions 1–3 are sufficient to describe any particle system as long as there are no frictional forces. For systems with friction one more condition is needed:

Condition 4: *Frictional forces act parallel to the contact plane, i.e., perpendicular to F^n. Let F^* be the tangential force that is necessary to prevent two particles from sliding (i.e., to keep the particles at rest with respect to each other). Then the tangential force is $|F| = \min(|F^*|, |\mu F^n|)$. If the particle slides (finite tangential velocity), the frictional force adopts its maximal value $|F| = \mu |F^n|$, in agreement with Coulomb's friction law. Its direction is opposite to the tangential relative acceleration (for static contacts) or opposite to the tangential relative velocity (for sliding contacts).*

To perform simulations, the forces acting on the particles in the normal and tangential directions have to be derived from these four conditions. The corresponding algorithm is explained in Sects. 5.3 and 5.4.

In our simple example (Fig. 5.1) there exists only one single contact point between the sphere and the plane. In more complex situations, e.g., for a resting cube, there are contact areas instead of points. Such contacts can always be reduced to point contacts. It will be shown that the described conditions are sufficient to determine the forces and torques acting on the particles, provided there are not too many contacts in the system. If the number of contacts is too large, only the total force and the total torque that act on a particle may be determined, but not each of the individual contact forces. For the computation of the particle trajectories, however, the total forces and torques are sufficient. We will return to this problem in Sect. 5.6. Additionally, once friction is incorporated, there are cases when there is no set of finite forces that can prevent mutual deformation of the particles (here we do not refer to collisions). These situations are resolved by applying impulses on the involved particles. Since these cases are very rare in most practical applications, they are not explained here.

Rigid-Body Dynamics has been intensively studied during the past two decades. Descriptions of important parts of this method can be found, e.g., in [162, 163]. The core of the algorithm is the numerical computation of the contact forces, which is a Linear Complementarity Problem [66, 67]. An efficient algorithm for this type of problems can be found in [16–19]. Rigid-Body Dynamics has also been applied to granular systems, e.g., [189, 247], where frictionless smooth spheres have been simulated. Systems of granular particles subject to friction have been studied, e.g., in [128]. A practical description of the application of Rigid-Body Dynamics to systems of spheres can be found in [281, 282].

An important example for the application of Rigid-Body Dynamics is the simulation of railway ballast [263]. Rigid-Body Dynamics is much better suited here than force-based Molecular Dynamics for the following reasons:

- Ballast particles are irregularly shaped and sharp-edged. Even if the bulk material properties were precisely known, the contact force law is unknown due to the complicated shape.
- Ballast particles are very stiff, which implies that the gradient of the interaction force is very steep. In this regime the numerical integration of Newton's equation is problematic. For typical loads the elastic deformation of the particles is negligible, i.e., the rigid-body assumption is well justified.
- Static friction, whose treatment in Molecular Dynamics simulations is problematic, is essential for the dynamics of the system. It is correctly modeled by Rigid-Body Dynamics.
- The long-term behavior of railway ballast is affected by abrasion and fragmentation of particles. Whereas the simulation of fragmentation using Molecular Dynamics requires a number of preconditions (see Sect. 2.9), the incorporation of fragmentation in Rigid-Body Dynamics is unproblematic.

5.2 Sketch of the Algorithm

The state of the granular system is described by the positions and orientations of its particles and by the corresponding time derivatives. Contacts between the particles may be classified as sticking and sliding contacts. The contact network is eventually modified by creation and breaking of contacts as well as by transformation of sticking into sliding contacts and vice versa. Whenever the contact network is modified, the state of the system changes qualitatively. The simulation proceeds in discrete time steps. Each of them consists of

1. **Contact detection:** All existing contacts are registered.
2. **Treatment of collisions:** A collision takes place if two contacting particles move at negative, normal relative velocity. In this case no *finite* contact force can avoid deformation of the particles since *any* force, however large it is, needs a short but finite time to decelerate the colliding particles. Hence, mutual deformation would be unavoidable. Therefore, collisions have to be considered separately (see Sect. 5.5).
3. **Clear-up of the contact list:** After a collision, in general, a number of contacting particles have a positive, normal relative velocity, i.e., these particles lose contact and are, thus, removed from the list of contacts.
4. **Formulation of the geometry equation:** The normal components of the relative velocities at all remaining contacts are zero. The geometry equation is established, which contains the information about the contact network of the system (see Sect. 5.3).
5. **Computation of the forces:** Section 5.4 deals with the computation of the forces and relative accelerations by means of the geometry equation.
6. **Integration of the equation of motion:** Finally, the equation of motion is integrated for all particles. When performing this operation, it may be necessary to update the geometry equation and to recompute the forces.

5.3 Mathematical Description

5.3.1 Frictionless Particles

First we restrict ourselves to systems of frictionless particles. When the mathematical framework is developed for this simplified case, frictional forces between the particles will be introduced.

The rigidity of the particles is enforced by means of mathematical motion constraints of the form

$$g\left(\tilde{q}\right) \geq 0 \,, \tag{5.2}$$

where $\tilde{q}$ is a vector whose components are the coordinate vectors $\vec{q}_i$ of *all* particles of the system,

$$\tilde{q} \equiv \begin{pmatrix} \vec{q}_1 \\ \vec{q}_2 \\ \dots \\ \vec{q}_N \end{pmatrix} . \tag{5.3}$$

The coordinate vectors contain the center of mass position and the orientation of the particles. For each possible contact there is a separate constraint function g with

$$g = \begin{cases} 0 & \text{if the particles are in contact} \\ > 0 & \text{otherwise.} \end{cases} \tag{5.4}$$

For spheres, which is the simplest case, the constraint function reads

$$g\left(\tilde{q}\right) = |\vec{r}_i - \vec{r}_j| - R_i - R_j . \tag{5.5}$$

If the spheres deformed each other ($|\vec{r}_i - \vec{r}_j| < R_i + R_j$), the function $g\left(\tilde{q}\right)$ would become negative; if the particles touch each other it is zero. To allow contact breaking we formally allow $g(\tilde{q}) > 0$ too, but since a contact is disregarded as soon as it breaks, only $g(\tilde{q}) = 0$ is encountered. For sharp-edged particles, such as particles described by polyhedrons or polygons (in two dimensions), the motion constraints are

$$g\left(\tilde{q}\right) = \vec{n}_j \cdot \left(\vec{r}_i + \vec{x}_i - \vec{r}_j\right) - d , \tag{5.6}$$

where $\vec{n}_j$ is the normal vector with respect to that surface of particle j whose contact with a vertex of particle i is considered (see Fig. 5.2, left). The vectors $\vec{r}_i$ and $\vec{r}_j$ are the center-of-mass positions of the particles; the vertex of particle i whose contact with particle j is considered, is at position $\vec{r}_i + \vec{x}_i$. The constant d is the distance of the corresponding face of j from the center of the particle. Face–face contacts can be described by two face-vertex contacts (see Fig. 5.2, right).

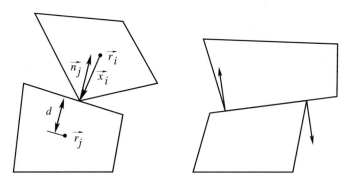

Fig. 5.2. *Left*: face–vertex contact, see text for explanation. *Right*: face–face contacts can be reduced to two face-vertex contacts. The arrows indicate the face normals of each of the contacts

Each motion constraint corresponds to a scalar contact force f. According to d'Alembert's principle, the direction of the contact force is given by the spatial derivative of g with respect to all components of the coordinate vector $\tilde{q}$, namely $\partial g / \partial \tilde{q}$. The contact force acting on a particle i is $f \partial g / \partial \vec{q}_i$, with $\vec{q}_i$ being the generalized coordinate vector of particle i,

$$\vec{q}_i \equiv \begin{pmatrix} x_i \\ y_i \\ z_i \\ \vartheta_i \\ \varphi_i \\ \psi_i \end{pmatrix} . \tag{5.7}$$

The equation of motion for the particles reads

$$\hat{M}_i \ddot{\vec{q}}_i = \vec{Q}_i + \sum_\alpha f_\alpha \frac{\partial g_\alpha}{\partial \vec{q}_i} , \tag{5.8}$$

$$g_\alpha (\tilde{q}) \geq 0 .$$

Here and in the following, particles are indexed by Latin letters, contacts by Greek letters. The mass matrix $\hat{M}_i$ of particle i is defined by

$$\hat{M}_i \equiv \begin{pmatrix} m_i & 0 & 0 & 0 \\ 0 & m_i & 0 & 0 \\ 0 & 0 & m_i & 0 \\ 0 & 0 & 0 & \hat{J}_i \end{pmatrix} , \tag{5.9}$$

where $\hat{J}_i$ is the moment of inertia tensor. In two dimensional systems, $\hat{J}_i$ reduces to a scalar J_i (there is only one orientation angle) and the mass matrix reduces to

$$\hat{M}_i^{(2D)} \equiv \begin{pmatrix} m_i & 0 & 0 \\ 0 & m_i & 0 \\ 0 & 0 & J_i \end{pmatrix} . \tag{5.10}$$

The term $\vec{Q}_i$ in (5.8) is the vector of the components of the external force and torque acting on particle i. In most cases of practical interest, $\vec{Q}_i$ is the gravitational force but other external forces and torques may be incorporated at this point as well.

The motion constraints are formulated as $g (\tilde{q}) \geq 0$ to allow for separation of particles; however, contact forces can only act if particles are actually in contact. Therefore, constraints which are strictly positive, i.e., $g (\tilde{q}) > 0$ (the particles are separated), can be disregarded. These constraints are said to be inactive. The remaining constraints—the active ones—are, thus, satisfying $g (\tilde{q}) = 0$. Since constraints $g (\tilde{q})$ are required to be non-negative at any time, the time derivatives of active constraints have to be non-negative too:

$$\dot{g}_\alpha = \frac{\partial g_\alpha}{\partial \vec{q}_k}\dot{\vec{q}}_k \geq 0 \;. \tag{5.11}$$

For simplicity of the notation, the Einstein convention is used, i.e., summation over the doubly occurring indices k and l is implied. Provided (5.11) is fulfilled as an equality, $\dot{g}_\alpha = 0$, the accelerations read[1]

$$\ddot{g}_\alpha = \frac{\partial g_\alpha}{\partial \vec{q}_k}\ddot{\vec{q}}_k + \frac{\partial^2 g_\alpha}{\partial \vec{q}_k \partial \vec{q}_l}\dot{\vec{q}}_k\dot{\vec{q}}_l \geq 0 \;. \tag{5.12}$$

The time derivatives $\dot{g}_\alpha$ and $\ddot{g}_\alpha$ are the relative velocity and relative acceleration of the particles at their contact points. Note that $\dot{g}_\alpha$ and $\ddot{g}_\alpha$ are not the relative velocity or acceleration of the centers of mass of the particles, but of the points of both particles which are actually in contact. By combination of translational motion and rotation, the relative velocity or acceleration of the contact point may be positive (the particles separate from each other) although their centers of mass approach each other. Inserting the equation of motion (5.8) into (5.12) yields

$$\ddot{g}_\alpha = \frac{\partial g_\alpha}{\partial \vec{q}_k}\left[\hat{M}_k^{-1}\left(\vec{Q}_k + \sum_\beta f_\beta \frac{\partial g_\beta}{\partial \vec{q}_k}\right)\right] + \frac{\partial^2 g_\alpha}{\partial \vec{q}_k \partial \vec{q}_l}\dot{\vec{q}}_k\dot{\vec{q}}_l$$
$$= \frac{\partial g_\alpha}{\partial \vec{q}_k}\left(\hat{M}_k^{-1}\vec{Q}_k\right) + \frac{\partial^2 g_\alpha}{\partial \vec{q}_k \partial \vec{q}_l}\dot{\vec{q}}_k\dot{\vec{q}}_l + \frac{\partial g_\alpha}{\partial \vec{q}_k}\hat{M}_k^{-1}\left(\sum_\beta f_\beta \frac{\partial g_\beta}{\partial \vec{q}_k}\right) \;. \tag{5.13}$$

The first term of the right hand side describes the action of the external forces, the second term describes the action of inertial forces such as centrifugal force and Coriolis force, and the third term finally describes the action of the contact forces. This equation may be written in the form

$$\ddot{g}_\alpha = b_\alpha + \sum_\beta A_{\alpha\beta} f_\beta \;, \tag{5.14}$$

with

$$A_{\alpha\beta} \equiv \frac{\partial g_\alpha}{\partial \vec{q}_k}\hat{M}_k^{-1}\frac{\partial g_\beta}{\partial \vec{q}_k}$$
$$b_\alpha \equiv \frac{\partial g_\alpha}{\partial \vec{q}_k}\left(\hat{M}_k^{-1}\vec{Q}_k\right) + \frac{\partial^2 g_\alpha}{\partial \vec{q}_k \partial \vec{q}_l}\dot{\vec{q}}_k\dot{\vec{q}}_l \;. \tag{5.15}$$

Also for simpler notation from now on we denote the relative acceleration of the contacting particles at the point of contact—the contact acceleration—by

[1] In a strict sense, if (5.11) holds true as an inequality, $\dot{g} > 0$, the acceleration $\ddot{g}$ may acquire any value. Positive $\dot{g}$, however, correspond to breaking contacts and is not considered in the first place. Hence, for any contact α which requires our attention, both conditions, $g_\alpha = 0$ and $\dot{g}_\alpha = 0$, must be fulfilled. Moreover, $\ddot{g} \geq 0$ is required in this case.

a_α instead of $\ddot{g}_\alpha$. With the Einstein convention, (5.14) turns into the *geometry equation*

$$a_\alpha = b_\alpha + A_{\alpha\beta} f_\beta \ , \tag{5.16}$$

where again summation over β is implied. By means of this equation and the consistency conditions introduced in Sect. 5.1, the contact forces f_β can be determined. The consistency conditions read

$$\begin{aligned} a_\alpha &\geq 0 \\ f_\alpha &\geq 0 \\ a_\alpha f_\alpha &= 0 \ . \end{aligned} \tag{5.17}$$

The first condition prevents deformation of the particles (see footnote on p. 218), the second one excludes attractive forces and the third one requires the contact forces acting only if the particles stay in contact, i.e., if $a_\alpha = 0$. These conditions together with the geometry equation (5.16) allow us to determine the unknown contact forces f_α. The system consisting of (5.16) and the conditions (5.17) is called a Linear Complementarity Problem. It can be solved by Dantzig's algorithm [66], as described in Sect. 5.4.

Note that the contact forces f are scalar quantities. They act in the direction given by the gradients $\partial g_\alpha / \partial q$. To avoid confusion with real forces we denote them in lower case.

5.3.2 Particle Systems with Friction

To incorporate friction, additional motion constraints which control the tangential motion of contacting particles are introduced. For polygonal particles they are of the form

$$g(\tilde{q}) = \vec{t}_j \cdot (\vec{r}_i + \vec{x}_i - \vec{r}_j - \vec{x}_j) \ . \tag{5.18}$$

This constraint has a similar form as the constraint for the normal motion (5.6), but instead of the normal unit vector of the contacting face of particle j the tangential unit vector $\vec{t}_j$ appears, thus ensuring that the vertex of particle i at position $\vec{r}_i + \vec{x}_i$ does not move along the face of particle j, away from the point $\vec{r}_j + \vec{x}_j$ of current contact.

The tangential motion constraints are, however, of different nature than the normal motion constraints. Whereas in the case of normal motion the constraints must *never* be violated, the constraints on the tangential motion may actually be violated, as it happens when contacting particles start to slide. This is due to the fact that the absolute values of the frictional forces are limited by Coulomb's law $f^t \leq \mu f^n$, with μ being the friction constant and f^n the corresponding normal force. As reflected by the consistency condition 4 (cf. Sect. 5.1), the frictional force adopts its maximum value if particles actually slide, i.e., if the tangential motion constraint is inactive (violated).

Thus, in this case the value of the frictional force is determined by Coulomb's law directly. In contrast to normal motion constraints, inactive constraints may not be disregarded because now the corresponding contact forces are non-zero and the constraint is necessary to determine the direction of the tangential force.

Since the incorporation of friction only causes further motion constraints, there is in principle, no need for further discussion of the problem because the geometry equation (5.16) can describe systems with friction as well. For simplicity of notation it is worth, however, to consider normal and frictional forces and their corresponding constraints separately. There are two classes of motion constraints,

$$
\begin{aligned}
g^{\mathrm{n}}(\tilde{q}) &\geq 0 \\
g^{\mathrm{t}}(\tilde{q}) &= 0 \,,
\end{aligned}
\tag{5.19}
$$

and the corresponding contact forces f^{n} and f^{t}. Now the equation of motion reads

$$
\hat{M}_k \ddot{\vec{q}}_k = \vec{Q}_k + \sum_\alpha \left(f_\alpha^{\mathrm{n}} \frac{\partial g_\alpha^{\mathrm{n}}}{\partial \vec{q}_k} + f_\alpha^{\mathrm{t}} \frac{\partial g_\alpha^{\mathrm{t}}}{\partial \vec{q}_k} \right) \,.
\tag{5.20}
$$

The second time derivative of the motion constraints are

$$
\begin{aligned}
\ddot{g}_\alpha^{\mathrm{n}} &= \frac{\partial g_\alpha^{\mathrm{n}}}{\partial \vec{q}_k} \left(\hat{M}_k^{-1} \vec{Q}_k \right) + \frac{\partial^2 g_\alpha^{\mathrm{n}}}{\partial \vec{q}_k \partial \vec{q}_l} \dot{\vec{q}}_k \dot{\vec{q}}_l + \frac{\partial g_\alpha^{\mathrm{n}}}{\partial \vec{q}_k} \hat{M}_k^{-1} \left(\sum_\beta f_\beta^{\mathrm{n}} \frac{\partial g_\beta^{\mathrm{n}}}{\partial \vec{q}_k} + f_\beta^{\mathrm{t}} \frac{\partial g_\beta^{\mathrm{t}}}{\partial \vec{q}_k} \right) \\
\ddot{g}_\alpha^{\mathrm{t}} &= \frac{\partial g_\alpha^{\mathrm{t}}}{\partial \vec{q}_k} \left(\hat{M}_k^{-1} \vec{Q}_k \right) + \frac{\partial^2 g_\alpha^{\mathrm{t}}}{\partial \vec{q}_k \partial \vec{q}_l} \dot{\vec{q}}_k \dot{\vec{q}}_l + \frac{\partial g_\alpha^{\mathrm{t}}}{\partial \vec{q}_k} \hat{M}_k^{-1} \left(\sum_\beta f_\beta^{\mathrm{n}} \frac{\partial g_\beta^{\mathrm{n}}}{\partial \vec{q}_k} + f_\beta^{\mathrm{t}} \frac{\partial g_\beta^{\mathrm{t}}}{\partial \vec{q}_k} \right) \,.
\end{aligned}
\tag{5.21}
$$

With $a_\alpha^{\mathrm{n}} \equiv \ddot{g}_\alpha^{\mathrm{n}}$, $a_\alpha^{\mathrm{t}} \equiv \ddot{g}_\alpha^{\mathrm{t}}$, the abbreviations

$$
\begin{aligned}
b_\alpha^{\mathrm{n}} &\equiv \frac{\partial g_\alpha^{\mathrm{n}}}{\partial \vec{q}_k} \left(\hat{M}_k^{-1} \vec{Q}_k \right) + \frac{\partial^2 g_\alpha^{\mathrm{n}}}{\partial \vec{q}_k \partial \vec{q}_l} \dot{\vec{q}}_k \dot{\vec{q}}_l \\
b_\alpha^{\mathrm{t}} &\equiv \frac{\partial g_\alpha^{\mathrm{t}}}{\partial \vec{q}_k} \left(\hat{M}_k^{-1} \vec{Q}_k \right) + \frac{\partial^2 g_\alpha^{\mathrm{t}}}{\partial \vec{q}_k \partial \vec{q}_l} \dot{\vec{q}}_k \dot{\vec{q}}_l
\end{aligned}
\tag{5.22}
$$

and

$$
\begin{aligned}
A_{\alpha\beta}^{\mathrm{nn}} &\equiv \frac{\partial g_\alpha^{\mathrm{n}}}{\partial \vec{q}_k} \hat{M}_k^{-1} \frac{\partial g_\beta^{\mathrm{n}}}{\partial \vec{q}_k} & A_{\alpha\beta}^{\mathrm{nt}} &\equiv \frac{\partial g_\alpha^{\mathrm{n}}}{\partial \vec{q}_k} \hat{M}_k^{-1} \frac{\partial g_\beta^{\mathrm{t}}}{\partial \vec{q}_k} \\
A_{\alpha\beta}^{\mathrm{tn}} &\equiv \frac{\partial g_\alpha^{\mathrm{t}}}{\partial \vec{q}_k} \hat{M}_k^{-1} \frac{\partial g_\beta^{\mathrm{n}}}{\partial \vec{q}_k} & A_{\alpha\beta}^{\mathrm{tt}} &\equiv \frac{\partial g_\alpha^{\mathrm{t}}}{\partial \vec{q}_k} \hat{M}_k^{-1} \frac{\partial g_\beta^{\mathrm{t}}}{\partial \vec{q}_k}
\end{aligned}
\tag{5.23}
$$

the modified geometry equation reads

$$a_\alpha^n = b_\alpha^n + \sum_\beta \left(A_{\alpha\beta}^{nn} f_\beta^n + A_{\alpha\beta}^{nt} f_\beta^t \right)$$

$$a_\alpha^t = b_\alpha^t + \sum_\beta \left(A_{\alpha\beta}^{tn} f_\beta^n + A_{\alpha\beta}^{tt} f_\beta^t \right) .$$

(5.24)

The full set of consistency conditions is then

$a_\alpha^n \geq 0$	Relative normal accelerations at contact points must not be negative to avoid mutual deformation of the particles.
$f_\alpha^n \geq 0$	Normal forces are always non-negative, i.e., there are no attractive forces.
$a_\alpha^n f_\alpha^n = 0$	If any normal acceleration is different from zero (breaking contact), the corresponding normal force vanishes.
$|f_\alpha^t| \leq \mu f_\alpha^n$	Tangential forces are limited by Coulomb's friction law. (5.25)
$a_\alpha^t \left(|f_\alpha^t| - \mu f_\alpha^n \right) = 0$	If the tangential acceleration is different from zero, the tangential force adopts its maximal value. Its sign is chosen such that the tangential force acts in the opposite direction to the relative tangential motion. For the case of actual sliding, i.e. $v_\alpha^t = \dot{g}_\alpha^t \neq 0$, we have $f_\alpha^t = -\mu \operatorname{sign}(v_\alpha^t)$.

Equations (5.24) together with the consistency conditions (5.25) can be solved using a modified Dantzig's algorithm [18], which will be described in the next section. Some of the tangential forces may be directly determined by the respective normal forces. This is the case when sliding at this contact occurs. The consistency conditions for these forces have to be fulfilled, nevertheless.

5.4 Dantzig's Algorithm for the Computation of the Forces

5.4.1 General Scheme

Contacts can be classified into breaking contacts ($a^n > 0$, thus $f^n = f^t = 0$), persisting static contacts ($a^n = 0$ and $a^t = 0$), and persisting sliding contacts ($a^n = 0$ but $a^t \neq 0$ or $v^t \neq 0$). If we knew the classification of all contact forces into these categories in advance, the contact forces could be determined by solving an inhomogeneous system of linear equations that consists of all equations for which either $a_\alpha^n = 0$ or $a_\alpha^t = 0$, with the corresponding f_α^n and

f_α^t as variables. All remaining normal forces are zero, while the remaining tangential forces assume their maximum values if the corresponding normal force is larger than zero. Unfortunately the contact classification is only known if we know the contact forces, i.e., the solution, as well.

The forces together with the corresponding contact classification can be determined using Dantzig's algorithm. It starts with considering a certain contact and disregarding all others, i.e., the contact forces of all other contacts are set to zero. After having found a solution for this contact, its classification is also known. The algorithm proceeds then with the next contact. Again the contact forces are determined, preserving the consistency of all contacts considered before. In this process the contact classifications of the already consistent contacts may be changed if necessary. The process is repeated until the last contact has been classified.

All contacts are assigned to one of four lists [18]:

- List NC contains breaking contacts. The corresponding normal force is zero.
- List C_F contains persisting static contacts ($a_\alpha^t = 0$). The absolute value of the frictional force f_α^t is smaller than μf_α^n.
- Lists $C^\pm$ contain sliding contacts (they are persisting as well). In list C^+ are all contacts where $f^t = \mu f^n$, in C^- are all contacts where $f^t = -\mu f^n$. These lists are specific for the two-dimensional simulation. In three dimensions one has to encode the direction of sliding by a different method.

The classification of the contacts is done successively for all contacts α using the following algorithm. Assume a consistent classification of the preceding $\alpha-1$ contacts into the above lists, i.e., all $\alpha-1$ contacts satisfy the consistency conditions. The contact α is then classified by the scheme:

1. Check if the normal force $f_\alpha^n = 0$ satisfies the consistency condition $a_\alpha^n \geq 0$. If this is the case the contact is already consistent and belongs to NC.
2. If $a_\alpha^n < 0$, the normal contact force f_α^n is increased in order to obtain a non-negative, normal acceleration (to avoid deformation of the particles). Increasing the normal force f_α^n changes the contact accelerations of the already classified contacts. Since for persisting contacts β the condition $a_\beta^n = 0$ holds (and for static contacts also $a_\beta^t = 0$), this would invalidate the classification of these contacts. The consistency of the already classified contacts is preserved by simultaneously adjusting the contact forces of the persisting contacts. For a given increase s of the new force f_α^n, we now determine the values by which these contact forces have to be changed in order to keep $a_\beta^n = 0$ and, if necessary (according to the current classification), $a_\beta^t = 0$. To this end a reduced set of geometry equations is formulated:

$$0 = a_\beta^n = b_\beta^n + \mathcal{A}_{\beta\gamma}^{nn} f_\gamma^n + \mathcal{A}_{\beta\gamma}^{nt} f_\gamma^t + \mathcal{A}_{\beta\alpha}^{nn} f_\alpha^n$$
$$0 = a_\beta^t = b_\beta^t + \mathcal{A}_{\beta\gamma}^{tn} f_\gamma^n + \mathcal{A}_{\beta\gamma}^{tt} f_\gamma^t + \mathcal{A}_{\beta\alpha}^{tn} f_\alpha^n \; .$$

$$(5.26)$$

The reduced set of geometry equations can be obtained from the original geometry equations (5.24) by disregarding all contacts in NC and replacing $f_\gamma^t = \pm \mu f_\gamma^n$ for all contacts γ from $C^\pm$ (sliding contacts). Since only persisting contacts remain, all contact accelerations in the reduced geometry equations are, thus, zero. If the new contact force $f_\alpha^n \to f_\alpha^n + s$ is changed, the previously determined contact forces, f_γ^n and f_γ^t, in the reduced set of geometry equations have to be varied by unknown quantities Δf_γ^n and Δf_γ^t in order to keep the contact accelerations at their required value zero:

$$0 = b_\beta^n + A_{\beta\gamma}^{nn}\left(f_\gamma^n + \Delta f_\gamma^n\right) + A_{\beta\gamma}^{nt}\left(f_\gamma^t + \Delta f_\gamma^t\right) + A_{\beta\alpha}^{nn}\left(f_\alpha^n + s\right)$$
$$0 = b_\beta^t + A_{\beta\gamma}^{tn}\left(f_\gamma^n + \Delta f_\gamma^n\right) + A_{\beta\gamma}^{tt}\left(f_\gamma^t + \Delta f_\gamma^t\right) + A_{\beta\alpha}^{tn}\left(f_\alpha^n + s\right)\ .$$

$$(5.27)$$

Inserting (5.26) into (5.27), a linear system of equations for the unknown Δf_γ^n and Δf_γ^t follows,

$$0 = A_{\beta\gamma}^{nn}\Delta f_\gamma^n + A_{\beta\gamma}^{nt}\Delta f_\gamma^t + A_{\beta\alpha}^{nn}s$$
$$0 = A_{\beta\gamma}^{tn}\Delta f_\gamma^n + A_{\beta\gamma}^{tt}\Delta f_\gamma^t + A_{\beta\alpha}^{tn}s\ ,$$

$$(5.28)$$

with the step size s being a parameter. The necessary variations Δf_γ^n and Δf_γ^t are proportional to the step size s, hence the solution is of the form

$$\Delta f_\gamma^n = F_\gamma^n\, s$$
$$\Delta f_\gamma^t = F_\gamma^t\, s\ ,$$

$$(5.29)$$

where F_γ^n and F_γ^t are the necessary variations for $s = 1$. By inserting the modified values back into the original reduced geometry equation, we obtain the values by which the accelerations of the breaking and sliding contacts change.

3. Proceeding with processing contact α, the value of the new force f_α^n is increased until either the acceleration $a_\alpha^n = 0$ (the contact is now consistent) or until the classification of any already consistent contact changes. The classification changes if any of the following occurs:

   (a) A normal acceleration $a_\beta^n > 0$ becomes zero: The contact becomes persisting and has to be moved from NC to either C_F or $C^\pm$ according to its present value of the corresponding tangential acceleration.

   (b) A normal force $f_\beta^n > 0$ (persisting contact) becomes zero: the contact breaks and has to be moved from C_F or $C^\pm$ to NC.

   (c) A tangential acceleration $a_\beta^t \neq 0$ becomes zero: if the corresponding tangential velocity is zero too the contact is now static and has to be moved from $C^\pm$ to C_F.

   (d) A frictional force reaches the maximum value $\pm \mu f_\beta^n$: the contact becomes sliding and has to be moved from C_F to $C^\pm$.

4. If no consistent values for a_α^n and f_α^n could be found yet, the algorithm proceeds with step 2.

If the contact α is persisting, the tangential component a_α^t of contact α is considered. The procedure is very similar to the calculation of a_α^n, with the only difference being that the frictional force is *decreased* until it assumes its negative maximum value, if $a_\alpha^t > 0$. Contrary, if $a_\alpha^t < 0$, the friction is increased until it adopts its positive maximum value.

5.4.2 Application of the Algorithm to a Simple Example

Danzig's algorithm appears to be rather complicated at first glance, therefore, we provide a basic example. The system consists of two particles, a fixed square of size L (particle 1), and a mobile rectangle of dimensions L by $L/2$ (particle 2), see Fig. 5.3. The upper particle is supported by the square at the length γL. If the upper particle is moved to the right, its position is stable until $\gamma = 1/2$ is reached. This section strictly follows the described algorithm to compute the forces acting on the upper particle.

First, the motion constraints are formulated. There are two contacts in the system (see Fig. 5.3). The first constraint prevents the vertex of particle 1 from penetrating the edge of particle 2. According to (5.6) this constraint reads

$$g_1\left(\tilde{q}\right) = \vec{n}_2 \cdot \left(\vec{r}_1 + \vec{x}_1 - \vec{r}_2\right) - d_1 \geq 0. \tag{5.30}$$

In the same way, the second constraint prevents the vertex of particle 2 from penetrating the edge of particle 1:

$$g_2\left(\tilde{q}\right) = \vec{n}_1 \cdot \left(\vec{r}_2 + \vec{x}_2 - \vec{r}_1\right) - d_2 \geq 0. \tag{5.31}$$

From the setup in Fig. 5.3 the unit vectors $\vec{n}_{1/2}$ and $\vec{x}_{1/2}$ can be specified:

$$\vec{n}_1 = \begin{pmatrix} 0 \\ 1 \end{pmatrix} , \quad \vec{n}_2 = \begin{pmatrix} 0 \\ -1 \end{pmatrix} , \quad \vec{x}_1 = \begin{pmatrix} L/2 \\ L/2 \end{pmatrix} , \quad \vec{x}_2 = \begin{pmatrix} -L/2 \\ -L/4 \end{pmatrix} . \tag{5.32}$$

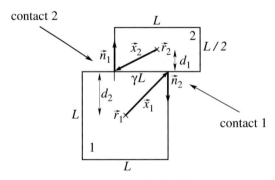

Fig. 5.3. Sketch of the model setup. Particle 1 is fixed, particle 2 is mobile

Without loss of generality the center of particle 1 is located at the origin of the coordinate system, hence,

$$\vec{r}_1 = \begin{pmatrix} 0 \\ 0 \end{pmatrix}, \qquad \vec{r}_2 = L \begin{pmatrix} 1 - \gamma \\ 3/4 \end{pmatrix}. \tag{5.33}$$

Next, the matrix $\hat{G}$ is constructed with the elements $G_{\alpha k}$ (with α the contact number and k the particle number) such that

$$G_{\alpha k} = \frac{\partial g_\alpha}{\partial q_k} \tag{5.34}$$

$$G_{11} = \frac{\partial g_1}{\partial q_1} = \frac{\partial}{\partial (\vec{r}_1, \varphi_1)} [\vec{n}_2 \cdot (\vec{r}_1 + \vec{x}_1 - \vec{r}_2) - d_1], \tag{5.35}$$

for example. Remember that the matrix elements $G_{\alpha k}$ are generalized coordinate vectors themselves. The derivative with respect to the center of mass position $\vec{r}_1$ is unproblematic:

$$\frac{\partial}{\partial \vec{r}_1} [\vec{n}_2 \cdot (\vec{r}_1 + \vec{x}_1 - \vec{r}_2) - d_1] = \vec{n}_2. \tag{5.36}$$

The differentiation with respect to the orientation angle φ_1 is less straightforward and will, therefore, be explained in detail. Given the value $g_1(\vec{r}_1, \varphi_1, \vec{r}_2, \varphi_2)$ of the constraint function for the current configuration $\tilde{q} = (\vec{r}_1, \varphi_1, \vec{r}_2, \varphi_2)$, the only relevant variable that changes when rotating particle 1 by a small angle $d\varphi$ is the vector $\vec{x}_1$:

$$\vec{x}_1' = \vec{x}_1 + d\vec{\varphi} \times \vec{x}_1, \tag{5.37}$$

where $d\vec{\varphi}$ is the vector of absolute value $d\varphi$ pointing into z-direction (perpendicular to the plane where the two-dimensional particles move). Correspondingly the constraint function (5.30) varies as

$$\begin{aligned} g_1(\vec{r}_1, \varphi_1 + d\varphi, \vec{r}_2, \varphi_2) &= \vec{n}_2 \cdot (\vec{r}_1 + \vec{x}_1 + d\vec{\varphi} \times \vec{x}_1 - \vec{r}_1) \\ &= g_1(\vec{r}_1, \varphi_1, \vec{r}_2, \varphi_2) + \vec{n}_2 \cdot (d\vec{\varphi} \times \vec{x}_1) = g_1(\vec{r}_1, \varphi_1, \vec{r}_2, \varphi_2) + d\varphi \vec{e}_z \cdot (\vec{x}_1 \times \vec{n}_2) \\ &= g_1(\vec{r}_1, \varphi_1, \vec{r}_2, \varphi_2) + d\varphi [\vec{x}_1 \times \vec{n}_2]_z, \end{aligned} \tag{5.38}$$

where $\vec{e}_z$ is the unit vector in z-direction. The symbol $[\cdot]_z$ in the last line denotes the z-component of a vector (see also p. 89). Therefore,

$$\frac{\partial g_1}{\partial \varphi_1} = [\vec{x}_1 \times \vec{n}_2]_z. \tag{5.39}$$

Combining (5.36) and (5.39), the first element of the matrix $\hat{G}$ reads

$$G_{11} = \frac{\partial g_1}{\partial \tilde{q}_1} = \begin{pmatrix} \vec{n}_2 \\ [\vec{x}_1 \times \vec{n}_2]_z \end{pmatrix}. \tag{5.40}$$

The same rules as for $\vec{x}_{1/2}$ also apply to the normal vectors $\vec{n}_{1/2}$. Consequently,

$$\hat{G} = \begin{pmatrix} G_{11} & G_{12} \\ G_{21} & G_{22} \end{pmatrix}$$

$$= \begin{pmatrix} \begin{pmatrix} \vec{n}_2 \\ [\vec{x}_1 \times \vec{n}_2]_z \end{pmatrix} & \begin{pmatrix} -\vec{n}_2 \\ -[(\vec{r}_1 + \vec{x}_1 - \vec{r}_2) \times \vec{n}_2]_z \end{pmatrix} \\ \begin{pmatrix} -\vec{n}_1 \\ -[(\vec{r}_2 + \vec{x}_2 - \vec{r}_1) \times \vec{n}_1]_z \end{pmatrix} & \begin{pmatrix} \vec{n}_1 \\ [\vec{x}_2 \times \vec{n}_1]_z \end{pmatrix} \end{pmatrix} . \tag{5.41}$$

The geometry matrix is constructed from $\hat{G}$ by

$$A_{\alpha\beta} = \sum_k G_{\alpha k} \hat{M}_k^{-1} G_{\beta k} . \tag{5.42}$$

To assure that particle 1 is fixed, as it was assumed in the system description at the very beginning of this section, it is assigned an infinite inertial mass and infinite moment of inertia, i.e., $\hat{M}_1^{-1} = 0$, thus, no finite force can move it away from its present position.[2] Therefore all $G_{\alpha 1}$ can be disregarded. In an implementation of this algorithm, of course, these elements would be skipped over from the beginning, they have been included in (5.41) here for the sake of a clear presentation. The result for $\hat{A}$ is

$$A_{\alpha\beta} = \sum_k G_{\alpha k}^{\mathrm{T}} \hat{M}_k^{-1} G_{\beta k} = G_{\alpha 2}^{\mathrm{T}} \hat{M}_2^{-1} G_{\beta 2} . \tag{5.43}$$

The mass matrix of the mobile particle 2 is[3]

$$\hat{M}_2 = \begin{pmatrix} m_2 & 0 \\ 0 & J_2 \end{pmatrix} , \quad \text{thus} \quad \hat{M}_2^{-1} = \begin{pmatrix} m_2^{-1} & 0 \\ 0 & J_2^{-1} \end{pmatrix} . \tag{5.44}$$

The elements of the geometry matrix can be expressed as

$$A_{11} = G_{12}^{\mathrm{T}} \hat{M}_2^{-1} G_{12}$$

$$= \begin{pmatrix} -\vec{n}_2 \\ -[(\vec{r}_1 + \vec{x}_1 - \vec{r}_2) \times \vec{n}_2]_z \end{pmatrix}^{\mathrm{T}} \begin{pmatrix} m_2^{-1} & 0 \\ 0 & J_2^{-1} \end{pmatrix} \begin{pmatrix} -\vec{n}_2 \\ -[(\vec{r}_1 + \vec{x}_1 - \vec{r}_2) \times \vec{n}_2]_z \end{pmatrix}$$

$$= \frac{\vec{n}_2^2}{m_2} + \frac{[(\vec{r}_1 + \vec{x}_1 - \vec{r}_2) \times \vec{n}_2]_z^2}{J_2} , \tag{5.45}$$

---

[2] Note that only the *inertial* mass is infinite but not the *gravitational* mass. Infinite gravitational mass would lead to an *infinite* downward force.
[3] Since the first two components in $G_{\alpha k}$ are combined to a two-dimensional vector, m_2 and m_2^{-1} appear only once.

$$A_{12} = G_{12}^{\mathrm{T}} \hat{M}_2^{-1} G_{22}$$

$$= \begin{pmatrix} -\vec{n}_2 \\ -[(\vec{r}_1 + \vec{x}_1 - \vec{r}_2) \times \vec{n}_2]_z \end{pmatrix}^{\mathrm{T}} \begin{pmatrix} m_2^{-1} & 0 \\ 0 & J_2^{-1} \end{pmatrix} \begin{pmatrix} \vec{n}_1 \\ [\vec{x}_2 \times \vec{n}_1]_z \end{pmatrix}$$

$$= -\frac{\vec{n}_1 \cdot \vec{n}_2}{m_2} - \frac{[(\vec{r}_1 + \vec{x}_1 - \vec{r}_2) \times \vec{n}_2]_z \, [\vec{x}_2 \times \vec{n}_1]_z}{J_2} \tag{5.46}$$

$$A_{21} = G_{22}^{\mathrm{T}} \hat{M}_2^{-1} G_{12} = A_{12} \tag{5.47}$$

$$A_{22} = G_{22}^{\mathrm{T}} \hat{M}_2^{-1} G_{22} = \begin{pmatrix} \vec{n}_1 \\ [\vec{x}_2 \times \vec{n}_1]_z \end{pmatrix}^{\mathrm{T}} \begin{pmatrix} m_2^{-1} & 0 \\ 0 & J_2^{-1} \end{pmatrix} \begin{pmatrix} \vec{n}_1 \\ [\vec{x}_2 \times \vec{n}_1]_z \end{pmatrix}$$

$$= \frac{\vec{n}_1^2}{m_2} + \frac{[\vec{x}_2 \times \vec{n}_1]_z^2}{J_2} \, , \tag{5.48}$$

using $\vec{n}_2^2 = n_2^2 = 1$. The components of the vector $\vec{b}$ of external forces is given by

$$b_\alpha = \sum_k G_{\alpha k}^{\mathrm{T}} \hat{M}_k^{-1} Q_k + \sum_k \frac{\partial}{\partial \dot{\vec{q}}_k} \left(\sum_l \frac{\partial g_\alpha}{\partial \dot{\vec{q}}_l} \dot{\vec{q}}_l \right) \dot{\vec{q}}_k \, . \tag{5.49}$$

Again $\hat{M}_1^{-1} = 0$ and $\dot{\vec{q}} = 0$ for the fixed particles 1. The variable $\dot{\vec{q}}$ is independent of $\vec{q}$, therefore

$$b_\alpha = G_{\alpha 2}^{\mathrm{T}} \hat{M}_2^{-1} Q_2 + \frac{\partial}{\partial \dot{\vec{q}}_2} \left(\frac{\partial g_\alpha}{\partial \dot{\vec{q}}_2} \dot{\vec{q}}_2 \right) \dot{\vec{q}}_2 \, . \tag{5.50}$$

Consider the second term for $\alpha = 1$. With (5.34) and (5.41), the term in brackets reads

$$\frac{\partial g_1}{\partial \dot{\vec{q}}_2} \dot{\vec{q}}_2 = G_{12} \dot{\vec{q}}_2 = -\vec{n}_2 \cdot \dot{\vec{r}}_2 - [(\vec{r}_1 + \vec{x}_1 - \vec{r}_2) \times \vec{n}_2]_z \, \omega_2 \, . \tag{5.51}$$

To evaluate the second term in (5.50) we need the derivatives of this expression with respect to $\vec{r}_2$ and to φ_2:

$$\frac{\partial}{\partial \vec{r}_2} G_{12} \dot{\vec{q}}_2 = -\omega_2 \frac{[(\vec{r}_1 + \vec{x}_1 - \vec{r}_2) \times \vec{n}_2]_z}{\partial \vec{r}_2} = \omega_2 \frac{[\vec{r}_2 \times \vec{n}_2]_z}{\partial \vec{r}_2}$$

$$= \omega_2 \frac{(\vec{r}_2 \times \vec{n}_2) \cdot \vec{e}_z}{\partial \vec{r}_2} = \omega_2 \frac{\vec{r}_2 \cdot (\vec{n}_2 \times \vec{e}_z)}{\partial \vec{r}_2} = -\omega_2 \vec{t}_2 \, , \tag{5.52}$$

with the definition $\vec{t}_2 \equiv \vec{e}_z \times \vec{n}_2$. To compute the derivative with respect to φ_2, we take advantage of the fact that the only relevant orientation-dependent quantity is the vector $\vec{n}_2$. Starting from (5.51) one obtains

$$G_{12} \dot{\vec{q}}_2 = -\vec{n}_2 \cdot \dot{\vec{r}}_2 - [(\vec{r}_1 + \vec{x}_1 - \vec{r}_2) \times \vec{n}_2] \cdot \vec{e}_z \omega_2$$

$$= -\vec{n}_2 \cdot \left\{ \dot{\vec{r}}_2 + [\vec{e}_z \times (\vec{r}_1 + \vec{x}_1 - \vec{r}_2)] \, \omega_2 \right\} \, . \tag{5.53}$$

The term in braces does not depend on the orientation, therefore as in (5.38),

$$
\begin{aligned}
\frac{\partial}{\partial\varphi_2}G_{12}\dot{\vec{q}}_2 &= -\left[\vec{n}_2 \times \left\{\dot{\vec{r}}_2 + [\vec{e}_z \times (\vec{r}_1 + \vec{x}_1 - \vec{r}_2)]\,\omega_2\right\}\right]_z \\
&= -\left[\vec{t}_2 \cdot \left\{\dot{\vec{r}}_2 + [\vec{e}_z \times (\vec{r}_1 + \vec{x}_1 - \vec{r}_2)]\,\omega_2\right\}\right] \\
&= -\vec{t}_2 \cdot \dot{\vec{r}}_2 - \left[\vec{t}_2 \times \vec{e}_z\right]_z (\vec{r}_1 + \vec{x}_1 - \vec{r}_2)\,\omega_2 \\
&= -\vec{t}_2 \cdot \dot{\vec{r}}_2 - \vec{n}_2 \cdot (\vec{r}_1 + \vec{x}_1 - \vec{r}_2)\,\omega_2 \ .
\end{aligned}
\tag{5.54}
$$

In the second line the identity $[\vec{n}_2 \times \vec{a}]_z = \vec{t}_2 \cdot \vec{a}$ for an arbitrary vector $\vec{a}$ is used and for the last expression $[\vec{t}_2 \times \vec{e}_z]_z = \vec{n}_2$. Combining (5.52) and (5.54), the second term of (5.50) reads

$$
\begin{aligned}
\left(\frac{\partial}{\partial\vec{q}_2}G_{12}\dot{\vec{q}}_2\right)\dot{\vec{q}}_2 &= \left(\frac{\partial}{\partial\vec{r}_2}G_{12}\dot{\vec{q}}_2\right)\dot{\vec{r}}_2 + \left(\frac{\partial}{\partial\varphi_2}G_{12}\dot{\vec{q}}_2\right)\omega_2 \\
&= -2\left(\vec{t}_2 \cdot \dot{\vec{r}}_2\right)\omega_2 - [\vec{n}_2 \cdot (\vec{r}_1 + \vec{x}_1 - \vec{r}_2)]\,\omega_2^2 \ .
\end{aligned}
\tag{5.55}
$$

In the same way the expression

$$
\left(\frac{\partial}{\partial\vec{q}_2}G_{22}\dot{\vec{q}}_2\right)\dot{\vec{q}}_2 = -(\vec{x}_2 \cdot \vec{n}_1)\,\omega_2^2
\tag{5.56}
$$

is derived. The external force is caused by gravity, i.e.,

$$
Q_2 = \begin{pmatrix} m_2\vec{g} \\ 0 \end{pmatrix} \ .
\tag{5.57}
$$

Inserting the latter three expressions into (5.49) yields the components of the vector $\vec{b}$ of external forces:

$$
\begin{aligned}
b_1 &= G_{12}^{\mathrm{T}}\hat{M}_2^{-1}Q_2 + \left(\frac{\partial}{\partial\vec{q}_2}G_{12}\dot{\vec{q}}_2\right)\dot{\vec{q}}_2 \\
&= \begin{pmatrix} -\vec{n}_2 \\ -[(\vec{r}_1 + \vec{x}_1 - \vec{r}_2) \times \vec{n}_2]_z \end{pmatrix}^{\mathrm{T}} \begin{pmatrix} m_2^{-1} & 0 \\ 0 & J_2^{-1} \end{pmatrix} \begin{pmatrix} m_2\vec{g} \\ 0 \end{pmatrix} \\
&\quad -2\left(\vec{t}_2 \cdot \dot{\vec{r}}_2\right) - [\vec{n}_2 \cdot (\vec{r}_1 + \vec{x}_1 - \vec{r}_2)]\,\omega_2^2 \\
&= \begin{pmatrix} -\vec{n}_2 \\ -[(\vec{r}_1 + \vec{x}_1 - \vec{r}_2) \times \vec{n}_2]_z \end{pmatrix}^{\mathrm{T}} \begin{pmatrix} \vec{g} \\ 0 \end{pmatrix} -2\left(\vec{t}_2 \cdot \dot{\vec{r}}_2\right)\omega_2 - [\vec{n}_2 \cdot (\vec{r}_1 + \vec{x}_1 - \vec{r}_2)]\,\omega_2^2 \\
&= -\vec{n}_2 \cdot \vec{g} - 2\left(\vec{t}_2 \cdot \dot{\vec{r}}_2\right)\omega_2 - [\vec{n}_2 \cdot (\vec{r}_1 + \vec{x}_1 - \vec{r}_2)]\,\omega_2^2 \ ,
\end{aligned}
\tag{5.58}
$$

and in the same way

$$
b_2 = \vec{n}_1 \cdot \vec{g} - (\vec{x}_2 \cdot \vec{n}_1)\,\omega_2^2 \ .
\tag{5.59}
$$

The centrifugal force in b_1 seems to have the wrong sign, i.e., it seems to pull the two particles together instead of separating them. In b_2 it has the correct sign. The solution to this puzzle is that there is another inertial force acting at the contact, the Coriolis force. This force appears if a particle moves relative to a rotating reference frame. In our contact problem the vertex of particle 1 moves relative to the edge of particle 2, which provides the rotating frame. It can be shown that the Coriolis force contains a term that is twice the centrifugal force but of opposite sign. In b_2 there is no Coriolis force since the reference particle is fixed.

With the abbreviation $\vec{\xi}_2 = \vec{r}_1 + \vec{x}_1 - \vec{r}_2$, the geometry equation (5.16) can now be formulated for our example:

$$
\begin{pmatrix} a_1 \\ a_2 \end{pmatrix} =
$$

$$
\begin{pmatrix}
\dfrac{\vec{n}_2^2}{m_2} + \dfrac{\left[\vec{\xi}_2 \times \vec{n}_2\right]_z^2}{J_2} & -\dfrac{\vec{n}_1 \cdot \vec{n}_2}{m_2} - \dfrac{\left[\vec{\xi}_2 \times \vec{n}_2\right]_z \left[\vec{x}_2 \times \vec{n}_1\right]_z}{J_2} \\[4mm]
-\dfrac{\vec{n}_1 \cdot \vec{n}_2}{m_2} - \dfrac{\left[\vec{\xi}_2 \times \vec{n}_2\right]_z \left[\vec{x}_2 \times \vec{n}_1\right]_z}{J_2} & \dfrac{\vec{n}_1^2}{m_2} + \dfrac{\left[\vec{x}_2 \times \vec{n}_1\right]_z^2}{J_2}
\end{pmatrix}
\begin{pmatrix} f_1 \\ f_2 \end{pmatrix}
$$

$$
+ \begin{pmatrix}
-\vec{n}_2 \cdot \vec{g} - 2\left(\vec{t}_2 \cdot \dot{\vec{r}}_2\right) w_2 - \left[\vec{n}_2 \cdot (\vec{r}_1 + \vec{x}_1 - \vec{r}_2)\right] w_2^2 \\[2mm]
\vec{n}_1 \cdot \vec{g} - (\vec{x}_2 \cdot \vec{n}_1) w_2^2
\end{pmatrix} . \tag{5.60}
$$

Inserting the vectors $\vec{n}_{1/2}$, $\vec{x}_{1/2}$, and $\vec{r}_{1/2}$ given in (5.32) and (5.33) as well as using

$$
\vec{g} = \begin{pmatrix} 0 \\ -G \end{pmatrix} , \qquad \vec{\xi}_2 = L \begin{pmatrix} -1/2 + \gamma \\ -1/4 \end{pmatrix} , \tag{5.61}
$$

the geometry equation turns into ($j_2 = J_2/m_2 L^2$)

$$
\begin{pmatrix} a_1 \\ a_2 \end{pmatrix} =
\begin{pmatrix}
\dfrac{1}{m_2} + \dfrac{\left[L\left(\frac{1}{2} - \gamma\right)\right]^2}{J_2} & \dfrac{1}{m_2} - \dfrac{\left(\frac{1}{2} - \gamma\right)\left(-\frac{L}{2}\right)}{J_2} \\[4mm]
\dfrac{1}{m_2} - \dfrac{L\left(\frac{1}{2} - \gamma\right)\left(-\frac{L}{2}\right)}{J_2} & \dfrac{1}{m_2} + \dfrac{\left(-\frac{L}{2}\right)^2}{J_2}
\end{pmatrix}
\begin{pmatrix} f_1 \\ f_2 \end{pmatrix}
$$

$$
+ \begin{pmatrix}
-G - 2v_{2x} - \dfrac{1}{4}Lw_2^2 \\[2mm]
-G - \dfrac{1}{4}Lw_2^2
\end{pmatrix} \tag{5.62}
$$

$$
= \dfrac{1}{m_2}
\begin{pmatrix}
1 + \dfrac{(1 - 2\gamma)^2}{4j_2} & 1 + \dfrac{1 - 2\gamma}{4j_2} \\[4mm]
1 + \dfrac{1 - 2\gamma}{4j_2} & 1 + \dfrac{1}{4j_2}
\end{pmatrix}
\begin{pmatrix} f_1 \\ f_2 \end{pmatrix}
+ \begin{pmatrix}
-G - 2v_{2x}w_2 - \dfrac{1}{4}Lw_2^2 \\[2mm]
-G - \dfrac{1}{4}Lw_2^2
\end{pmatrix} .
$$

Suppose particles 1 and 2 are at rest. Let us study the qualitatively different cases $\gamma < 1/2$ (particle 2 is unstable) and $\gamma > 1/2$ (particle 2 is stable). With $m_2 = 1$ the simplified geometry equation reads

$$\begin{pmatrix} a_1 \\ a_2 \end{pmatrix} = \begin{pmatrix} 1 + \dfrac{(1-2\gamma)^2}{4j_2} & 1 + \dfrac{1-2\gamma}{4j_2} \\ 1 + \dfrac{1-2\gamma}{4j_2} & 1 + \dfrac{1}{4j_2} \end{pmatrix} \begin{pmatrix} f_1 \\ f_2 \end{pmatrix} + \begin{pmatrix} -G \\ -G \end{pmatrix} . \tag{5.63}$$

Let us start with all forces equal to zero and consider the first contact. Hence $a_1 = -G < 0$, which violates the consistency requirements since the particles would deform each other. This must be prevented by increasing the force f_1. The reduced geometry equation (all forces except f_1 are zero) then reads

$$a_1 = \left(1 + \frac{(1-2\gamma)^2}{4j_2}\right) f_1 - G . \tag{5.64}$$

The force f_1 is increased until a_1 becomes zero, i.e.,

$$f_1 = \frac{4j_2 G}{4j_2 + (1-2\gamma)^2} > 0 . \tag{5.65}$$

Since $a_1 = 0$ and $f_1 > 0$ the first contact is consistent (for the moment). Now the second contact is considered. Checking $f_2 = 0$ yields

$$a_2(f_2 = 0) = \left(1 + \frac{1-2\gamma}{4j_2}\right) \frac{4j_2 G}{4j_2 + (1-2\gamma)^2} - G = \frac{2\gamma(1-2\gamma)}{4j_2 + (1-2\gamma)^2} G . \tag{5.66}$$

If $\gamma \leq 1/2$, we obtain $a_2 \geq 0$, i.e., the contact is consistent and the calculation is finished. As expected, contact 2 breaks and particle 2 topples over the edge of particle 1. If $\gamma > 1/2$, the contact is not yet consistent since $a_2 < 0$. Therefore, f_2 is increased to prevent deformation *without* destroying the consistency of contact 1. In order to preserve the consistency of contact 1 simultaneously, f_1 is readjusted while changing the value of f_2. To this end the effect of changing both forces on a_1 is computed:

$$\begin{aligned} a_1 &= \left(1 + \frac{(1-2\gamma)^2}{4j_2}\right)(f_1 + \Delta f_1) + \left(1 + \frac{1-2\gamma}{4j_2}\right)(f_2 + \Delta f_2) - G \\ &= \left(1 + \frac{(1-2\gamma)^2}{4j_2}\right) f_1 + \left(1 + \frac{1-2\gamma}{4j_2}\right) f_2 - G \\ &\quad + \left(1 + \frac{(1-2\gamma)^2}{4j_2}\right) \Delta f_1 + \left(1 + \frac{1-2\gamma}{4j_2}\right) \Delta f_2 . \end{aligned} \tag{5.67}$$

Before adjusting f_1 and f_2 we had $a_1 = 0$, therefore,

$$a_1 = \left(1 + \frac{(1-2\gamma)^2}{4j_2}\right) \Delta f_1 + \left(1 + \frac{1-2\gamma}{4j_2}\right) \Delta f_2 . \tag{5.68}$$

The consistency condition $a_1 = 0$ should stay fulfilled, hence,

$$\Delta f_1 = -\frac{4j_2 + 1 - 2\gamma}{4j_2 + (1 - 2\gamma)^2}\Delta f_2 . \tag{5.69}$$

Increasing f_2 decreases f_1 if $4j_2 > 2\gamma - 1$. For such large values of j_2, particle 2 is so inert that lifting it at the left end (contact 2) also lifts the other contact point 1, thus reducing the contact force f_1. Equations (5.65) and (5.69) yield the inequality

$$0 < \frac{4j_2 G}{4j_2 + (1 - 2\gamma)^2} - \frac{4j_2 + 1 - 2\gamma}{4j_2 + (1 - 2\gamma)^2}\Delta f_2 \tag{5.70}$$

for the marginal value by which f_2 may be increased without violating $f_1 = 0$, i.e., the consistency of contact 1. Its solution reads

$$\Delta f_2 < \frac{4j_2}{4j_2 + (1 - 2\gamma)}G \tag{5.71}$$

if $4j_2 > 2\gamma - 1$, otherwise $\Delta f_1 > 0$, i.e., Δf_2 can grow unlimited. If f_2 is changed from zero to Δf_2 while at the same time changing, f_1 by Δf_1, according to (5.69) to keep the first contact consistent, a_2 assumes the value

$$
\begin{aligned}
a_2 &= \left(1 + \frac{1 - 2\gamma}{4j_2}\right)\left(\frac{4j_2 G}{4j_2 + (1 - 2\gamma)^2} + \Delta f_1\right) + \left(1 + \frac{1}{4j_2}\right)\Delta f_2 - G \\
&= \frac{2\gamma(1 - 2\gamma)G}{4j_2 + (1 - 2\gamma)^2} - \left(1 + \frac{1 - 2\gamma}{4j_2}\right)\frac{4j_2 + 1 - 2\gamma}{4j_2 + (1 - 2\gamma)^2}\Delta f_2 + \left(1 + \frac{1}{4j_2}\right)\Delta f_2 \\
&= \frac{2\gamma(1 - 2\gamma)G}{4j_2 + (1 - 2\gamma)^2} + \frac{4\gamma^2}{4j_2 + (1 - 2\gamma)^2}\Delta f_2 \\
&= \frac{2\gamma(1 - 2\gamma)G + 4\gamma^2\Delta f_2}{4j_2 + (1 - 2\gamma)^2} .
\end{aligned}
\tag{5.72}
$$

Hence, contact 2 becomes consistent ($a_2 = 0$) when adjusting f_2 by

$$f_2 = \Delta f_2 = \frac{2\gamma - 1}{2\gamma}G . \tag{5.73}$$

Since $\gamma > 0.5$ this value is positive in agreement with the consistency requirements. It can be checked that this value always satisfies (5.71). The corresponding value of f_1 follows from (5.65) and (5.69):

$$f_1 = \frac{G}{2\gamma} . \tag{5.74}$$

Summarizing the result of this example, for $\gamma \leq 1/2$ (unstable case) we obtain

$$f_1 = \frac{4j_2G}{4j_2 + (1 - 2\gamma)^2} > 0 \,, \qquad f_2 = 0 \,,$$

$$a_1 = 0 \,, \qquad\qquad a_2 = \frac{2\gamma(1 - 2\gamma)G}{4j_2 + (1 - 2\gamma)^2} > 0 \,,$$

(5.75)

which means that contact 2 breaks and particle 2 flips over the vertex of particle 1. For $\gamma > 1/2$ (stable case) the result reads

$$f_1 = \frac{G}{2\gamma} \,, \qquad f_2 = \frac{2\gamma - 1}{2\gamma}G \,,$$

$$a_1 = 0 \,, \qquad a_2 = 0 \,,$$

(5.76)

i.e., both contacts persist. Consequently the result is in agreement with the hypothesis from the beginning of this section, hence, Rigid-Body Dynamics indeed yields correct results.

The presented example is one of the most simple (non-trivial) examples. The aim of this example is to apply the described algorithm in a very regular way to a mechanical system to elucidate its basic steps. All parts of the computation can be directly implemented in a computer simulation program.

5.5 Collisions

In the framework of Rigid-Body Dynamics, collisions occur if two contacting particles have a negative, normal relative velocity at their contact point, i.e., if not stopped, the particles would deform each other. Obviously no *finite* contact force can prevent this deformation since no matter how large the force is, it always takes a finite, however small, period of time to stop the approaching particles. Thus, to prevent deformation of the particles, an *infinite* repulsive force of infinitesimal duration must be applied.

Collisions of grains of large (but finite) elastic material constant (Young modulus) at a given initial relative velocity are of short (but finite) duration, where the deformation is small as well. If the elastic constant is increased, the duration of the collision decreases, i.e., the interaction force increases since the particles collide in less time at smaller deformation.

It turns out that the total momentum transfer Δp between the two particles is finite. It can be obtained by time integration of the interaction force:

$$\Delta p = \lim_{t_c \to 0} \int_0^{t_c} f \, \mathrm{d}t = \text{finite} \,.$$

(5.77)

Therefore, instead of forces we discuss the momentum transfer. The duration t_c of the collision is only formally the parameter of the limit in the equation above. This limit is actually achieved by increasing the elastic constant to infinity. In this limiting process, the duration of collision approaches zero.

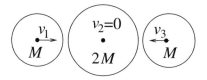

Fig. 5.4. Three-particle collision. The outer particles impact the middle one

Due to the infinite stiffness of the particles, or equivalently, the infinite speed of sound in the particle material, the velocities of particles after a multi-particle collision are not uniquely determined. In the limit of infinite speed of sound, all information on the exact collision mechanism is lost, in particular on the sequence of collisions. Colliding particles of finite stiffness do not exhibit this feature since all processes here take finite time.

Let us give an example to illustrate this important feature. Figure 5.4 shows the collision of two particles of mass M with a third particle of mass $2M$. The velocities of the small particles are of equal absolute value, while the large particle is at rest. Observed with a finite time resolution (time intervals smaller than a certain size cannot be resolved), all three particles collide simultaneously. If the particles are very stiff, this finite resolution is insufficient to distinguish different (equally possible) collision scenarios. There are three different possibilities of what may actually happen:

1. The collisions take place sequentially, particle 1 collides first. In this case three collisions occur:
   i. Particles 1 and 2:
   $$v_1' = -\frac{v}{3}, \qquad v_2' = \frac{2}{3}v, \qquad v_3' = -v . \tag{5.78}$$

   ii. Particles 2 and 3:
   $$v_1'' = -\frac{v}{3}, \qquad v_2'' = -\frac{4}{9}v, \qquad v_3'' = \frac{11}{9}v . \tag{5.79}$$

   iii. Particles 1 and 2:
   $$v_1''' = -\frac{13}{27}v, \qquad v_2''' = -\frac{10}{27}v, \qquad v_3''' = \frac{11}{9}v . \tag{5.80}$$

   This is the final state where particles 1 and 2 move to the left and particle 3 moves to the right.

2. The collisions take place sequentially with particle 3 colliding first with particle 2. Here the same sequence of collisions occurs as in case 1 but with particles 1 and 3 exchanged. The final velocities are
   $$v_1''' = -\frac{11}{9}v, \qquad v_2''' = \frac{10}{27}v, \qquad v_3''' = \frac{13}{27}v . \tag{5.81}$$

3. The collisions take place precisely at the same time, leading to the post-collision velocities

$$v_1 = -v , \qquad v_2 = 0 , \qquad v_3 = v . \tag{5.82}$$

If the collisions are not completely separated in time due to finite stiffness, the result is undetermined. The final velocities may assume any value that is in agreement with the conservation of momentum and (for elastic particles) with the conservation of energy.

These results are independent of the absolute time lag Δt (positive if particle 1 collides first). Hence, the limit $\Delta t \to 0$ does not exist. Likewise, for infinitely stiff particles (elastic constant $E \to \infty$) the limits $\lim_{E \to \infty} v_i$ ($i = 1, 2, 3$) do not exist either and collisions of rigid particles are inherently ambiguous. Thus with the model assumption of vanishing collision time (rigid-body assumption), many-particle collisions cannot be considered consistently.

The above given example is not critical since in a three-particle system the perfect coincidence $t_1 = t_3$ occurs with vanishing probability, i.e., never. There are, however, situations where multiple-particle contacts occur inevitably, as sketched in Fig. 5.5. In the table-like arrangement, three particles are already in mutual (persistent) contact. As soon as the sphere contacts the rectangle in the middle, a four-particle collision takes place.

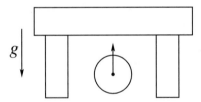

Fig. 5.5. Example of a realistic multiple-particle collision [16]

According to these arguments the information on the detailed sequence of collisions (the system's history) is not accessible. The following set of assumptions turn out to yield realistic results, although they do not follow from first principles:

1. All individual pair collisions occur at once.
2. The transfer of momentum at contact points is finite. There is no momentum transfer which corresponds to attractive forces.
3. With ε being the coefficient of restitution and v the impact velocity we require that the final velocity $v' \geq -\varepsilon v$. ($v < 0$, therefore $-\varepsilon v$ is positive.)
4. If the velocity after the collision is strictly larger than $-\varepsilon v$, there is no momentum transfer at this contact.

The first two assumptions have been introduced already. The remaining two assumptions deserve further discussion. Two-particle collisions can be described by means of the coefficient of restitution, which relates the impact velocity v and the final velocity v' after the collision:

$$v' = -\varepsilon v \, . \tag{5.83}$$

For a detailed discussion of the coefficient of restitution see Sect. 3.5. In the case of multi-particle collisions, however, two particles that are initially at rest relative to each other may separate after a collision (see the example sketched in Fig. 5.5). Therefore, the final velocity may indeed be larger than the value $-\varepsilon v$ (remember that $v \leq 0$, therefore $-\epsilon v \geq 0$.) The opposite case $v' < -\varepsilon v$ is not allowed. In particular, two particles colliding with a finite impact velocity must not remain at rest relative to each other after the collision.

To illustrate this requirement consider again the situation sketched in Fig. 5.5. After the collision of the sphere with the central rectangle, the latter particle moves upward. Moreover, this four-particle collision changes the relative velocity of the central rectangle and the left and right rectangles. Before the collision this relative velocity is zero, but after the collision it is positive. Consequently, (5.83) is violated for this pair of particles involved in the collision. Assumption 3 takes care of this fact.

According to the fourth assumption, two particles departing from each other after a collision at velocity larger than $-\varepsilon v$ may not be accelerated further by an additional momentum transfer. Let us formulate these assumptions mathematically. The excess velocity

$$\Delta v \equiv v' + \varepsilon v \, , \tag{5.84}$$

is zero if $v' = -\varepsilon v$. The source of any velocity change is a momentum transfer between the colliding particles. The excess velocities at the contacts can be related to the momentum transfers by means of the collisional geometry equation, which is derived in a similar way as the geometry equation of the force algorithm:

$$v'_\alpha = v_\alpha + A^{col}_{\alpha\beta} \Delta p_\beta \tag{5.85}$$

or, using the excess velocities instead:

$$\Delta v_\alpha = (1 + \varepsilon) \, v_\alpha + A^{col}_{\alpha\beta} \Delta p_\beta \, . \tag{5.86}$$

The collision matrix $\hat{A}^{col}$ is computed in the same way as the geometry matrix, but with two major differences. First we consider the collision process as frictionless. Therefore the collision matrix is similar to the term $\hat{A}^{nn}$ of the geometry matrix. Second, when computing the geometry matrix all contacts with non-zero relative velocity at the contact point are disregarded. These contacts have to be included now. Otherwise the same expressions to determine the matrix elements apply. A simple way to convince oneself of this fact is to integrate the geometry equation over time:

$$\int\limits_t^{t+t_c} dt \, a_\alpha = \int\limits_t^{t+t_c} dt \, b_\alpha + A_{\alpha\beta}^{\text{nn,ext}} \int\limits_t^{t+t_c} dt f_\beta \; . \tag{5.87}$$

The matrix $\hat{A}^{\text{nn,ext}}$ (ext for extended) contains *all* contacts, not only those with vanishing, normal relative velocity. Performing the limit $t_c \to 0$ the term $\int dt \, b_\alpha$ vanishes since the terms represented by b_α are finite. With $v'_\alpha = v_\alpha + \int dt \, a_\alpha$ and $\Delta p_\beta = \int dt f_\beta$ we obtain

$$v'_\alpha = v_\alpha + A_{\alpha\beta}^{\text{nn,ext}} \Delta p_\beta \; . \tag{5.88}$$

Calling the extended matrix $\hat{A}^{\text{nn,ext}} = A^{\text{col}}$ for simplicity's sake, we arrive at (5.85).

Note that the components v_α are *relative* velocities of the contact points of two contacting particles, but not the velocities of the particles themselves. In mathematical terms, the above discussed assumptions can be expressed as

$$\Delta v_\alpha \geq 0$$
$$\Delta p_\alpha \geq 0 \tag{5.89}$$
$$\Delta v_\alpha \Delta p_\alpha = 0 \; .$$

where Δp_α is the exchange of momentum at the contact point α. The first condition assures that the final velocity is larger than $-\varepsilon v$, the second condition excludes attractive interactions between particles, and finally, the third condition excludes momentum transfer if $\Delta v \neq 0$. These conditions together with the geometry equations (5.86) establish a Linear Complementarity Problem that can be solved by Dantzig's algorithm, as explained in Sect. 5.4.

5.6 Resolution of Static Indeterminacy

If the number of contacts in the system is too large, the contact forces cannot be uniquely determined by the force algorithm. This is the case if the number of free variables in the system, i.e., the number of contact forces, is larger than the number of mechanical degrees of freedom, $3N$ in two dimensions or $6N$ in three dimensions, with N being the number of particles. However, the total forces and torques acting on the particles and, hence, their trajectories are unique. This drawback restricts the applicability of Rigid-Body Dynamics since for some applications the knowledge of the contact forces is crucial for understanding the system's properties.

So far the contact forces have been considered as independent of each other. This assumption is the reason for the force indeterminacy. In realistic systems, however, the forces are not independent as shown for the example in Fig. 5.6. If an external force is applied on the central particle (directed to the right), the contact force with the particle to its right is increased while

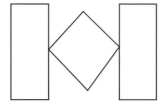

Fig. 5.6. The central particle is in contact with two other particles. The forces at both contacts are not independent of each other

at the same time the contact force with the particle to its left is decreased. In reality, both contact forces in this example depend on a single parameter, which is the applied external force.

One can mimic this behavior by introducing infinitesimal displacements of the particles. Each particle has a set of macroscopic coordinates, $\vec{r}$ and $\vec{\varphi}$, and a set of microscopic coordinates, $\delta\vec{r}$ and $\delta\vec{\varphi}$. We define the vector $\delta\tilde{q}$ of all microscopic (infinitesimal) coordinates in analogy to the vector $\tilde{q}$ of all (macroscopic) coordinates of all particles. The contact network and the kinematic state of the particles are determined solely by the macroscopic coordinates, hence they can be considered the actual coordinates of the particles. The microscopic coordinates describe a displacement from the macroscopic coordinates. They are the sole reason for deformations of the contacting particles. By definition they are of *infinitesimal* size, which allows us to employ a linear approximation in $\delta\tilde{q}$ for the computation of the deformations. The vector $\tilde{\xi}$ of all deformations in the system

$$\tilde{\xi} = \hat{D}\delta\tilde{q} \tag{5.90}$$

is defined by the deformation matrix $\hat{D}$ and the microscopic coordinates. The dependence of $\hat{D}$ on the geometric properties of the systems is straightforward but lengthy, therefore it is not given in explicit expressions here. For presenting the idea it is sufficient to state that the dependence of $\tilde{\xi}$ on $\delta\tilde{q}$ is linear, as in (5.90). We define a force law to relate the deformations at the contact points with the contact forces

$$f = f(\tilde{\xi}) , \tag{5.91}$$

which are functions of the displacements $\delta\tilde{q}$. To calculate the forces, the geometry equation (5.24) is used together with the consistency conditions (5.25). The solutions to this set of equations yield the microscopic displacements $\delta\tilde{q}$. Hence, there are as many variables as degrees of freedom, i.e., the system is unique.

To determine the microscopic coordinates, an overdamped relaxation method is applied. We start with a set of inconsistent coordinates $\delta\tilde{q}$. The term inconsistent means that the consistency conditions for the resulting forces and accelerations are not fulfilled. Now we let the system relax. This way new microscopic coordinates $\tilde{\xi}' = \tilde{\xi} - h\tilde{a}$ appear, with h being the step size and $\tilde{a}$ the vector of all contact accelerations. If the contact acceleration is negative the displacement is larger, yielding a larger force to stop the approaching motion of the contacting particles. Since the adjustment step for the displacements is proportional to the acceleration, the sequence of microscopic coordinates of the particles can be understood as overdamped motion.

The adjusted microscopic coordinates are the solution of the linear system of equations

$$\tilde{\xi} - h\tilde{a} = \hat{D}\delta\tilde{q}' \tag{5.92}$$

or, equivalently,

$$\hat{D}\left(\delta\tilde{q}' - \delta\tilde{q}\right) = -h\tilde{a} \ . \tag{5.93}$$

In cases with fewer degrees of freedom than contact accelerations, the system of equations is overdetermined. There may be vectors $h\tilde{a}$ that are not representable by any vector $\delta\tilde{q}' - \delta\tilde{q}$. In this case the projection of the vector $-h\tilde{a}$ into the image space of the operator $\hat{D}$ is needed before solving the system of equations.

The adjustment of microscopic coordinates is repeated until the consistency conditions are met. To further improve the speed of this method we can store the microscopic coordinates, which yielded a consistent system, in the previous time step. If the system did not change too much, this set of microscopic coordinates is very close to the new solution and only a few iteration steps are needed to arrive at the new solution.

The presented method combines the advantage of Rigid-Body Dynamics, namely the ability to simulate very stiff particles, with the advantage of Molecular Dynamics, namely uniquely-defined contact forces. The method of small displacements may also be applied to simulate certain degradation mechanisms. As an example, abrasion of edges of the particles can be simulated by gradually changing the force law (5.91), which describes changing edge properties of the particles.

5.7 Integration of the Equation of Motion

For the integration of Newton's equation of motion, a Runge–Kutta scheme of fourth order has been applied. This integration algorithm requires four force computations per time step. Since we use discrete time steps, we frequently face the problem that after progressing by one time step some particles deform each other due to a collision – a situation to be avoided for Rigid-Body Dynamics. Therefore, the time of the next contact is determined by means of a nested intervals approach: Starting with a certain time step, the system is advanced accordingly and checked for particle deformations. If there are no deformations the system is advanced further. In case that any deformations occur, the system state before this time step is restored, the time step is decreased, and the system is advanced again. This process is repeated until the distance of the particle surfaces drop below a certain threshold while not deforming each other. This new state of the system is then accepted and the other steps of the algorithm are performed. This procedure ensures that there are no particle deformations in the system at the beginning of any accepted time step.

On the contrary, if there are no collisions in the systems for a certain number of time steps the time step is increased until it reaches an upper limit, which assures that particles cannot move across each other (thus missing a collision) in a single time step.

Due to the finite accuracy of the integration scheme, particle deformations cannot be excluded completely. There exist sophisticated integration methods that can avoid this problem, although these algorithms are complicated and numerically expensive. A simpler way to avoid mutual deformations is called *post-stabilization*, where after applying the integration scheme the particles are moved by tiny distances until the deformations are corrected. There are several methods for post-stabilization; for an in-depth discussion see [64].

With the integration method presented here, the discussion of the simulation algorithm is now complete.

5.8 Simple Examples

In this section, Rigid-Body Dynamics is demonstrated by means of some simple examples. Animated sequences of these examples can be found on the book's web site. The sequence of snapshots in Fig. 5.7 visualizes the action of static friction. The system consists of a fixed surface (lower rectangle), a

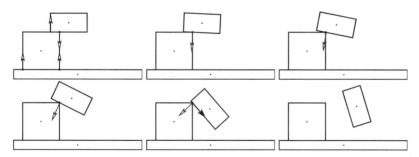

Fig. 5.7. A mobile particle topples from an unstable position on top of two fixed particles. At the fifth picture of the sequence the particle starts to slide (*filled arrow*) due to the increasing tangential force. The last picture shows the mobile particle after its collision with the surface. The normal vectors of the contacts are drawn with hollow arrows

fixed square, and a mobile particle lying on top. Initially the mobile particle is at rest. According to the position of its center of mass, the particle tilts to the right and eventually starts to slide. Sliding occurs only after reaching a certain inclination due to static friction at the contact point, as visualized by the velocity vectors.

The vector of velocity points opposite to the direction in which the contact point (the corner of the fixed particle) moves with respect to the involved surface of the mobile particle. Consequently, here the velocity vector points in the direction of the motion of the mobile particle.

Figure 5.8 shows the transition to sliding motion for a system of three mobile particles.

Figure 5.9 shows three toppling dominos. After a certain time the particles come to rest due to the action of static friction.

The last example, Fig. 5.10, shows a sediment of 50 irregular pentagons. The container has been filled sequentially, particle by particle. Just as in the experiment we obtain a loose packing of grains. The granular material can be further densified by recurrent tapping of the container. Experimentally

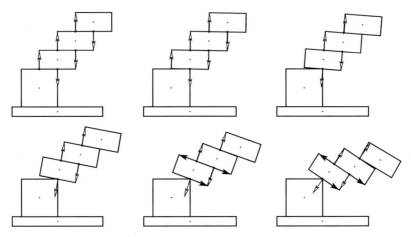

Fig. 5.8. Three mobile particles topple. The fifth snapshot shows the transition to sliding motion (velocity unit vectors drawn as filled arrows)

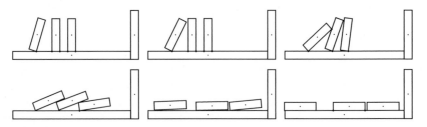

Fig. 5.9. Three toppling dominos come to rest after a finite time (no vectors drawn)

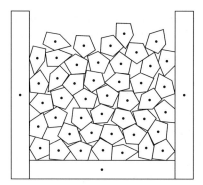

Fig. 5.10. Sedimentation of 50 irregular pentagons

one observes that the densification due to tapping is an extraordinarily slow process which has attracted large scientific interest, e.g., [27, 58, 123, 129, 158, 196, 210]

5.9 Discussion of the Model

The idea of Rigid-Body Dynamics is complementary to the idea of (force-based) Molecular Dynamics: While in Molecular Dynamics the forces that govern the motion of the particles are determined from the (small) mutual deformation of the particles, in Rigid-Body Dynamics the forces are determined by motion constraints such that mutual deformation of particles is avoided. Thus Rigid-Body Dynamics is intended to describe the motion of very stiff particles. Molecular Dynamics simulations become numerically problematic for very stiff particles when the forces are steep functions of the deformation. Moreover, the force must be known as a function of the deformation and the deformation rate. These problems, which are discussed in more detail in the beginning of this chapter, are unavoidable within the Molecular Dynamics approach, but are avoided in a natural way in Rigid-Body Dynamics. Therefore we believe that Rigid-Body Dynamics is much better suited than Molecular Dynamics to simulate very hard materials, such as hard stones under moderate load in railway ballasts. Most notable, in our opinion, is the fact that we may obtain correct trajectories of the particles without knowing the force law governing the particle interactions.

However, from the presentation of the algorithm it becomes obvious that one time step in Rigid-Body Dynamics is a lot more complicated than one time step in Molecular Dynamics. Each time step requires by far more computing time than a time step in Molecular Dynamics simulations. Fortunately this disadvantage is offset at least partly by the fact that in Rigid-Body Dynamics, we can choose a larger time step than in Molecular Dynamics. Whereas in Molecular Dynamics the time step is limited by the typical length scale of the interaction force, divided by the typical velocity (usually on the order $\sim 10^{-7}$ sec), the time step in Rigid-Body Dynamics is determined by the characteristic *time* in which the forces in the system actually change.[4] The latter quantity is usually much larger ($\sim 10^{-3}$ sec). Thus, the computer time requirements for both methods are comparable when simulating mainly static problems, like heaps of granular particles or railway ballasts.

An additional advantage of Rigid-Body Dynamics is that fracture of particles can be introduced in a very natural way (avoiding serious problems which may occur in Molecular Dynamics simulations when particle fragmentation is considered, see Sect. 2.9). In Rigid-Body Dynamics particle fragmentation is simulated by introducing new edges into a polygonal particle, thus splitting a polygon into a number of smaller polygons. Unlike the method described in Sect. 2.9, the fragment particles always fit into the same space as the original particle before fragmenting.

Rigid-Body Dynamics and Molecular Dynamics represent two complementary approaches to modeling particle systems. While Molecular Dynamics works best when dealing with *dynamical* systems of soft to moderately stiff particles of simple shape (although many tricks have been developed to overcome this limitations), Rigid-Body Dynamics is particularly well suited to handle the opposite case of systems comprising very stiff particles of complicated shape, which exhibit slow dynamics. However, the approximations that allowed for simulating rigid bodies came at a price—we had to sacrifice the uniqueness of the contact forces. This is a direct consequence of disregarding material properties. Thus, the algorithm does not necessarily compute the physically-correct sets of contact forces of a many-particle system, but one out of many (in general out of an infinite set of contact forces). Therefore, it cannot be assured that the system behaves in a physically-realistic way.

However, it is our firm believe that Rigid-Body Dynamics represents an interesting complement to Molecular Dynamics, which could provide new insights into many systems that are notoriously hard to model using Molecular Dynamics methods.

---

[4] To formulate the difference more pointedly: In Molecular Dynamics we have to use the time in which the force *may* change as time step, in Rigid-Body Dynamics we use the time in which it actually changes, which is larger.

6

Cellular Automata

6.1 Overview

The study of granular systems had been disregarded almost completely for decades among physicists when in 1987 Bak, Tang, and Wiesenfeld [14] defined a cellular automaton model to simulate avalanches on the surface of a sand heap. With this example they introduced the concept of *self-organized criticality*, which applies to a much wider class of problems [13]. Although there are doubts about the validity of this model for the description of avalanches, it has been this particular work that, among other important contributions, inspired the rapid development of the field. A general introduction into the theory of cellular automata can be found, e.g., in [305].

The literature on cellular automata simulations of granular materials can be classified into 3 topics:

1. Heap growth and avalanche statistics
2. Structure formation (dunes and ripples) caused by air, water flows, or vibrated containers
3. Lattice gas models for flows of granular material

The underlying models and some results obtained with the respective algorithms will be discussed in the following sections.

6.2 Heap Formation and Avalanches

The paradigm of self-organized criticality (SOC) states that [14, 15]: *There are driven systems which evolve by their internal dynamics, independent of any control parameters, toward a critical state. This state is characterized by large-scale spatial or spatio-temporal correlations.*

This idea has been adopted immediately by physicists and scientists of other fields. At the present time there exists an enormous body of scientific literature in many fields, which supports the postulates of SOC for certain

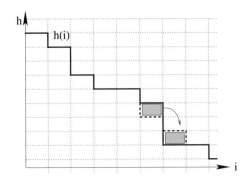

Fig. 6.1. Sand pile automaton described by Bak, Tang and Wiesenfeld

systems or which tries to disprove them. SOC systems invariably seem to show arbitrarily large fluctuations, a fact that has been the focus of many publications. An introduction to this interesting subject is given in the book by Bak [13]. Attempts to formulate *necessary* conditions for systems that lead to the emergence of SOC have been unsuccessful so far.

The first system for which the SOC-idea was developed was a cellular automaton model for avalanches which occur on a heap of sand if one puts sand grains individually, one by one, on top of the pile. Figure 6.1 shows a sketch of the one-dimensional[1] automaton. The height of the heap at (horizontal) position i is denoted by $h(i)$. The dynamics of the heap is governed by the local gradient

$$\Theta_i = h_i - h_{i+1} . \tag{6.1}$$

If a grain is added at position i the gradients change according to

$$\begin{aligned}
\Theta_i &\to \Theta_i + 1 \\
\Theta_{i-1} &\to \Theta_{i-1} - 1 .
\end{aligned} \tag{6.2}$$

If the local gradient Θ_i at position i exceeds a certain critical threshold Θ^c, one grain falls from position i to the neighboring site $i+1$. The corresponding transition rules for the local gradients read

$$\begin{aligned}
\Theta_i &\to \Theta_i - 2 \\
\Theta_{i+1} &\to \Theta_{i+1} + 1 \\
\Theta_{i-1} &\to \Theta_{i-1} + 1 .
\end{aligned} \tag{6.3}$$

If $\Theta_N > \Theta^c$ (boundary condition) the particle at position N is removed from the system. The corresponding rule reads

---

[1] This is the dimension of the base on which the heap grows. Realistic heaps in this sense are, hence, two-dimensional.

$$\Theta_N \to \Theta_N - 1$$
$$\Theta_{N-1} \to \Theta_{N-1} + 1 \,.$$

$$(6.4)$$

A heap is stable if

$$\Theta_i \le \Theta^c \quad \text{for all} \quad i = 1, 2, \ldots, N \,. \tag{6.5}$$

In this version of the sand heap model grains are always added at the same position ($i = 1$). It is sufficient to restrict ourselves to the simulation of one half of the heap because particles on the right half of the heap always move to the right, particles on the left always move to the left. Thus, both sides cannot interact and can, therefore, be treated independently.

Let us start the gedankenexperiment with an empty plane of length N. If particles are placed on the plane little happens for a certain time: the heap grows and occasionally rule (6.3) makes particles move from one site to another. At some point in time the local gradient at one site, which received a particle from a neighboring site, is also larger than the threshold value Θ^c, i.e., this site in its turn has to give a particle to its downhill neighbor. This may cause a domino effect. A one-dimensional system quickly evolves into a state where all gradients just adopt their critical values $\Theta = \Theta^c$. Once this state is reached, adding one particle turns the first site supercritical, thus giving a particle to its neighbor downhill, which now in its turn becomes supercritical and hands a particle downhill and so on. A practical definition of the size of an avalanche is the number of sites that become supercritical after adding one particle, thus distributing particles to their neighbors. The present model reveals avalanches that span the entire system from the point of origin to the end of the chain of sites. The avalanches would therefore have size (length) N.

Although this system is very simple it shows an essential property of a self-organized system: the heap develops *by its intrinsic dynamics* to its critical state.

The dynamics of the equivalent two-dimensional automaton for which Bak, Tang, and Wiesenfeld developed the theory of self-organized criticality is significantly more interesting. In what follows we describe this two-dimensional automaton and discuss some selected results obtained with this model. In its simple form where all particles are deposited at the same position ($i = 1, j = 1$), thus, in analogy to the one-dimensional case, the simulation area can be reduced to one quadrant of the heap. On this quadrant a quadratic lattice with sites (i, j), where $1 \le i \le N$ and $1 \le j \le N$, is constructed. The local gradient of the heap is defined by

$$\Theta_{i,j} = 2h_{i,j} - h_{i+1,j} - h_{i,j+1} \,. \tag{6.6}$$

The automaton rules (6.2) and (6.3) can be generalized to the two-dimensional case: If a grain is placed on the lattice site (i, j), the local gradients change according to

$$\Theta_{i-1,j} \to \Theta_{i-1,j} - 1$$
$$\Theta_{i,j-1} \to \Theta_{i,j-1} - 1 \tag{6.7}$$
$$\Theta_{i,j} \to \Theta_{i,j} + 2 \,.$$

If site (i, j) is critical, i.e., $\Theta_{i,j} > \Theta^c$, one particle is moved to each $(i + 1, j)$ and $(i, j + 1)$, i.e., in the positive x and y-directions. The automaton rule analogous to (6.3) reads

$$\Theta_{i,j} \to \Theta_{i,j} - 6$$
$$\Theta_{i+1,j} \to \Theta_{i+1,j} + 2$$
$$\Theta_{i-1,j} \to \Theta_{i-1,j} + 2$$
$$\Theta_{i,j+1} \to \Theta_{i,j+1} + 2 \tag{6.8}$$
$$\Theta_{i,j-1} \to \Theta_{i,j-1} + 2$$
$$\Theta_{i+1,j-1} \to \Theta_{i+1,j-1} - 1$$
$$\Theta_{i-1,j+1} \to \Theta_{i-1,j+1} - 1 \,.$$

There are also alternative ways of modeling a growing sand heap (see, e.g., [135]). The dynamics of this system turns out to be significantly more complex than that of the one-dimensional heap. In particular, the sizes of the avalanches follow a power-law distribution. For a detailed discussion of automata for modeling granular systems see, e.g., [14, 135].

Since the described automaton has been repeatedly implemented by several authors and since these programs are well documented on the internet, we present here a slightly different cellular automaton which has been inspired by Puhl [226]. As the main difference to the model discussed above, each direction is considered separately, i.e., in this model sand flows only in the direction of a gradient that is actually steeper than a given threshold value Θ^c. In the model above the material flew in all directions regardless of the gradient in this direction, occasionally even uphill. Since in the present model particles may cross quadrant boundaries, we cannot limit ourselves to a single quadrant but have to consider the entire heap.

The gradients in all four directions are checked sequentially. If one of the four inequalities

$$h(i, j) - h(i \pm 1, j \pm 1) \geq \Theta^c \tag{6.9}$$

is satisfied, there is a flow of material in the corresponding direction. The amount m of grains to be displaced is a random number from the interval $1 \leq m \leq \Theta^c/2$, thus assuring that the gradient cannot be reversed by the material displacement.

First the program includes some standard headers and defines the necessary parameters: L is the lattice size, N is the number of particles to be deposited, tps is the interval between Postscript snapshots of the system, and threshold is the critical threshold Θ^c defined in (6.9). Additionally, ranf() is defined for generating random numbers in the interval $[0, 1)$.

```
                          ━━━━━━━━━ heap.cpp ━━━━━━━━━
 1  #include <iostream>
 2  #include <fstream>
 3  #include <stdlib.h>
 4  #include <vector>
 5
 6  const int L = 201, N = 100000000, tps = 1000000, threshold = 11;
 7  const char *psfile = "heap.ps", *avfile = "avalanches.dat";
 8
 9  int H[L+2][L+2];
10
11  float ranf(){ return float(rand())/(float(RAND_MAX)+1); }
12  int equil();
13  void pssnap(ostream &,int);
                       ━━━━━━━━━ (continued) ━━━━━━━━━
```

After opening the output files **heap.ps** and **avalanches.dat**, the program
starts with an empty N×N-lattice H. In each iteration step a grain is placed at
the center site and the function **equil()** is called to put the heap into a stable
state. The return value of **equil()** is the number of particles displaced in this
iteration step, i.e., the size of the avalanche. This value is recorded in the file
avalanches.dat. Finally, in intervals of **tps** iteration cycles, a snapshot of
the heap is taken in Postscript format.

```
                          ━━━━━━━━━ heap.cpp ━━━━━━━━━
14  int main()
15  {
16    ofstream psout, avout;
17    psout.open(psfile);
18    avout.open(avfile);
19
20    for(unsigned int i=0; i<L+2; i++){
21      for(unsigned int j=0; j<L+2; j++){
22        H[i][j] = 0;
23      }
24    }
25    for(int it=0; it<N; it++){
26      H[int(L/2)+1][int(L/2)+1]++;
27      avout << equil() << endl;
28      if( !(it % tps)){
29        cout << "snapshot "<< int(it/tps) << endl;
30        pssnap(psout,int(it/tps));
31      }
32    }
33    psout.close();
34    avout.close();
35  }
                       ━━━━━━━━━ (continued) ━━━━━━━━━
```

The rules of the automaton are implemented in **equil()**. For each direction
condition (6.9) is checked. If it is satisfied, a random number of grains is moved
accordingly.

The coordinates of the lattice sites that have changed their height H in
the current iteration step are stored in the vectors Li and Lj. After entering the procedure **equil()**, these vectors contain only the coordinate of
the particle that is newly placed onto the heap, i.e., Li[0]=int(N/2)+1 and
Lj[0]=int(N/2)+1. If the local gradient at this position is too large according to the automaton rules, some particles are distributed to the neighboring
sites. Thus, those neighboring sites may become supercritical too. Therefore,

we have to check the stability of these sites as well. This is done in the `while`-loop starting at line 42. We check the most recent elements of the vectors `Li` and `Lj` to see if the local gradients exceed the threshold. In the process the number `ncrit` of supercritical gradients is counted. If there are supercritical gradients, a random number `mov` of particles is moved to the neighboring site in the corresponding direction. The number of particles moved depends on the number `ncrit`, and must be chosen such that the gradient can never change its sign. The coordinates of the destination site are stored in the vectors `Li` and `Lj` for later checking. After checking for all four directions that the current site is stable, its coordinates are removed from `Li` and `Lj`. Once all sites in `Li` and `Lj` are visited, i.e., once `Li` and `Lj` are empty, the relaxation step is completed.

```
                        ———————— heap.cpp ————————
36  int equil()
37  {
38    int avcount = 0, si, sj, mov;
39    vector<int> Li, Lj;
40    Li.push_back(int(L/2)+1);
41    Lj.push_back(int(L/2)+1);
42    while(Li.size()){
43      si=Li.back();
44      sj=Lj.back();
45      Li.pop_back();
46      Lj.pop_back();
47      int oldH = H[si][sj];
48      int ncrit=0;
49      for(int i=-1; i<=1; i+=2) {
50        if(oldH >= H[si+i][sj]+threshold) ncrit++;
51        if(oldH >= H[si][sj+i]+threshold) ncrit++;
52      }
53      if(ncrit>0){
54        for(int i=-1; i<=1; i+=2){
55          if(oldH >= H[si+i][sj]+threshold) {
56            mov = int(threshold*ranf()/(ncrit+1));
57            H[si][sj]-=mov;
58            H[si+i][sj]+=mov;
59            avcount += mov;
60            Li.push_back(si+i); Lj.push_back(sj);
61          }
62          if(oldH >= H[si][sj+i]+threshold) {
63            mov = int(threshold*ranf()/(ncrit+1));
64            H[si][sj]-=mov;
65            H[si][sj+i]+=mov;
66            avcount += mov;
67            Li.push_back(si); Lj.push_back(sj+i);
68          }
69        }
70      }
71      if((Li.back()==1) || (Li.back()==L) || (Lj.back()==1) || (Lj.back()==L)){
72        for(int i=0; i<L+2; i++){
73          H[0][i] = H[i][0] = H[L+1][i] = H[i][L+1] = 0;
74        }
75      }
76    }
77    return avcount;
78  }
                        ———————— (continued) ————————
```

If a particle reaches the boundary of the simulation area, it disappears from the system, i.e., the height H of the heap is set to zero at the boundary ($i = 0$, $i = N + 1$, $j = 0$, $j = N + 1$) to simulate open boundary conditions.

Finally `psplot()` produces the Postscript output. The height H is depicted as a wireframe grid. On the first call of this procedure the Postscript header is written.

```
                             ─── heap.cpp ───
79  void pssnap(ostream & psout, int page){
80    if(!page){
81      psout << "%!PS-Adobe-2.0" << endl;
82      psout << "%%BoundingBox: 0 0 700 500" << endl;
83      psout << "/fx {3 -1 roll exch moveto" << endl;
84      psout << "0.5 mul dup rmoveto 1 0 rlineto} def" << endl;
85      psout << "/fy {3 -1 roll exch moveto 0.5 mul dup rmoveto 0.5 0.5 rlineto}";
86      psout << "def" << endl;
87      psout << "/lx {0 exch rlineto 1 0 rlineto} def" << endl;
88      psout << "/ly {0 exch rlineto 0.5 0.5 rlineto} def" << endl;
89      psout <<  "/s {stroke} def" << endl;
90    }
91    psout << "%%Page: " << page << " " << page << endl;
92    psout << 460/float(L) <<" "<<100/float(L)<<" scale 5 5 translate 0.1";
93    psout << " setlinewidth" << endl;
94    for(unsigned int i=1; i<L+1; i++){
95      psout << i << " 1 " << H[i][1]/3.0 << " fy" << endl;
96      for(unsigned int j=2; j<L+1; j++){
97        psout << (H[i][j]-H[i][j-1])/3.0 << " ly" << endl;
98      }
99      psout << " s" << endl;
100   }
101   for(unsigned int j=1; j<L+1; j++){
102     psout << " 1 " << j << " " << H[1][j]/3.0 << " fx" << endl;
103     for(unsigned int i=2; i<L+1; i++){
104       psout << (H[i][j]-H[i-1][j])/3.0 << " lx" << endl;
105     }
106     psout << " s" << endl;
107   }
108   psout << "showpage" << endl;
109 }
```

Figure 6.2 shows a series of snapshots of the heap's evolution for lattice size 201×201. After a certain simulation time the heap adopts the (quadratic) form of the supporting plane.

Aside from the heap itself the size distribution of the avalanches has often been of interest (Fig. 6.3). After about 9×10^7 iterations the shape of the heap does not change noticeably anymore. Starting from this moment, another 10^7 particles are deposited. The size of the resulting avalanches for all depositions are recorded in the file **avalanches.dat**. However, for the statistics the initial 9×10^7 entries are disregarded. In contrast to the algorithm by Bak, Tang, and Wiesenfeld [14], it turns out that the size distribution of the avalanches does not obey the law $f(l) \propto 1/l$. Additional discrepancies arise from the fact that the base area is not circular but quadratic. Therefore the heap growth along the diagonals is different from the growth in parallel to the edges. The implementation of a circular base area is straightforward.

More sophisticated algorithms, especially algorithms for parallel computers, can be found in [226].

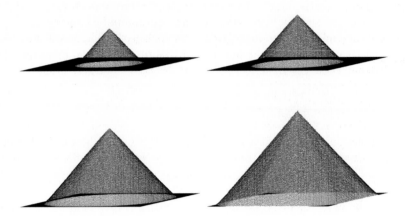

Fig. 6.2. Snapshot of the growing sand heap

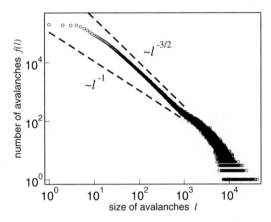

Fig. 6.3. Size distribution of the avalanches as a log-log plot. The *dashed lines* show the curves $f(l) \propto 1/l$ and $f(l) \propto 1/l^{3/2}$, respectively

6.3 Formation of Ripples

The formation of structures in sand beneath wind or fluid flow has been of scientific interest for a long time, especially in geology. An overview of the structures observed can be found in the books by Bagnold [12] and Pye and Tsoar [227]. For modeling sand beneath wind flow, four mechanisms of particle motion are assumed:

- Sand raised by the wind covers large distances as dust cloud.
- Individual grains may hop (driven by the wind) over distances of a few to several hundreds or thousands particle diameters (saltation).
- Blown sand particles crawl along the surface. This form of motion is observed especially in the formation of dunes on beaches or in deserts. The characteristic ripple structure of the sea floor is largely influenced by this type of motion.
- Particles move downhill according to local gradients.

Cellular automata have been applied to model ripple structures as well. Examples are the two-dimensional automaton by Ouchi and Nishimori [198, 199, 202] and Caps and Vandewalle [60], who studied the formation of ripples in wind-overblown sand and the automaton by Strassburger et al. [268] to describe structures that can be observed in shallow, horizontally shaken sand. Since the methods are very similar, we restrict ourselves to the description of the model by Ouchi and Nishimori [199].

The automaton is two-dimensional, with a horizontal extension of $0 \leq i <$ Lx and a vertical extension of $0 \leq j < $ Ly. The dynamics is governed by freely flying, wind-driven particles and by flows along steep gradients of the overblown surface.

Each site is characterized by the (real valued) difference $h(i, j)$ between the height of the sand at this position and the average height. The simulation is started at homogenous height distribution disturbed by small random fluctuations.

In the first lines of the program the standard headers are included, the function ranf() for drawing random numbers and the system parameters are defined. The size of the simulation area is given by Lx and Ly, nstep is the number of the automaton steps to be simulated, and tps is the interval between Postscript snapshots (file ripple.ps). The parameters threshold, L0, qmax and b will be explained below.

```
                          ──────── Ripples.cpp ────────
 1 │ #include <iostream>
 2 │ #include <fstream>
 3 │ #include <stdlib.h>
 4 │
 5 │ const int Lx=500, Ly=200;
 6 │ const float qmax = 0.15, L0=10, b=1, threshold=0.3;
 7 │ float H[Lx][Ly], dH[Lx][Ly];
 8 │ const int nstep = 1000, nps = 20;
 9 │
10 │ ofstream fps("ripple.ps");
11 │
12 │ void blow();
13 │ void drift();
14 │ void psplot(ofstream & psout, int page);
15 │ float ranf() {return float(rand())/(float(RAND_MAX)+1);}
                          ──────── (continued) ────────
```

The array H (representing $h(i, j)$) is initialized with random numbers. The dynamics of the automaton consists of two steps, blow() and drift().

```
                              Ripples.cpp
16  int main()
17  {
18    for(int i=0;i<Lx;i++){
19      for(int k=0;k<Ly;k++){
20        H[i][k]=ranf()-0.5;
21      }
22    }
23    psplot(fps,0);
24    for(int i=0;i<nstep;i++){
25      blow();
26      drift();
27      if((i+1)%nps == 0) psplot(fps,(i+1)/nps);
28    }
29  }
                              (continued)
```

The first step, `blow()`, describes lift, flight, and deposition of sand caused by the wind. The wind is assumed to blow in the positive x-direction. At each site (x, y), a random amount `q` of sand from the interval $0 \le q_{max} < 1$ is lifted and is deposited at $(x + L, y)$, conserving the total mass of sand:

$$h(x, y) \rightarrow h(x, y) - q$$
$$h(x + L, y) \rightarrow h(x + L, y) + q \ . \tag{6.10}$$

The saltation length L depends on the intensity of the wind and the height $h(x, y)$ from where the sand was lifted. The strength of the wind is considered to be constant and it enters the calculation as an additive parameter. Linear dependence on the height is assumed:

$$L = L_0 + b \cdot h(x, y) \ . \tag{6.11}$$

The sites are updated simultaneously, i.e., the changes in height are computed in the second loop (lines 39–47) and stored in `dH`. In the third loop (lines 48–52) the array `dh` is used to update `H`. Periodic boundary conditions in the x-direction are implied.

```
                              Ripples.cpp
30  void blow()
31  {
32    for(int i=0;i<Lx;i++){
33      for(int k=0;k<Ly;k++){
34        dH[i][k]=0;
35      }
36    }
37    int L,iL;
38    float q;
39    for(int i=0;i<Lx;i++){
40      for(int k=0;k<Ly;k++){
41        q=qmax*ranf();
42        dH[i][k]-=q;
43        L=int(L0+b*H[i][k]);
44        iL = (i+L+Lx) % Lx;
45        dH[iL][k]+=q;
46      }
47    }
48    for(int i=0;i<Lx;i++){
49      for(int k=0;k<Ly;k++){
```

```
50        H[i][k]+=dH[i][k];
51      }
52    }
53 }
```

───────────── (continued) ─────────────

The action of the wind simulated by `blow()` may lead to steep gradients of the surface structure which are smoothed by the second step, `drift()`. The drift of sand follows the local gradients,[2] i.e., the sand flows downhill in the direction of steepest descent. Suppose the direction of steepest descent points toward site (x', y') from (x, y). If the gradient in this direction exceeds a given critical value `threshold`, a random amount `mov` moves to the corresponding neighboring site:

$$h(x, y) \rightarrow h(x, y) - \text{mov}$$
$$h(x', y') \rightarrow h(x', y') + \text{mov} .$$
(6.12)

The value of `mov` has to be smaller than half of the threshold, thus ensuring that the gradient may not be reversed by the drift of particles. In `drift()`, again we first temporarily store the new values of h in the array `dH` to ensure simultaneous update of h.

─────────────── Ripples.cpp ───────────────

```
54 void drift()
55 {
56   for(int i=0;i<Lx;i++){
57     for(int k=0;k<Ly;k++){
58       dH[i][k]=0;
59     }
60   }
61   int im1, ip1, km1, kp1, ll, kk;
62   float slopemax, s, q;
63   for(int i=0;i<Lx;i++){
64     for(int k=0;k<Ly;k++){
65       im1 = (i-1+Lx)%Lx;
66       ip1 = (i+1)%Lx;
67       km1 = (k-1+Ly)%Ly;
68       kp1 = (k+1)%Ly;
69       slopemax=0;
70       ii=i;
71       kk=k;
72       if((s=H[i][k]-H[im1][k])>slopemax) {slopemax=s;ii=im1;}
73       if((s=H[i][k]-H[ip1][k])>slopemax) {slopemax=s;ii=ip1;}
74       if((s=H[i][k]-H[i][km1])>slopemax) {slopemax=s;kk=km1;}
75       if((s=H[i][k]-H[i][kp1])>slopemax) {slopemax=s;kk=kp1;}

77       if(slopemax>threshold){
78         q=0.25*slopemax*(1+ranf());
79         dH[i][k]-=q;
80         dH[ii][kk]+=q;
81       }
82     }
83   }
84   for(int i=0;i<Lx;i++){
85     for(int k=0;k<Ly;k++){
86       H[i][k]+=dH[i][k];
```

───

[2] Here we deviate from the model by Ouchi and Nishimori in which the drift is modeled as isotropic diffusion.

```
87 |     }
88 |   }
89 | }
```
——————————————— (continued) ———————————————

Finally, in intervals of **tps** time steps a Postscript snapshot of the system is recorded. The local heights are gray-scale encoded. The output function is straightforward. Note that the gray scale (the values in line 95) has to be adapted when the parameters of the automaton are changed.

———————— Ripples.cpp ————————
```cpp
90  | void psplot(ofstream & psout, int page)
91  | {
92  |   if(!page){
93  |     psout << "%!PS-Adobe-2.0" << endl;
94  |     psout << "%%BoundingBox: 30 30 " << 60+Lx << " " << 60+Ly << endl;
95  |     psout << "/h {0.62 mul 0.31 add setgray 1 0 rlineto "
96  |           << "currentpoint stroke moveto} def" << endl;
97  |     psout << "/sc {40 40 translate 1 setlinewidth} def" << endl;
98  |   }
99  |   psout << "%%Page: " << page << " " << page << endl;
100 |   psout << "sc" << endl;
101 |   for(int j=0; j<Ly; j++){
102 |     psout << "newpath 0 " << j << " moveto" << endl;
103 |     for(int i=0; i<Lx; i++){
104 |       psout << H[i][j] << " h" << endl;
105 |     }
106 |   }
107 |   psout << "showpage" << endl;
108 | }
```

A sequence of snapshots of forming ripples is assembled in Fig. 6.4. The ripple structure is clearly visible, where elevated ridges (light gray) alternate with valleys (dark gray).

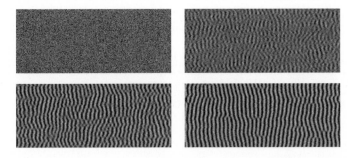

Fig. 6.4. Snapshots of the automaton by Ouchi and Nishimori. Dark gray corresponds to small height h. *Top, left:* homogenous initial configuration. *Right:* snapshot after 2 iterations. *Bottom:* snapshots after 5 (*left*) and 50 iterations (*right*)

The presented simple automaton can reproduce characteristic features of ripples qualitatively:

2. While individual ripples are subject to fluctuations, the characteristic properties of the structure are time-invariant.

In animated sequences of snapshots one can see that the ripples move in opposite direction to the wind, in agreement with many experiments. A detailed discussion of the properties of this automaton can be found in [199]. Later, Ouchi and Nishimori generalized the automaton such that it models size segregation that can be observed during ripple formation [202]. They introduced two types of particles that differ in size and have different flight and drift properties.

We wish to remark that this simple model is not a suitable model for quantitative investigations. There is a wide range of publications on models and simulations of dunes and ripples. Much more realistic systems can be found, e.g., in [7, 8, 116–118, 126, 143, 144, 157, 205, 250, 295–297].

6.4 Lattice Gas Simulations

6.4.1 Lattice Gas Automaton

Lattice gas models have been introduced first in 1986 by Frisch, Hasslacher, and Pomeau [81] to study fluid flows by means of numerical simulations. Lattice gases use a triangular lattice (in two dimensions), i.e., each lattice node has 6 neighbors. This type of neighborhood relation is also called Moore neighborhood. In the following example granular pipe flow is simulated (see [207–209]). The corresponding hexagonal lattice is shown in Fig. 6.5.

In contrast to the simple deterministic sand heap automaton as described in Sect. 6.2, the lattice gas automaton operates probabilistically, i.e, the state of the lattice at time $t+1$ is not completely determined by the state of the lattice at time t. Instead there is a certain *probability* to transit from a given state to another. The probabilities are determined by the automaton rules.

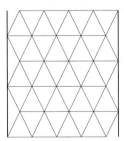

Fig. 6.5. The triangular lattice for simulating granular pipe flow with periodic boundary conditions in the vertical coordinate. In the horizontal direction the pipe is confined by rough walls

The state of each lattice site i is described by seven binary numbers $v_i^{(j)}$, with $j = (0, 1, 2, .., 6)$. The variable $v_i^{(0)} = 1$ indicates a particle at rest at site i, whereas $v_i^{(k)} = 1$ indicates a particle moving toward the kth neighbor of the site. The numbering of the neighbors is a matter of convention and will be discussed below when the implementation of the algorithm is presented. All states of a site can be occupied by at most one particle, thus, a site can be populated by up to seven particles. The state of a site can be, therefore, described by a 7-bit binary number, where each of the variables $v_i^{(j)}$ is represented by one bit.

Each automaton step consists of two phases, motion and collision. The first phase is deterministic: the particles move in the direction of their present velocities. If, e.g., $v_i^{(3)} = 1$, the particle moves to neighboring site number three. In the collision phase the particles may change their velocities according to the probabilistic collision rules.

The collision rules are shown in Fig. 6.6. An arrow pointing away from site i in direction j (i.e., $v_i^{(j)} = 1$) represents a particle in site i that moves toward site j. An incoming arrow from site j stands for a particle from site j moving toward site i. A dot represents a particle at rest ($v_i^{(0)} = 0$). The collision rules have been defined differently by different authors, those in Fig. 6.6 have been taken from [115]. Due to the stochastic nature of the collision rules, a certain pre-collision state may result in different post-collision states. The values in square brackets in Fig. 6.6 are the probabilities of the corresponding alternative states. Except for the collision type sketched in the first line of the table, the probabilities depend on a parameter p. For $p = 0$ the particles collide elastically. Choosing $p > 0$ the particles collide inelastically.[3]

At the boundary walls of the system the lattice structure is different from the interior as the sites close to the wall have fewer neighbors. For the pipe lattice sketched in Fig. 6.5 this concerns the left and right vertical walls. The automaton rules for these sites differ from those of the regular (internal) sites. Their rules have to take into account the boundary conditions of the physical system. For simulating granular pipe flow the rules are sketched in Fig. 6.7, which are taken from [207]. The parameter p_{wall} models the roughness of the wall: A particle colliding with the wall is scattered back with probability p_{wall} and it is reflected by the wall with a probability of $1 - p_{\text{wall}}$ (first line in Fig. 6.7). Thus, $p_{\text{wall}} = 0$ represents a frictionless smooth wall.

Finally, gravity has to be modeled. Again we follow [207] where it is suggested that a particle at rest starts to move downward with probability p_{grav} if there is no particle already occupying this direction. Since there is no vertical lattice line, it is decided stochastically whether the particle moves left-downward or right-downward.

---

[3] Strictly speaking, for $p > 0$ collisions may occur elastically as well, but there is a certain probability for losing energy. Thus, the expectation value of the energy after a collision is smaller than the initial energy.

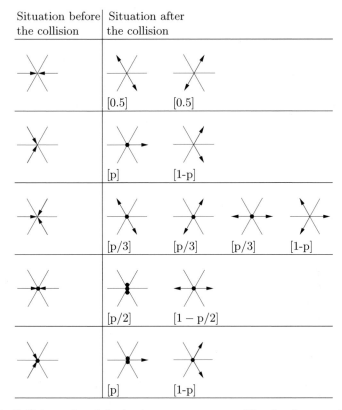

Situation before the collision	Situation after the collision

Fig. 6.6. Collision rules of the lattice gas automaton. The sketches may be rotated through multiples of $60°$. In two of the rules there reside more than one resting particles in the site, which seems to contradict the proposition that there is at most one resting particle. The resolution of this situation will be described below (see functions `collide_3()` and `distribute()`)

We remark that there were early attempts to model lattice gases on rectangular lattices [110]. However, it turned out that it is impossible to ensure momentum conservation on such lattices, i.e., in the thermodynamical limit the Navier–Stokes equations are violated [81].

Similar models as well as further enhancements of the simple lattice gas described here can be found, e.g, in [70, 137, 138, 178, 195, 197].

6.4.2 Granular Pipe Flow

The pipe is modeled as a lattice of size $nx \times ny$, assuming periodic boundary conditions in the vertical direction y, i.e., particles leaving the system downward enter it again from above and vice versa. Due to the hexagonal lattice structure the coordinate system is somewhat more complicated than

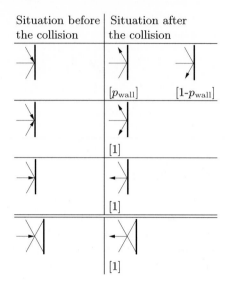

Situation before the collision	Situation after the collision	
	$[p_{\text{wall}}]$	$[1\text{-}p_{\text{wall}}]$
	$[1]$	
	$[1]$	
	$[1]$	

Fig. 6.7. Rules for collisions with the wall

a normal Cartesian. The x-axis points in the horizontal direction. The y-axis is in the upward right direction (see Fig. 6.8). The non-Cartesian co-ordinate system is needed to enumerate the lattice sites uniquely by their (x, y) coordinates. We have chosen this unusual non-rectangular coordinate system to simplify the data structure. It does not reflect any physical peculiarities and has significant advantages over a Cartesian system in that both axes are parallel to lattice lines. The only complication is that the side walls are not parallel to the physical vertical axis (the direction of gravity). Thus we are faced with the situation in which, e.g., at height $y = 0$ the walls are at position $x = 0$ and $x = n_x - 1$, whereas at height $y = 10$ they are at position $x = -5$ and $x = n_x - 6$ (see Fig. 6.8).

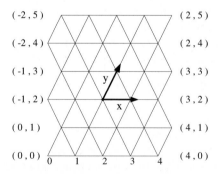

Fig. 6.8. Sketch of the coordinate system. The numbers in brackets are the coordinates of the wall sites in (x, y) format

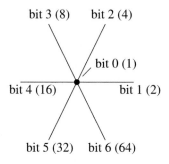

Fig. 6.9. Encoding of the directions

The state of a lattice site is encoded bitwise. Bit 0 (least significant) is set to one if there is a particle at rest in the site. The other bits represent particles moving in one of the six directions emanating from the site (the numbering of the directions can be seen in Fig. 6.9).

The collision rules are implemented by mapping each possible initial state to a final state, both represented by 7-bit binary numbers. The collision rule from the second row in Fig. 6.6 will serve as an example. Two particles, coming from the upper and lower left neighboring sites, collide. After the collision there is either a particle at rest and one moving to the right, or two particles moving to the upper and lower right neighbors. One should note that the direction *from the upper left* is identical to the direction *to the lower right*, from the point of view of the site where the particle comes from. Both are represented by setting bit 6 to one (decimal 64).[4] In the same way both directions *from the lower left* and *to the upper right* are represented by bit 2 set to one (decimal 4). Thus, the initial state is encoded by (decimal) $4 + 64 = 68$. The first possible final state is encoded by 3 (1 resting particle plus 2 particles moving to the right), the second again by 68. Hence, the considered collision rule is represented by: *state 68 transforms with probability p into 3 and with 1–p into 68.* Each collision rule sketched in Fig. 6.6 stands for 6 equivalent rules which can be constructed from the given one by rotating through 60°, 120°, 180°, 240°, and 300° (the derived rules may be identical in some cases). All these individual rules may be written explicitly, although all equivalent (with respect to rotation) rules may be represented in a compact notation:

$$F = (I\,N) \quad \mathrm{mod} \ \left(2^1 + 2^2 + 2^3 + 2^4 + 2^5 + 2^6\right)$$
$$= (I\,N) \quad \mathrm{mod} \ 126\,. \tag{6.13}$$

The initial state *without* considering a possible resting particle is encoded by the even number I. The final state is encoded by F, again disregarding a possible resting particle. Finally, N is a constant which is specific for the particular collision rule. For our example we obtain $N = 13$ for the first part

---

[4] The bits encode *directions*, not lattice vertices.

of the rule (i.e., $68 \to 3$), which can be checked by substituting the values in (6.13). For the second part of the rule ($68 \to 68$) obviously no special encoding is needed. The specific values of N have to be separately determined for each rule.

Let us discuss the simulation program in detail: The constant pgrav is the probability for a particle at rest starting to move downward to mimic gravity. Damping is modeled by p and pwall is the probability for a particle colliding with the wall to scatter back into the incoming direction. The constant npart is the number of particles and tmax is the number of desired automaton steps, while tps is the number of steps between two successive Postscript snapshots. The header further contains the definition of the class Cell, which describes the state of a lattice site. Although the simulation area is two-dimensional, internally the sites are organized as a one-dimensional vector cell. Furthermore, the header contains the prototypes of all functions that are used below.

```
──────────────── latticegas.cc ────────────────
1  #include <fstream>
2  #include <vector>
3
4  const int nx=50, ny=5000;
5  const double p=1.0, pgrav=0.7, pwall=0.8;
6  const int npart=250000, tmax=1000000, tps=1000;
7
8  class Cell {
9  public:
10    Cell(){_nrest=0;_status=0;}
11    int & nzero() {return _nrest;}
12    int nzero() const {return _nrest;}
13    unsigned char & status() {return _status;}
14    unsigned char status() const {return _status;}
15  private:
16    int _nrest;
17    unsigned char _status;
18  };
19
20  vector<Cell> cell(nx*ny);
21  ofstream fps("plot.ps");
22  int nc,nneglect;
23
24  void init(vector<Cell> & c);
25  void random_init(vector<Cell> & cell);
26  int index(int x,int y);
27  void step(vector<Cell> & c);
28  void gravity(vector<Cell> & c);
29  void propagate(vector<Cell> & c);
30  void collide(vector<Cell> & c);
31  bool distribute(vector<Cell> & c);
32  void psplot(ostream & os, const vector<Cell> & c);
33  void cell_gravity(Cell & c);
34  void wall_collide(Cell & c, int deflect);
35  void cell_collide(Cell & c);
36  int count_ones(int s);
37  void collide_2(Cell & c);
38  void collide_3(Cell & c);
39  int no_of_particles(const vector<Cell> & c);
──────────────── to be continued ────────────────
```

In the main program first `random_init()` is called to initialize the sites with random values. We have to make sure that the wall sites do not contain particles moving toward the wall since those particles would leave the simulation area. Thus, the wall sites have to be considered separately in `random_init()`. After generating a snapshot of the initial state of the system by `psplot()`, `tmax` automaton steps are performed. Snapshots of the automaton state are generated for every `tps` step.

```
                           ─── latticegas.cc ───
40  int main()
41  {
42    nc=nneglect=0;
43    random_init(cell);
44    psplot(fps,cell);
45    for(int i=0;i<tmax;i++){
46      if(i%10==0) cout << "\r" << i << flush;
47      step(cell);
48      if(i%tps==0) {
49        cout << "\rstep: " << i << " #p: " << no_of_particles(cell)
50             << " #c: " << nc << " #>3: " << nneglect << " = "
51             << int(1000*nneglect/(double)nc+0.5)/10.0 << "%\n";
52        nc=0;
53        nneglect=0;
54        psplot(fps,cell);
55      }
56    }
57    cout << "\n" << flush;
58  }
59
60  void random_init(vector<Cell> & cell)
61  {
62    for(int i=0;i<npart;i++){
63      bool allowed;
64      int mask;
65      int ix,iy;
66
67      do{
68        allowed=true;
69        iy=int(ny*drand48());
70        ix=int(nx*drand48())-int(iy/2);
71        Cell c=cell[index(ix,iy)];
72        int n=int(7*drand48());
73        mask=1;
74        for(int i=0;i<n;i++) mask*=2;
75        if(c.status()&mask) allowed=false;
76        if(iy%2==0){
77          if((ix==-int(iy/2)) && (mask&57)) allowed=false;
78          if((ix==nx-1-int(iy/2)) && (mask&3)) allowed=false;
79        } else {
80          if((ix==-int(iy/2)) && (mask&17)) allowed=false;
81          if((ix==nx-1-int(iy/2)) && (mask&71)) allowed=false;
82        }
83      } while(!allowed);
84      cell[index(ix,iy)].status()|=mask;
85    }
86  }
                           ─── to be continued ───
```

The automaton step is computed in `step()` which, in its turn, calls `gravity()`, `propagate()`, `collide()`, and `distribute()`. These functions compute the new state of the automaton at time $t+1$ from the state at time t. The function `distribute()` ensures that at the end of each step there is at most one resting particle at each lattice site.

```
                          latticegas.cc
87  void step(vector<Cell> & c)
88  {
89    int count=0;
90
91    gravity(c);
92    propagate(c);
93    collide(c);
94    while((count<10000) && distribute(c)) count++;
95    if(count==10000){
96      cerr << "distribute failed\n";
97      abort();
98    }
99  }
100
101 int index(int x,int y)
102 {
103   if(y<0){
104     y=(y+ny)%ny;
105     x-=ny/2;
106   }
107   if(y>=ny){
108     y=y-ny;
109     x+=ny/2;
110   }
111   int realx=x+(y/2);
112   return y*nx+realx;
113 }
                          to be continued
```

If `distribute()` is unable to find such a legal state after a certain number of trials (10,000 in our case), the program aborts with an error message. This may happen if the number of particles in the simulation area is too large. To map the internal representation of the sites, the one-dimensional array (`vector`), to the non-Cartesian coordinates x and y, the function `index()` is called.

In the function `gravity()`, resting particles are moved downward with probability `pgrav`, according to the rules modeling the action of gravity. Since there are two options, another random number is drawn to decide whether it moves left-downward or right-downward. If the chosen direction is occupied, the particle stays at its position. The described procedure is performed for each site in the function `cell_gravity()`, which is a member of the class `Cell`. The function `gravity()` serves only to call this function for all sites.

```
                          latticegas.cc
114 void gravity(vector<Cell> & c)
115 {
116   for(unsigned int i=0;i<c.size();i++)
117     if(c[i].status()%2==1) cell_gravity(c[i]);
118 }
119
120 void cell_gravity(Cell & c)
121 {
122   if((c.status() & 96) == 96) return;
123   if(drand48()<pgrav){
124     if(drand48()<0.5){
125       if((c.status() & 32) == 0){
126         c.status() |= 32;
127         c.status() -= 1;
```

```
128        }
129      } else {
130        if((c.status() & 64) == 0){
131          c.status() |= 64;
132          c.status() -= 1;
133        }
134      }
135    }
136 }
```
───── to be continued ─────

The function `propagate()` moves the particles from the source to the destination sites. For each site, it is tested which directions are occupied and the particles present are moved, i.e., the corresponding bits in the affected neighboring sites are set to one. The encoding of the directions can be seen in Fig. 6.9.

───── latticegas.cc ─────
```
137 void propagate(vector<Cell> & c)
138 {
139   vector<Cell> tmp(nx*ny);
140
141   for(int iy=0;iy<ny;iy++){
142     for(int ix=-int(iy/2);ix<nx-int(iy/2);ix++){
143       int s=c[index(ix,iy)].status();
144       if(s&1)  tmp[index(ix,iy)].status()   |= 1;
145       if(s&2)  tmp[index(ix+1,iy)].status() |= 2;
146       if(s&4)  tmp[index(ix,iy+1)].status() |= 4;
147       if(s&8)  tmp[index(ix-1,iy+1)].status() |= 8;
148       if(s&16) tmp[index(ix-1,iy)].status() |= 16;
149       if(s&32) tmp[index(ix,iy-1)].status() |= 32;
150       if(s&64) tmp[index(ix+1,iy-1)].status() |= 64;
151     }
152   }
153   c=tmp;
154 }
```
───── to be continued ─────

In the function `collide()`, we consider the particle–particle and particle–wall collisions according to the rules shown in Figs. 6.6 and 6.7. This function is thus the core of the simulation. First, `wall_collide()` is called for collisions of particles with the wall, then particle–particle collisions are considered by calling `cell_collide()` for each site.

───── latticegas.cc ─────
```
155 void collide(vector<Cell> & c)
156 {
157   for(int iy=0;iy<ny;iy++){
158     if(iy%2==0){
159       wall_collide(c[index(-iy/2,iy)], 56);
160       wall_collide(c[index(nx-1-iy/2,iy)], 2);
161     } else {
162       wall_collide(c[index(-iy/2,iy)], 16);
163       wall_collide(c[index(nx-1-iy/2,iy)], 70);
164     }
165   }
166   for(int iy=0;iy<ny;iy++)
167     for(int ix=1-iy/2;ix<nx-1-iy/2;ix++) cell_collide(c[index(ix,iy)]);
168 }
169
170 void wall_collide(Cell & c, int deflect)
171 {
172   int s=c.status();
```

```
173    int newstatus=0;
174    int atrest = (c.status()&1);
175    newstatus=s-(s&deflect);
176    s&=deflect;
177    if(s&2)  newstatus|=16;
178    if(s&16) newstatus|=2;
179    if((s & 68) == 68) newstatus |= 40;
180    if((s & 40) == 40) newstatus|=68;
181    if(s&4)  newstatus |= ((drand48()<pwall) ? 32 : 8);
182    if(s&8)  newstatus |= ((drand48()<pwall) ? 64 : 4);
183    if(s&32) newstatus |= ((drand48()<pwall) ? 4 : 64);
184    if(s&64) newstatus |= ((drand48()<pwall) ? 8 : 32);
185    c.status() =atrest+newstatus;
186  }
187
188  void cell_collide(Cell & c)
189  {
190    int s=c.status(),n=count_ones(s);
191    if(n==2) nc++;
192    if(n>3) nneglect++;
193    if(n==2) collide_2(c);
194    if(n==3) collide_3(c);
195  }
```
──────────────── to be continued ────────────────

In `cell_collide()`, the number of colliding particles is first determined by calling `count_ones()`. If less than 2 particles are present there is no collision at the concerned site. If there reside more than three particles no collision occurs either. In order to allow collisions of more than three particles we would have to formulate more collision rules in the spirit of the ones given in Fig. 6.6. Tests revealed that for moderate particle density, less than 5% of all collisions involve more than 3 particles. Therefore, neglecting those collisions is justified. To provide a means to check the validity of this assumption, the number of neglected collisions is recorded in **nneglect**.

──────────────── latticegas.cc ────────────────
```
196  void collide_2(Cell & c)
197  {
198    int s=c.status();
199
200    if(s%2==1) return;
201    if((s==6) || (s==12) || (s==24) || (s==48) || (s==96) || (s==66)) return;
202    if((s==18) || (s==36) || (s==72)){
203      c.status() = (drand48()<0.5) ? ((4*s) % 126) : ((16*s) % 126);
204      return;
205    }
206    if((s==10) || (s==20) || (s==40) || (s==80) || (s==34) || (s==68)){
207      if(drand48()>p) return;
208      c.status() = (13*s) % 126 + 1;
209      return;
210    }
211  }
```
──────────────── to be continued ────────────────

The function `collide_2()` performs the two-particle collisions. According to the collision rules, the situation does not change for some particle configurations (see, e.g., the $[1-p]$ branch in row 2 of Fig. 6.6). Thus, we first check for such configurations. For all other cases the state is computed according to the automaton rules.

```
                              latticegas.cc
212  void collide_3(Cell & c)
213  {
214    int s=c.status();
215
216    if((s==7) || (s==13) || (s==25) || (s==49) || (s==97) || (s==67)) return;
217    if((s==14) || (s==28) || (s==56) || (s==112) || (s==98) || (s==70)) return;
218    if((s==11) || (s==21) || (s==41) || (s==81) || (s==35) || (s==69)){
219      if(drand48()>p) return;
220      c.status() = ((13*(s-1)) % 126) + 1;
221      c.nzero()+=1;
222      return;
223    }
224    if((s==19) || (s==37) || (s==73)){
225      if(drand48()>p/2) return;
226      c.status() = 1;
227      c.nzero()+=2;
228      return;
229    }
230    if((s==42) || (s==84)){
231      if(drand48()>p) return;
232      int z=int(3*drand48())+1;
233      if(z==3) z=4;
234      c.status()=(18*z)+1;
235      return;
236    }
237    if((s==22) || (s==44) || (s==88) || (s==50) || (s==100) || (s==74)){
238      if(drand48()>p) return;
239      c.status() = (52*s)%126+1;
240      return;
241    }
242    if((s==26) || (s==52) || (s==104) || (s==82) || (s==38) || (s==76)){
243      if(drand48()>p) return;
244      c.status() = (25*s)%126+1;
245      return;
246    }
247  }
                             to be continued
```

The function `collide_3()` operates in a similar way for three-particle collisions. First it is checked whether the configuration at the site requires any action. If this is the case we compute the final state. After a three-particle collision there may reside more than one resting particle at a lattice site. The fourth row in Fig. 6.6 gives an example. In the context of the lattice gas automaton as described here, this state is illegal. Studies in [208] suggested distributing such resting particles stochastically to their neighboring sites. This step is performed in `distribute()` after all collisions are processed and repeated until a legal state is found.

```
                              latticegas.cc
248  bool distribute(vector<Cell> & c)
249  {
250    bool illegal_sites=false;
251
252    for(int iy=0;iy<ny;iy++){
253      for(int ix=-iy/2;ix<nx-iy/2;ix++){
254        int i=index(ix,iy);
255        int nz=c[i].nzero();
256        for(int k=0;k<nz;k++){
257          illegal_sites=true;
258          int z;
259          bool weiter;
260          do{
261            weiter=false;
```

```
262        z=int(6*drand48());
263        if(ix==nx-2-iy/2){
264            if(z==0) weiter=true;
265            if(iy%2==1){
266                if((z==1) || (z==5)) weiter=true;
267            }
268        }
269        if(ix==1-iy/2){
270            if(z==3) weiter=true;
271            if(iy%2==0){
272                if((z==2)||(z==4)) weiter=true;
273            }
274        }
275        } while(weiter);
276        int dx,dy;
277        switch(z){
278        case(0): dx=1;dy=0;break;
279        case(1): dx=0;dy=1;break;
280        case(2): dx=-1;dy=1;break;
281        case(3): dx=-1;dy=0;break;
282        case(4): dx=0;dy=-1;break;
283        case(5): dx=1;dy=-1;break;
284        default:
285            { cerr << "unrecognized direction: " << z << endl;
286            abort();
287            }
288        }
289        int ii=index(ix+dx,iy+dy);
290        if(c[ii].status() & 1) c[ii].nzero()++; else c[ii].status()|=1;
291    }
292    c[i].nzero()=0;
293        }
294    }
295    return illegal_sites;
296 }
```
——— to be continued ———

First the Boolean variable `illegal_sites` is initialized with `false`. This variable is true if illegal states have been found. `distribute()` is called until this variable stays at `false`. In the outer loops (indices `ix` and `iy`) we iterate over all sites and test whether there are more than one resting particle (`nz>0`). If there are, `illegal_sites` is set to `true` and the particles are distributed to the neighboring sites. To this end a random number between 0 and 5 is drawn, representing a direction to a neighboring site for each superfluous particle. If there is an admittable site in that direction (which may be not the case close to the walls) a particle is moved to the selected site. Obviously, resting particles are not admitted at the wall sites.

The value of `illegal_sites` is returned to indicate whether `distribute()` has been successful (`illegal_sites=false`). If not, it is called again.

——— latticegas.cc ———

```
297 int count_ones(int s)
298 {
299    int n=0;
300    int mask=1;
301
302    for(int i=0;i<7;i++){
303        n+=((s&mask) ? 1 : 0);
304        mask*=2;
305    }
306    return n;
```

```
307  }
308
309  int no_of_particles(const vector<Cell> & c)
310  {
311    int n=0;
312
313    for(unsigned int i=0;i<c.size();i++){
314      Cell cell=c[i];
315      n+=count_ones(cell.status())+cell.nzero();
316    }
317    return n;
318  }
```
———————— to be continued ————————

The auxiliary function **count_ones()** counts the number of legal particles at a site. The function **no_of_particles()** counts the total number of particles in the simulation area.

Finally, the output functions **psplot()** and **density_plot()** are provided, which generate snapshots of the system.

———————— latticegas.cc ————————
```
319  void psplot(ostream & os, const vector<Cell> & c)
320  {
321    static int npage=0;
322    const double scale=2;
323
324    if(npage==0){
325      os << "%!PS-Adobe-2.0" << endl;
326      os << "%%BoundingBox: 0 0 " << scale*nx+20 << " "
327         << scale*ny*0.5*sqrt(3)+20 << endl;
328      os << "/c {0 360 arc stroke} def\n";
329      os << "/cf {0 360 arc fill stroke} def\n";
330    }
331    os << "%%Page: " << npage << " " << npage << endl;
332    os << "0.01 setlinewidth\n";
333    os << "10 10 translate\n";
334    os << scale << " " << scale << " scale\n";
335    os << "0 0 moveto\n";
336    int maxval=4;
337    for(int i=1;i<=maxval;i++){
338      os << (maxval-i)/double(maxval) << " setgray\n";
339      for(int iy=0;iy<ny;iy++){
340        for(int ix=-(iy/2);ix<nx-(iy/2);ix++){
341          double x=ix+iy*0.5;
342          double y=iy*sqrt(0.75);
343          int n=count_ones(c[index(ix,iy)].status());
344          if(((i<maxval) && (n==i)) || ((i==maxval) && (n>i))) {
345            os << x << " " << y << " 0.4 cf\n";
346          }
347        }
348      }
349    }
350    os << "showpage\n" << flush;
351    npage++;
352  }
```

6.4.3 Density Inhomogeneities in Granular Pipe Flow

As an example of the application of lattice gas automata to granular systems, we simulated granular flow through a narrow pipe. Density inhomogeneities were observed, which are undesired in technical systems. Figure 6.10 shows, on the left, a snapshot of an experiment where fine sand of typical particle diameter 0.18 mm flowed through a pipe that measured 2 mm in diameter [212, 213], and a close-up of this snapshot. To the right, a snapshot of a lattice gas automaton simulation of 250,000 particles and a close-up thereof are shown. The lattice consisted of 50 × 5000 sites.

Despite the drastically simplified model, the simulation yields results that are in qualitative agreement with the experiment. If the snapshot is enlarged sufficiently (Fig. 6.11) the hexagonal lattice structure becomes visible. An animation of this simulation can be found at the book's web site.

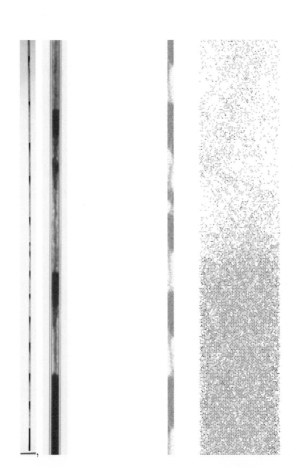

Fig. 6.10. Density inhomogeneities in granular pipe flow with a close-up. *Left*: experiment. *Right*: simulation

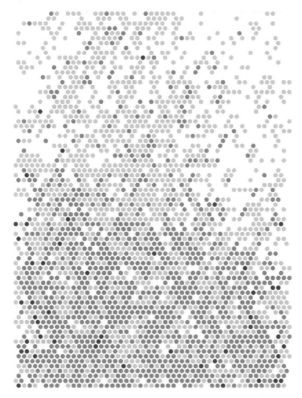

Fig. 6.11. High resolution image of the simulation snapshot shown in Fig. 6.10. Here the lattice sites becomes visible. The *grayscale* of the sites encodes the number of particles at that site (resting and moving)

7

Bottom-to-Top Reconstruction

7.1 Idea of the Method

Visscher and Bolsterli [286] suggested an algorithm that allows for the fast simulation of large static granular packings. While in Molecular Dynamics simulations, the coupled system of Newton's equations of motion is solved for all particles simultaneously, the fundamental idea of their method is to consider the motion of the particles sequentially. The deposition of the particles of a granular packing, e.g., a heap on the plane $(x, y, 0)$, proceeds as follows: the first particle is inserted at position $(x_1^{\text{init}}, y_1^{\text{init}}, z_1^{\text{init}})$. The coordinate z_1^{init} should be larger than the expected final height of the heap, the coordinates x_1^{init} and y_1^{init} can be chosen at random (particles are scattered over a certain area) or can be fixed, e.g., $(x_1^{\text{init}}, y_1^{\text{init}}) = (0, 0)$, to simulate the build-up of a heap from a point source. The particle then falls until it touches the ground at $(x_1^{\text{init}}, y_1^{\text{init}}, R_1)$. At this position the particle remains fixed. Then the second particle is inserted at position $(x_2^{\text{init}}, y_2^{\text{init}}, z_2^{\text{init}})$. It falls down until it touches either the ground at $(x_2^{\text{init}}, y_2^{\text{init}}, R_2)$ or the first particle, whatever happens first. If it touches the ground it remains fixed there just like the first particle. If it, however, touches the first particle, it rolls down its surface in the downslope direction until it either touches the ground (if $R_2 \geq R_1$) where it remains fixed or until it loses contact at $z_2 = R_1$ when both centers are at the same height (if $R_2 < R_1$). From there it falls to the ground where it remains fixed. The next particles are treated likewise. A particle remains fixed if it either touches the ground or if it attains a local minimum where it is supported by two (in two dimensions) or three (in three dimensions) other already fixed particles.

The algorithm is sketched below for a two dimensional system. The moving particle is drawn with a dotted line, fixed particles are drawn with a solid line.

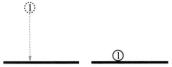

Particle 1 is inserted and moves downward until it touches the ground.

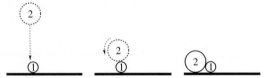

Particle 2 is inserted and moves downward until it touches particle 1. Then it rolls on the surface of particle 1 until it touches the ground.

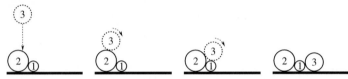

Particle 3 is inserted and moves downward until it touches particle 2. It then rolls to the right until it touches particle 1. This position is unstable (not a local minimum), thus, the particle continues to roll on particle 1 until it touches the ground.

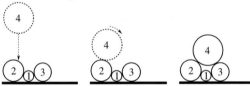

Particle 4 first touches particle 2 and starts to roll to the right until it touches particle 3. This position is stable (local minimum), thus, particle 4 is fixed at this position.

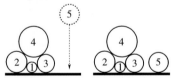

Particle 5 moves analogously to particle 1 and is fixed on the ground.

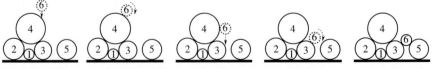

Particle 6 first touches particle 4 and rolls on it until it loses contact when the centers of both particles are at the same height. Then the particle falls again until it touches particle 3. It continues rolling until it touches particle 5 too where it is fixed (position is stable since it is a local minimum).

In general the motion of each particle i is computed with the wall and the other already-deposited particles $j = 0, \ldots, i-1$ considered as fixed obstacles. The positions of the particles $j = 0 \ldots i - 1$ are not influenced by the motion of the new particle i. Thus, the trajectory of each particle is computed while taking gravity as the only driving force into account. The other particles $j = 0 \ldots i-1$ and the wall establish the (complicated) boundary conditions to this motion. This way, the system of Newton's equation for the N-body system is decoupled and N single-particle equations are solved instead. Moreover, the time dependence of the particles' motion is disregarded. These simplifications increase the efficiency of the simulation significantly. Jullien et al. [134] were able to simulate packings of up to 10^8 particles using this algorithm.

However, for many situations this method is not applicable: since only one particle moves at a time, the trajectories are different from the ones obtained by solving Newton's equation of motion. Dynamical systems such as granular flows cannot be simulated. The limitations of the algorithm are discussed in Sect. 7.4.

The algorithm by Visscher and Bolsterli [286] was later improved and generalized, e.g., [131–133]. First we describe a simple variant to simulate heaps and static packings and then we present a generalization that has been applied by Baumann et al. [21, 22] to simulate certain dynamical systems.

Using the bottom-to-top reconstruction algorithm, one does not solve Newton's equation of motion but, as described, each particle moves according to a set of rules from one state to the next. In this sense the algorithm belongs to the class of event-driven algorithms.

7.2 Simulating a Heap

The formation of a heap on a plane by sequentially depositing particles is the simplest application for the bottom-to-top reconstruction scheme. Here we restrict ourselves to the simulation of a two-dimensional system of polydisperse particles.

To compute the deposition of the ith particle, we do not need the positions of all $i - 1$ previously deposited particles. Instead, it is enough to know the positions of particles that can come into contact with particle i, i.e., particles at the surface of the heap. This simplification is always justified for two-dimensional systems, whereas for three-dimensional systems it is only justified if the ratio between the largest and smallest radii of the particles in the system does not exceed the Apollonian ratio $r_A \equiv R_{\max}/R_{\min} = \sqrt{3}/(2 - \sqrt{3}) \approx$ 6.46. Otherwise the smaller particles can fall into the gaps between the larger particles and thus, in principle, interact with all particles of the heap. In this case, we cannot exclude any pair interactions à priori.

The program for simulating heaps in two dimensions starts with the standard header files and the definition of the system parameters. The coordinates of the particles are contained in the arrays x and y (of STL-type

vector<double>). Aside from the parameters describing the particle radii
Rmin and Rmax and the number of particles N, the name of the output file
"v.ps" is defined and the function prototypes are declared.

The container surface of type map deserves some extra explanation: As its
name suggests, this container describes the exposed surface of the heap, i.e., it
contains the surface particles sorted by their x-coordinate. A container of type
map may be thought of as a generalization of an array: Whereas in an array
data of equal type are accessible via an index of integer type, the index of a
map may be of any type. The only restriction is that there must be a function
to compare different indices to allow for sorting. In our case, the map surface
contains the particle number of the surface particles (of type int), indexed
by their x-position (of type double). Thus we declare the container surface
as map<double,int>, the first template parameter being the index type and
the second parameter being the data type. For iterating through the data (in
ascending index order) of a map or a part thereof, iterators that are used are
declared as variables of the type map<double,int>::iterator. Actually such
an iterator points to a variable of type pair, which contains two data elements
named first and second. Element first is the index, element second is the
corresponding data element. If, for example, i is an iterator of surface, the
x-coordinate of the particle to which i points can be accessed, via i->first.
Its particle number index is obtained by i->second. The iterator pointing
to the first element of the map, i.e., the element with the smallest index (x-
coordinate), is provided by the member function begin(). For the case of the
map surface, it reads surface.begin(). The member function end() (i.e.,
surface.end()) provides an iterator pointing to the element *after* the last
element (largest x-coordinate), thus pointing to an illegal element.

```
                         ────── VisscherHeap.cc ──────
 1  #include <math.h>
 2  #include <fstream>
 3  #include <list>
 4  #include <vector>
 5  #include <map>
 6
 7  typedef map<double,int>::iterator iter;
 8
 9  const double Rmin=3, Rmax=10, N=1000000;
10
11  vector<double> x(N), y(N), R(N);
12  map<double,int> surface;
13  ofstream psout("v.ps");
14
15  iter fall(int);
16  bool fall_height(int, int, double &);
17  bool find_new_partner(int, iter, iter &);
18  bool wcontact(int,int,double &,double &);
19  bool pcontact(int,int,double &, double &);
20  void psplot(ostream &);
21  void update_surface(int, iter, iter);
22  double ranf(double x){ return double(rand())*x/(double(RAND_MAX));}
                         ────── (continued) ──────
```

The main program initializes first the radii and positions of the particles. The
radii are chosen randomly from the interval (R_{min}, R_{max}) in such a way that

the total mass of particles from the interval $(R, R+dR)$ is constant regardless of R. If $M(R)dR$ denotes the total mass of all particles in the interval dR around R and $p(R)dR$ denotes the number of particles in this interval, it is required to have

$$M(R)dR = \rho\pi R^2 p(R)dR = \text{const}\, dR \tag{7.1}$$

and, therefore,

$$p(R) \sim \frac{1}{R^2}\,. \tag{7.2}$$

With the normalization condition

$$\int_{R_{\min}}^{R_{\max}} p(R)\,dR = 1 \tag{7.3}$$

for the distribution function of the radii, $p(R)$ follows

$$p(R) = \frac{R_{\min} R_{\max}}{R_{\max} - R_{\min}} \frac{1}{R^2} \equiv \frac{K}{R^2}\,. \tag{7.4}$$

Since the problem with generating random numbers according to a given probability distribution occurs rather frequently, this procedure is described in detail here. Given a monotonous function $R(z)$, we assume $z \in [0, 1)$ is a random variable that obeys an equiprobability distribution. The probability to find R in the interval $[R_1, R_2)$ is then

$$P\left(R_1 \leq R \leq R_2\right) = P\left(z_1 \leq z \leq z_2\right)\,; \quad R_1 = R\left(z_1\right),\ R_2 = R\left(z_2\right)\,. \tag{7.5}$$

With $p(R)$ being the probability density of R and the density of the equidistribution, $f(z) = 1$, this equation is equivalent to

$$\int_{z_1}^{z_2} f(z)dz = \int_{R_1}^{R_2} p(R)dR = \int_{z(R_1)}^{z(R_2)} \pm p[R(z)]\frac{dR}{dz}dz\,, \tag{7.6}$$

where in the last expression the positive sign applies to a monotonously increasing function $R(z)$ and the negative sign applies to a monotonously decreasing function. For the limit $z_2 \to z_1$ implying $R_2 \to R_1$, it follows that

$$f(z) = \pm\frac{dR}{dz}p[R(z)]\,. \tag{7.7}$$

The solution of the differential equation

$$\frac{dR}{dz} = \pm\frac{f(z)}{p(R)} = \pm\frac{1}{p(R)} \tag{7.8}$$

is the transformation $R(z)$ to generate random numbers according to the distribution $p(R)$ from equidistributed random numbers z. Choosing the positive sign ($R(z)$ as an increasing function), we obtain from (7.4)

$$\frac{\mathrm{d}R}{\mathrm{d}z} = \frac{R^2}{K} \tag{7.9}$$

with the solution

$$\frac{1}{R} = -\frac{z}{K} + C = -\frac{z}{K} + \frac{1}{R_{\min}} \; . \tag{7.10}$$

The integration constant C has been determined from the conditions $R(0) = R_{\min}$ or $R(1) = R_{\max}$, equivalently. Equation (7.10) can be rewritten as

$$R = \frac{R_{\min} R_{\max}}{R_{\max} - z \left(R_{\max} - R_{\min} \right)} \; , \tag{7.11}$$

which is the desired mapping $R(z)$ to compute the radii according to (7.4) from equidistributed random numbers z. The presented method for generating random numbers can be generalized (see e.g., [224]). The program code for the initialization of the radii according to the described algorithm is shown in lines 28 and 29.

In the for-loop starting from line 32, all N particles are sequentially deposited on the heap. The described relaxation of a particle into the next local minimum takes place in the do-loop starting at line 34. This loop is repeated for each particle until it has found a stable position, i.e., until it is located in a local minimum, touching two other particles, or until it touches the ground.

———— VisscherHeap.cc ————

```
23  int main()
24  {
25    for(int i=0;i<N;i++){
26      x[i]=0.01*Rmin*(ranf()-0.5);
27      y[i]=2*Rmax*i;
28      double z=ranf();
29      R[i]=Rmin*Rmax/(Rmax-z*(Rmax-Rmin));
30    }
31    iter partner, newpartner;
32    for(int i=0;i<N;i++){
33      bool ok,stable=false;
34      do{
35        newpartner=fall(i);
36        if(newpartner!=surface.end()){
37          do{
38            partner=newpartner;
39            ok=find_new_partner(i,partner, newpartner);
40            if((partner!=surface.end()) && (newpartner!=surface.end())){
41              if(ok)stable=(x[partner->second]<x[i])^(x[newpartner->second]<x[i]);
42            }
43          } while(ok && (!stable) && (newpartner!=surface.end()));
44        } else {
45          ok=true;
46          partner=newpartner;
47        }
48      } while(!ok);
49      update_surface(i,partner,newpartner);
50      if((i+1)%10000==0) {
51        cout << i << endl;
52        psplot(psout);
53      }
54    }
55  }
```

———— (continued) ————

The function `fall()` moves a particle vertically downward until it touches a particle on the surface of the heap or the ground and it returns an iterator pointing to this particle (`newpartner`). If the particle touches the ground, `fall()` returns `surface.end()`. This return value signals to the calling function that the particle has reached a valid position (at the ground) and needs no further consideration.

If, however, the particle touches another (`newpartner!=surface.end()`), it continues its motion by rolling downward on the other particle. This process is modeled in `find_new_partner()`. The rolling motion continues until either both particles have the same vertical coordinate (`find_new_partner` returns `false`, no new partner has been found) or the rolling particle comes into contact with another particle—its new partner (`find_new_partner` returns `true` in this case). The return parameter `newpartner`, which is an iterator to the container `surface`, points then to the new partner particle. If a new partner is found, the stability of the new position is checked in line 41. If the position is unstable (`stable==false`), the process continues until a stable position is found. Thus, the main program implements precisely the algorithm sketched on p. 272.

Since the new particle in its stable position just found is now part of the surface, `update_surface()` inserts it into the container `surface`. Finally a snapshot of the growing heap is produced once every 10,000 deposited particles (line 52).

If a particle i is not in contact with any other particle it moves downward until it touches the ground or another particle (function `fall()`). Apart from the ground plane all particles on the surface of the heap whose centers belong to the interval $(x[i] - R[i] - R_{\max}, x[i] + R[i] + R_{\max})$ are potential contact partners. The two iterators `start` and `stop` define an index range of the container `surface` that contains at least all particles in this interval (see Fig. 7.1). The iterators `start` and `stop` point to the first and last surface

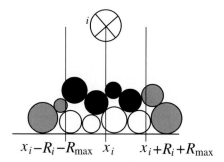

Fig. 7.1. The centers of the contact candidates of the falling particle i (circle drawn with a *cross*) belong to the interval $(x[i] - R[i] - R_{\max}, x[i] + R[i] + R_{\max})$ (*solid black circles*). The other surface particles are *gray*, particles which are not part of the surface are *hollow*

particles of the interval and allow us to access all candidates of the interval in an elegant way.[1]

```
                    ──────── VisscherHeap.cc ────────
56  iter fall(int n)
57  {
58    if(y[n]==R[n]) return surface.end();
59    iter start=surface.lower_bound(x[n]-R[n]-Rmax);
60    iter stop=surface.upper_bound(x[n]+R[n]+Rmax);
61    if(start==surface.end()) start=surface.begin();
62
63    double ymax=0, yy=0;
64    iter ii=surface.end();
65
66    for(iter i=start;i!=stop; i++){
67      if(y[i->second]<y[n]){
68        if(fall_height(n,i->second,yy)){
69          if(yy>ymax){
70            ymax=yy;
71            ii=i;
72          }
73        }
74      }
75    }
76    if(R[n]>=ymax){
77      ymax=R[n];
78      ii=surface.end();
79    }
80    y[n]=ymax;
81    return ii;
82  }
                    ──────── (continued) ────────
```

For all contact candidates that are located below the particle to be deposited, i.e., y[i->second]<y[n], the function fall_height() checks if and at which height yy both particles touch, disregarding for the moment possible interference from other particles. The maximum of those heights is the desired contact point (see Fig. 7.2). Finally, a possible contact with the ground is taken into account. The downward motion of the particle is executed and an

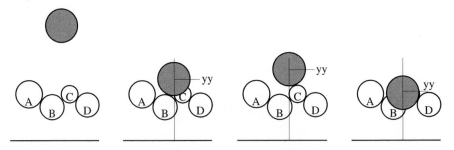

Fig. 7.2. For each of the candidates A-D (drawn in *black* in Fig. 7.1) the function fall_height() returns the y coordinate yy of a possible contact. The highest contact is with candidate C, therefore, C is the contact partner we are looking for

---

[1] The functions upper_bound() and lower_bound() (lines 59 and 60) are methods of the container map and are, thus, provided by the Standard Template Library (STL).

iterator pointing to the contact partner (or `surface.end()` in case of contact with the ground) is returned.

The function `fall_height()` computes the height h at which a falling particle n would come into contact with candidate i, disregarding the presence of other particles. The parameter h is declared as a reference (`double &`) since it is used to return the height to the calling function. The function itself returns a Boolean value that is `true` if there is a possible contact between the particles or `false` otherwise (e.g., candidate A in Fig. 7.2). In the latter case the equation

$$h = y_i + \sqrt{(R_n + R_i)^2 - (x_i - x_n)^2} \, , \tag{7.12}$$

has no real solution since the argument of the square root (variable `tmp` in the function) is negative.

```
────────────────── VisscherHeap.cc ──────────────
83 │ bool fall_height(int n, int i, double & h)
84 │ {
85 │   double tmp=(R[n]+R[i]-x[n]+x[i])*(R[n]+R[i]+x[n]-x[i]);
86 │   if(tmp>=0){
87 │     h=y[i]+sqrt(tmp);
88 │     return true;
89 │   }
90 │   return false;
91 │ }
────────────────── (continued) ──────────────
```

The particle to be deposited is now in contact with another particle or on the ground. If the particle falls all the way down to the ground its new position is stable by definition and its deposition is accomplished. If, however, the particle meets another particle first, its new position cannot be stable. Instead it continues its motion by rolling down the surface of the contact partner until it comes into contact with the wall or with another particle, or it continues to fall down. This part of the motion is modeled in the function `find_new_partner()`. Figure 7.3 illustrates the operation of this function.

The function `find_new_partner()` expects three arguments: the index n of the particle to be deposited, the iterator p pointing to the partner, and the iterator np pointing to the additional partner that is yet to be found. The last argument is returned to the calling function. Therefore, it is declared as a reference to an iterator. The return value of the function is a Boolean value that indicates whether another partner has been found.

Similar to `fall()`, all possible candidates for the additional contact partner are determined first, i.e., all particles that may be contacted by particle n while maintaining contact to its present partner p. Knowing the coordinates of the particles n and p, `x[n]` and `x[p->second]`, it is also known whether the new particle rolls to the left ($x_n < x_p$) or to the right ($x_n > x_p$). The candidate list is again an index range between the iterators `start` and `stop` of the container `surface` (see Fig. 7.4). We define particles whose x-coordinate is in the interval $x_p \ldots x_p \pm (R_p + 2R_n + R_{\max})$ as candidates, adjusting the sign according to the direction of motion.

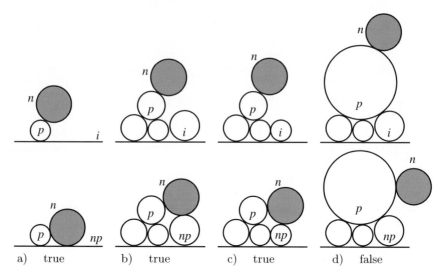

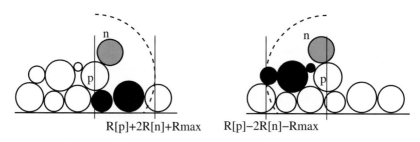

Fig. 7.3. Illustration of find_new_partner(). Depending on the particle configuration different scenarios take place (*left* to *right*): (**a**) Particle n rolls on p, touches the wall and is stable. (**b**) Particle n touches particle i and is stable. (**c**) Particle n is unstable on i. (**d**) Particle n does not find a new contact. It moves to the position $(x_p + R_p + R_n, y_p)$ and continues falling. The last line gives the return value of the function for the individual cases

Fig. 7.4. The centers of the candidates for the second partner particle are located within the circle of radius $r = R_p + 2R_n + R_{\max}$ around x_p. Depending on the relative position of the particles n and p the circle can be reduced to a half circle. The particles represented by *filled* circles are the candidates

In the loop starting at line 107, each candidate particle of the surface (iterator i) is tested to determine whether the new particle can come into contact with it. The function pcontact() is called (except for $i = p$), which returns false if no contact is possible between i, p, and n (details see below), otherwise it returns true. In the latter case the return arguments xx and yy are set to the coordinates of the new particle in contact with the new partner. As in the case of fall(), we need to find the candidate for which the deposited particle attains its maximum vertical coordinate yy. This new neighbor np is found (provided it exists) by iterating through all candidates.

Finally, wcontact() checks whether the contact of the new particle with the wall results in a larger vertical component, i.e., whether the new particle hits the wall before hitting any other particle. In this case find_new_partner() returns np=surface.end(), indicating that the ground is the new partner.

If there is a new partner (including the ground), the variables xmax and ymax contain the coordinates for which the particle n adopts its largest height while touching two other particles, or one particle and the ground. This is then the new position that the particle attains by rolling down the surface of its partner p. If no valid position has been found, ymax keeps its initial value ymax=0.

Finally, we test if the new position is above the center of the partner particle p. If this is not the case, the particle cannot move to this position by rolling on the surface of its partner. Instead, the particle moves along the surface of its partner until both particles have the same vertical coordinate. The particle moves, thus, to the coordinates

$$x_n = x_p \pm (R_n + R_p)$$
$$y_n = y_p \, ,$$

$$(7.13)$$

and the Boolean variable found has the value false, which indicates that no new partner has been found. From this position the particle continues falling (case d in Fig. 7.3).

If, however, ymax is larger than the center of particle p, the new partner is found and the particle is moved to its new position (x_{max}, y_{max}). The Boolean variable found has then the value true.

```
                              VisscherHeap.cc
92  bool find_new_partner(int n, iter p, iter & np)
93  {
94    bool found=false;
95    iter start=p,stop=p;
96    double xmax=x[n],ymax=0,xx,yy;
97
98    if(x[n]<=x[p->second]) {
99      double left=x[p->second]-R[p->second]-2*R[n]-Rmax;
100     start=surface.lower_bound(left);
101     if(start==surface.end()) start=surface.begin();
102     stop++;
103   } else {
104     double right=x[p->second]+R[p->second]+2*R[n]+Rmax;
105     stop=surface.upper_bound(right);
106   }
107   for(iter i=start;i!=stop;i++){
108     if(i!=p){
109       if(pcontact(n,p->second,i->second,xx,yy)){
110         found=true;
111         if(yy>ymax){
112           ymax=yy;
113           xmax=xx;
114           np=i;
115         }
116       }
```

```
117 │      }
118 │    }
119 │    if(wcontact(n,p->second,xx,yy)){
120 │       if(yy>ymax){
121 │          ymax=yy;  xmax=xx;
122 │          np=surface.end();
123 │       }
124 │    }
125 │    if(ymax>y[p->second]){
126 │       x[n]=xmax;  y[n]=ymax;
127 │    } else {
128 │       y[n]=y[p->second];
129 │       if(x[n]>x[p->second]) x[n]=x[p->second]+R[n]+R[p->second];
130 │       else x[n]=x[p->second]-R[n]-R[p->second];
131 │       found=false;
132 │    }
133 │    return found;
134 │ }
```
———————— (continued) ————————

The function `find_new_partner()` uses `pcontact()` and `wcontact()`, which will be discussed now.

The function `pcontact()` tests whether there may be a contact between the new particle n and the already deposited particles n_1 and n_2. The position vectors $\vec{r}$, $\vec{r}_{n_1}$ and $\vec{r}_{n_2}$ denote the coordinates of these particles. A contact is possible if the equations

$$|\vec{r}_n - \vec{r}_{n_1}| = R_n + R_{n_1}$$
$$|\vec{r}_n - \vec{r}_{n_2}| = R_n + R_{n_2} \tag{7.14}$$

have a real solution for $\vec{r} \equiv (x, y)$. With the abbreviations

$$A = -\frac{y_{n_1} - y_{n_2}}{x_{n_1} - x_{n_2}}$$
$$B = \frac{(R_n + R_{n_2})^2 - (R_n + R_{n_1})^2 + x_{n_1}^2 + y_{n_1}^2 - x_{n_2}^2 - y_{n_2}^2}{2(x_{n_2} - x_{n_1})}$$
$$C = \frac{A(B - x_{n_1}) - y_{n_1}}{1 + A^2} \tag{7.15}$$
$$D = \frac{(B - x_{n_1})^2 + y_{n_1}^2 - (R_n + R_{n_1})^2}{1 + A^2}$$
$$E = C^2 - D ,$$

the solution is[2]

$$y = -C + \sqrt{E}$$
$$x = B + Ay , \tag{7.16}$$

provided $E \geq 0$. In this case the function `pcontact()` returns `true` and the parameters `xx` and `yy` contain the resulting coordinates x and y. If $E < 0$ the return value is `false`, indicating that there is no contact possible.

_____
[2] The system (7.14) has a second solution that is irrelevant here.

```
_____ VisscherHeap.cc _____
135  bool pcontact(int n,int n1,int n2,double & xx, double & yy)
136  {
137     double Rn1=R[n]+R[n1];
138     double Rn2=R[n]+R[n2];
139     double A=(y[n1]-y[n2])/(x[n2]-x[n1]);
140     double B=(Rn1+Rn2)*(Rn1-Rn2)/(x[n2]-x[n1]) + x[n1]+x[n2] - A*(y[n1]+y[n2]);
141     B/=2;
142     double Bxn1=B-x[n1];
143     double C=(A*Bxn1-y[n1])/(A*A+1);
144     double D=(Bxn1*Bxn1+(y[n1]-Rn1)*(y[n1]+Rn1))/(A*A+1);
145     double E=C*C-D;
146
147     if(E<0){
148        return false;
149     } else{
150        yy=-C+sqrt (E);
151        xx=A*yy+B;
152        return true;
153     }
154  }
_____ (continued) _____
```

The function `wcontact()` tests whether particle n and the fixed particle n_1 contact each other and the ground simultaneously:

$$|\vec{r}_n - \vec{r}_{n_1}| = R_n + R_{n_1}$$
$$y_n = R_n \ . \tag{7.17}$$

The solution of (7.17) reads

$$x_n = x_{n_1} \pm \sqrt{(R_n + R_{n_1})^2 - (R_n - y_{n_1})^2} \ , \tag{7.18}$$

provided the argument of the square root is non-negative. If $x_n - x_{n_1} > 0$ the positive sign is chosen since the particle moves in positive x direction, otherwise the negative sign is valid.

```
_____ VisscherHeap.cc _____
155  bool wcontact(int n,int n1, double & xx, double & yy)
156  {
157     double A=(2*R[n]+R[n1]-y[n1])*(R[n1]+y[n1]);
158
159     if(A<0) return false;
160     if(x[n]<x[n1]) xx=x[n1]-sqrt(A); else xx=x[n1]+sqrt(A);
161     yy=R[n];
162     return true;
163  }
_____ (continued) _____
```

After completing the do-while-loop in the main program (lines 34–48), the stable position of the new particle is found and, hence, it becomes a member of the surface. Therefore, this particle is inserted into the container `surface` by `update_surface()`. After depositing a particle other particles may be screened, i.e., no further particles may come in contact with them. Consequently, these particles are not members of the surface anymore and are removed from `surface` (Fig. 7.5).

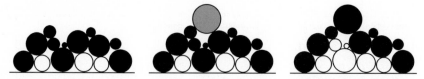

Fig. 7.5. *Left*: surface particles before the deposition of the new particle (*drawn gray*). *Center*: a new particle is deposited. *Right*: corrected list of surface particles

```
                                    VisscherHeap.cc
164  void update_surface(int n, iter p1, iter p2)
165  {
166    if(p1==p2) {
167      surface[x[n]]=n;
168      return;
169    }
170    if(p1==surface.end()) {
171      p1=p2;
172      p2=surface.end();
173    }
174    if(p2==surface.end()) {
175      if(x[n]>x[p1->second]){
176        p2=surface.upper_bound(x[n]);
177      } else {
178        p2=surface.lower_bound(x[n]);
179        if(p2==surface.end()) p2=surface.begin();
180      }
181    }
182    if(p2!=surface.end()){
183      if(x[p1->second]>x[p2->second]){
184        iter h=p1; p1=p2;p2=h;
185      }
186    }
187    if(p1!=p2){
188      iter pp=p1;
189      pp++;
190      if(pp!=p2){
191        surface.erase(pp,p2);
192      }
193    }
194    surface[x[n]]=n;
195  }
                                    (continued)
```

The program is completed with **psplot()** to output Postscript snapshots of the heap. The action of this function is straightforward and therefore will not be discussed further.

```
                                    VisscherHeap.cc
196  void psplot(ostream & psout)
197  {
198    static int page=0;
199    int lx=600, ly=600;
200
201    if(page==0){
202      psout << "%!PS-Adobe-2.0" << endl;
203      psout << "%%BoundingBox: 0 0 "<<lx<<" "<<ly << endl;
204      psout << "%%EndComments" << endl;
205      psout << " 1 setlinewidth\n";
206      psout << "/red {1 0 0 setrgbcolor stroke} def" << endl;
207      psout << "/c { 0 360 arc stroke} def" << endl;
208      psout << "/cf { 0 360 arc fill stroke} def" << endl;
209    }
210    psout << "%%Page: " << page << " " << page << endl;
211    double xm=0, ym=0;
```

```
212    for (iter i = surface.begin(); i != surface.end(); i++ ){
213        xm=max(xm,fabs(x[i->second]));
214        ym=max(ym,fabs(y[i->second]));
215    }
216    xm+=Rmax;
217    ym+=Rmax;
218    double sc=min((lx-10)/2/xm, (ly-20)/ym);
219    psout << lx/2 <<" 10 translate newpath" << endl;
220    psout << sc << " "<< sc <<" scale" << endl;
221    for(unsigned int p=0; p<R.size(); p++){
222        if(y[p]<ym){
223            psout << x[p] << " " << y[p] << " " << R[p];
224            if(R[p]<2*Rmax*Rmin/(Rmax+Rmin)) psout << " cf\n";
225            else psout << " c\n";
226        }
227    }
228    psout << "red"<< endl;
229    for (iter i = surface.begin(); i != surface.end(); i++ )
230        psout << "newpath "<< x[i->second] <<" "<< y[i->second] <<" "
231            << R[i->second] <<" cf"<< endl;
232    psout << "stroke showpage " << endl;
233    page++;
234 }
```

Figure 7.6 shows snapshots of a growing heap. The particles which are part of the surface are drawn filled.

When a heap of particles is created by sequentially depositing particles of different size, size segregation (stratification) is observed, which is caused by different angles of repose for large and small particles [46, 160, 298, 299]. The particles form stripes as shown in Fig. 7.7. In three dimensional dunes and ripples, other more complex stratification patterns are observed [4]. This effect has been studied and modeled extensively, e.g., in [31, 32, 99–101, 130, 140, 168–174, 182]. Similar structures have been observed in sand overflown by wind or water [267].

Figure 7.8 shows a heap of $N = 10^6$ particles of two different radii. The effect of stratification is visible in the close-ups. To generate the bimodal distribution of particle sizes, line 29 in `main()` has to be replaced by "`if(z>0.8) R[i]=Rmax; else R[i]=Rmin;`". The small particles are shown as filled circles, the large ones are drawn as hollow circles.

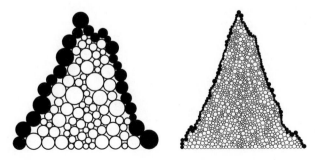

Fig. 7.6. Snapshots of the growing heap. The number of particles is $N = 100$ and $N = 1400$

Fig. 7.7. Formation of stripes in a heap consisting of particles of different properties (experimental result). The figure was taken from [102]

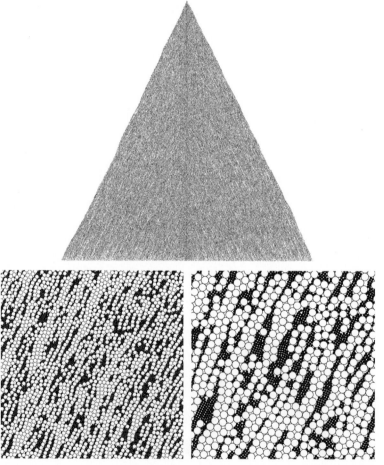

Fig. 7.8. A heap consisting of $N = 10^6$ particles of two different radii. Size segregation (stratification) can be seen in the close-ups

7.3 Dynamic Simulations

At first glance it might appear strange to apply the bottom-to-top reconstruction technique to the simulation of dynamic processes since due to the main principle of this algorithm, deposited particles cannot leave their positions anymore. But if we partition the dynamics of the system into alternating steps of collective motion and sequential deposition, certain dynamical systems can be simulated. As discussed below, this idealization is justified only for a restricted class of problems.

This simulation technique was first applied to simulate size segregation [133]. Consider a vertically shaken container ($z = A\cos(\omega t)$). Each oscillation period is subdivided into two parts: for the first interval when the container moves upward ($(2i - 1)\pi \leq \omega t < 2i\pi$), the material is assumed to be compact and, therefore, the particles do not move relative to each other. This part of the motion is modeled by shifting all particles upward by $2A$. For the second part of the oscillation ($2i\pi \leq \omega t < (2i + 1)\pi$), the material is assumed to be sufficiently dilute such that the interaction between the particles can be neglected. This part of the period is simulated by depositing the particles sequentially, starting with the lowest. Although this algorithm can indeed describe segregation [133], the quantitative interpretation of the results is problematic: the simulation result is independent of the amplitude and frequency of the vibration. Moreover, additional assumptions are needed to consider the influence of the vertical walls whose properties (in particular their roughness) are known to be essential to size segregation.

The algorithm has also been applied to model mixing and demixing processes in a partially filled cylinder that rotates slowly around its horizontal axis (e.g., [21, 22]). In the simulation this continuous, externally driven motion is subdivided into alternating steps of collective rotation and deposition. In the rotation step all particles move collectively through a small angle according to the rotation of the cylinder. This step is performed without regard to mechanical stability. It is, therefore, possible to form overhangs (which are small due to the small rotation angle). In the deposition step, each of the particles finds its new stable position, again starting with the lowest particle. Hence, the simulation proceeds as follows:

1. **Initialization:** Place the particles at random inside of the container.
2. **Motion of the container:** The positions of all particles follow the motion of the container for a small time step Δt.
3. **Preparation of the current time step:** Increase the y-coordinate of all particles by a constant, e.g., $R_{\mathrm{cyl}}/10$.
4. **Relaxation of the particles $i = 1 \ldots N$ in sequence of increasing y-coordinate:** Move particle i downward until it touches the wall or attains a local minimum, as described in Sect. 7.1.
5. **Measurement:** Once all particles are deposited, data on the system's state can be recorded.

6. **Loop:** Increase the system time by Δt and continue with step 2.

The algorithm is sketched in Fig. 7.9.

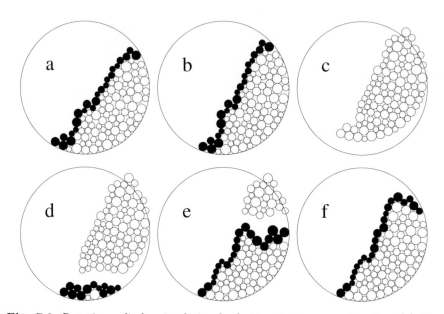

Fig. 7.9. Rotating cylinder simulation by bottom-to-top reconstruction. (**a**) The system at time t. (**b**) The cylinder is rotated (angle of rotation appears exaggerated). (**c**) All particles are lifted by $R_{\mathrm{cyl}}/10$. (**d, e**) The particles are deposited sequentially. (**f**) Situation at time $t + \Delta t$ when all particles have been deposited

The relaxation step is the core of the algorithm. To implement it properly one has to overcome a number of algorithmic and numeric difficulties whose detailed description would not contribute much to the understanding of the simulation method. Instead we refer to the web site where the source code for this simulation can be obtained.

Figure 7.10 shows a snapshot of a simulation of $N = 10^6$ particles of different radii $R_i \in (0.1, 1)$ cm. The radii are chosen randomly in such a way that the total mass of all particles from the interval $(R, R + \mathrm{d}R)$ is constant regardless of R (see p. 275). Initially the particles are distributed at random in the cylinder of radius 70 cm. Figure 7.10 shows the system after approximately one rotation, which corresponds to 100 deposition steps. We notice that the small particles are concentrated close to the center of the cylinder. This effect, which is observed also experimentally, was found in simulations in [21, 23].

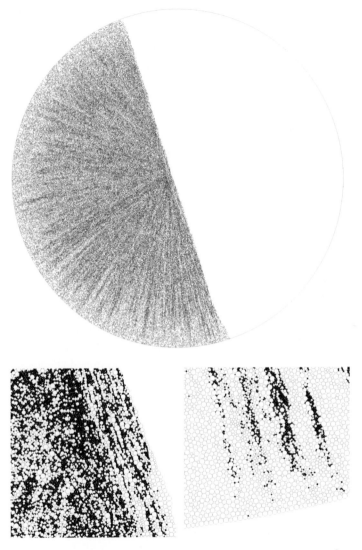

Fig. 7.10. A snapshot of a slowly rotating cylinder filled with $N = 10^6$ granular particles and close-ups. After one revolution consisting of 100 deposition steps, size segregation is clearly visible. For better presentation in the *top* figure only 30% of the particles have been drawn

7.4 Critical Analysis of the Model

The bottom-to-top reconstruction method, as first suggested in [286], is much faster than regular Molecular Dynamics as presented in Chap. 2. It was successfully applied to the simulation of various large granular systems. However, as already briefly discussed in the beginning of Sect. 7.3, this method is not universal. There are cases where physically incorrect behavior of the simulated granular system was observed [20].

A deposited particle does not move under the influence of particles deposited later, even if the particle suffers (in the realistic system) violent collisions with them. The motion of the particles is, thus, not governed by Newton's equation of motion. Instead, each single grain performs an overdamped motion in a complicated potential landscape comprising the already deposited grains. This is equivalent to disregarding inertial forces and moments. Even from a very basic and intuitive concept of classical mechanics it is clear that this algorithm cannot describe the granular many-body problem in a general way. It should be regarded as a compromise between computing power requirements and realistic description of physical reality. Two simple examples will demonstrate that the algorithm can lead to non-physical description of the motion of particles: Figure 7.11 shows a situation that can occur during real sedimentation processes but cannot be described by the algorithm. None of the gray drawn particles is in a stable position without the others, i.e., according to the described algorithm none of them can have been deposited first. On the contrary, Fig. 7.12 shows a sequence that is physically impossible since the configuration on the right hand side is mechanically unstable, although it may appear in the simulation.

While Figs. 7.11 and 7.12 show rather artificial systems, Fig. 7.13 demonstrates that unrealistic configurations are indeed generated by the algorithm. The figure shows a snapshot of a simulation of a growing heap. The

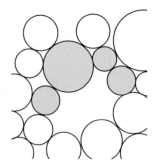

Fig. 7.11. There are realistic situations which cannot be described by the *bottom-to-top* reconstruction algorithm. The configuration shown here is physically possible but cannot be generated by the algorithm

Fig. 7.12. This sequence may be generated by the algorithm although it is not realistic since the configuration on the *right* hand side is unstable

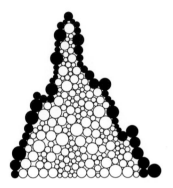

Fig. 7.13. A snapshot from the same sequence as shown in Fig. 7.6. The heap consists of 220 particles

significantly more realistic pictures (although the angle of repose is too large) of Fig. 7.6 originate from the same simulation.

The highly efficient algorithm presented allows us to simulate heaps of up to 10^8 particles in two dimensions [183] whereas classical Molecular Dynamics presently only allows for simulations of 10^4–10^5 particles. In Molecular Dynamics the motion of each particle does, in principle, affect the motion of each other particle, i.e., a complicated many-particle problem is numerically solved. The bottom-to-top reconstruction scheme treats this system as a sequence of single particle problems. Moreover the performance of the algorithm is independent of an integration time step—the new position of a particle is a function of the positions of the other particles, but not of their velocities.

The algorithm may yield satisfactory results when treating systems where the boundary conditions change only very slowly. In such cases Molecular Dynamics is frequently inefficient. In particular, for problems where the computation of the trajectories of individual particles is less important, the algorithm can be seen as a good compromise between efficiency and precision of the result.

Brownian Dynamics for the Simulation of Granular Flows

8.1 Langevin Equation for Pipe Flow

Langevin dynamics has been used for modeling several granular systems, such as the flattening of a vibrated heap [186], the motion of a particle on a rough inclined plane, and the properties of size segregation mechanisms [237]. The Langevin dynamics of granular flows was worked out theoretically in [65, 251], based on early work by Litwiniszyn [159, 160]. In this section we will apply the paradigm of Langevin random forces for modeling the flow of granular material through a narrow pipe [234–236].

Experiments show that such flows are not homogeneous but tend to form clusters, e.g., [125, 212, 228, 255]. Certainly the presence of air has much effect on clustering, e.g., [228]—in the case of pneumatic conveyors (where clusters are observed) air is even used to haul material through pipes, e.g., [55, 57, 142, 203, 211, 307]. Nevertheless, the formation of density inhomogeneities has also been reproduced using models which do not consider the effect of air, e.g., [207, 208, 212]. Here the surrounding air is disregarded, i.e., granular pipe flow is simulated in vacuum. It turns out that inhomogeneous pipe flow can be realistically modeled using simple and plausible assumptions on the interactions of the particles with the wall.

The stochastic theory of Langevin random forces will be presented only briefly. A more detailed description of this mathematical method can be found, e.g., in [206].

To derive the Langevin equation of interacting particles in a pipe, we need to obtain the Boltzmann equation first. Consider the motion of a *single* particle i moving in a vertical pipe under the action of gravity mg. Positive x-direction points downward, i.e., in the direction of gravity. The Langevin equation reads

$$\dot{x}_i = v_i$$
$$m\dot{v}_i = mg - \gamma v_i + \sqrt{2E\gamma}\xi_i(t) \ . \tag{8.1}$$

The frictional force γv_i and the Langevin random force $\sqrt{2E\gamma}\xi_i(t)$ describe the interaction of the particle with the wall. For the random force, Gaussian white noise is assumed:

$$\langle \xi_i(t) \rangle = 0$$
$$\langle \xi_i(t)\xi_j(t') \rangle = \delta_{ij}\delta(t - t') .$$
(8.2)

The thermal energy $E = k_B T$ is determined by the *granular temperature* T which characterizes the random motion of granular particles (see (8.9)). The expression for the noise intensity $\sqrt{2E\gamma}$ (Einstein relation) is exact only in thermal equilibrium and is, hence, an approximation. Equation (8.1) is a single-particle equation, i.e., it does not account for the interaction of particles. In the stationary state we expect a Maxwell distribution for the particle velocity of mean value $v^0 = mg/\gamma$.

The distribution of the probability $p(x, v, t)$ to find the (non-interacting) particles at time t, at position x, and with velocity v is governed by the collisionless Fokker–Planck equation

$$\frac{\partial p}{\partial t} + \frac{\partial}{\partial x}(vp) + \frac{\partial}{\partial v}\left[\left(g - \frac{\gamma}{m}v\right)p\right] - \frac{E\gamma}{m^2}\frac{\partial^2 p}{\partial v^2} = 0 .$$
(8.3)

For the incorporation of particle collisions into the model consider first a related problem: In 1971 Prigogine and Herman [225] studied vehicular traffic on a one lane road and investigated the spontaneous formation traffic jams. For the interaction of the cars they assumed that the faster car i in the rear adopts the velocity of the slower car j in front when the car behind catches up:

$$\bigcirc \xrightarrow{v_i} \bigcirc \xrightarrow{v_j} \quad \Rightarrow \quad \bigcirc\bigcirc \xrightarrow{v_j}$$

After the interaction both cars move at common velocity v_j. The modeling of traffic flows by means of Langevin equations and hydrodynamics has been elaborated to a high level by now (see, e.g., [111, 112]).

Although the close relationship between traffic flow and flow of granular matter has been pointed out by several authors, e.g., [114, 153, 212], the direct application of the vehicle interaction law to granular particles is problematic since it violates momentum conservation for the particle–particle collision. Nevertheless, it may be justified for the description of a narrow pipe which is not wider than few particle diameters since it can be expected that most of the particle–particle collisions are immediately followed by a collision with the wall. Therefore, the momentum of a particle is mainly determined by its interaction with the walls and the violation of momentum conservation for particle–particle collisions is insignificant. The pipe itself definitely does not belong to the system described by (8.1), i.e., it does *not* change its momentum via collisions with particles. Therefore, for the stationary flow it may be assumed that the increase of the momentum due to gravity is absorbed by the wall via inelastic collisions.

To derive the Boltzmann equation we need the collision integral. The probability density $p(x, v, t)$ of particles at position x and moving at velocity v is increased by collisions with faster particles traveling at velocity $v' > v$:

$$\left[\frac{\partial p(x, v, t)}{\partial t}\right]_{\text{gain}} \propto \int_v^\infty p(x, v, t)p(x, v', t)(v' - v) \, dv' \,, \tag{8.4}$$

since the faster particle adopts the velocity v of the slower one. The rate of change of the probability density caused by collision with a particle at velocity v' is proportional to the probability densities at v and v' (i.e., the numbers of collision partners) and, furthermore, to the relative velocity of the collision particles $v' - v$. Integration is performed over the interval v to ∞ since a collision occurs only if $v' > v$, otherwise the distance between both particles increases and no collision can occur. On the other hand, the probability $p(x, v, t)$ decreases due to collisions with slower particles, i.e., $v > v'$:

$$\left[\frac{\partial p(x, v, t)}{\partial t}\right]_{\text{loss}} \propto - \int_{-\infty}^v p(x, v, t)p(x, v', t)(v - v') \, dv' \,. \tag{8.5}$$

Combining (8.4) and (8.5), the Boltzmann equation for the one particle probability density $p(x, v, t)$ follows:

$$\frac{\partial p}{\partial t} + \frac{\partial}{\partial x}(vp) + \frac{\partial}{\partial v}\left[\left(g - \frac{\gamma v}{m}\right)p\right] - \frac{E\gamma}{m^2}\frac{\partial^2 p}{\partial v^2}$$

$$= C \int_{-\infty}^\infty p(x, v, t)p(x, v', t)(v' - v) \, dv' = C\, p(x, v, t)\, n(x, t)(u - v) \,, \tag{8.6}$$

where the reduced particle number density $n(x, t)$ (normalized to 1), the material velocity $u(x, t)$, and the granular temperature are defined by

$$n(x, t) = \int_{-\infty}^\infty p(x, v, t) \, dv \tag{8.7}$$

$$u(x, t) = \frac{1}{n(x, t)} \int_{-\infty}^\infty v \, p(x, v, t) \, dv \tag{8.8}$$

$$T(x, t) = \frac{m}{2k_B n(x, t)} \int_{-\infty}^\infty [v - u(x, t)]^2 \, p(x, v, t) \, dv \,. \tag{8.9}$$

The effective cross section C is a complicated constant whose value is determined by the pipe geometry and the properties of the granular particles, which must be determined experimentally [234].

The Boltzmann equation (8.6) describes the dynamics of the probability density of granular particles in the pipe when collisions are taken into consideration. It has the stationary solution

$$p^0(v) = \sqrt{\frac{m}{2\pi\,k_\mathrm{B}T^0}}\,n^0\,\exp\left[-\frac{m}{2\,k_\mathrm{B}T^0}\left(v-u^0\right)^2\right]$$

$$u^0 = \left(mg - Ck_\mathrm{B}T^0 n^0\right)/\gamma \tag{8.10}$$

$$T^0 = E/k_\mathrm{B}\ .$$

Assuming local thermodynamic equilibrium, i.e., $n^0 \to n(x,t)$, $u^0 \to u(x,t)$, and $T^0 \to T(x,t)$, (8.10) can be inserted into (8.6) to obtain, again, a Fokker–Planck equation

$$\frac{\partial p}{\partial t} + \frac{\partial}{\partial x}\,(vp) + \frac{\partial}{\partial v}\left[\left(\frac{F(n,T)}{m} - \frac{\gamma}{m}v\right)p\right] - \frac{E\gamma}{m^2}\frac{\partial^2 p}{\partial v^2} = 0\ . \tag{8.11}$$

Equation (8.11) has the same form as (8.3) if gravity is replaced by a self-consistent local force $F(n,T)$

$$F(n,T) = mg - Ck_\mathrm{B}T(x,t)\,n(x,t)\ . \tag{8.12}$$

By comparing with (8.1), we finally obtain the Langevin equation for a granular particle flowing through a narrow pipe under the action of gravity and with consideration of particle collisions:

$$\dot{x}_i = v_i$$

$$m\dot{v}_i = F\left(n\left(x_i,t\right),T\left(x_i,t\right)\right) - \gamma\,v_i + \sqrt{2E\gamma}\,\xi_i(t)\ . \tag{8.13}$$

For a detailed discussion on the derivation of (8.13) see [236].

8.2 Simulation of the Langevin Equation

Equations (8.13) are the starting point of a hydrodynamic theory [234, 236]. The analysis of their stability shows that the homogeneous flow becomes unstable if the density increases above a critical value.

    This equation was derived to simulate granular pipe flow using Langevin dynamics. The discretization of stochastic differential equations such as (8.13) differs from the discretization of ordinary differential equations: The action of the deterministic part is proportional to the time step Δt, whereas the action of the stochastic force is proportional to $\sqrt{\Delta t}$, reflecting the fact that the solution of stochastic differential equations have no time derivative. Hence, in discretized form the Langevin equation reads

$$x_i(t + \Delta t) = x_i(t) + v_i(t)\Delta t$$

$$v_i(t + \Delta t) = v_i(t) + \frac{1}{m}\left\{F\left[n\left(x_i,t\right),T\left(x_i,t\right)\right] - \gamma v_i(t)\right\}\Delta t + \frac{1}{m}\sqrt{2E\gamma\Delta t}\ z\ ,$$

$$\tag{8.14}$$

with z being a Gaussian random number of standard deviation 1. This type of integration scheme and extensions are described in much detail in [175].

The set of equations (8.14) can be directly used as an iteration rule for a numerical simulation [234, 236] (see also [256] for the simulation method). When choosing the constants one has to bear in mind that the time step Δt has to be small as compared to the relaxation time m/γ. The computation of field quantities $n(x,t)$, $T(x,t)$, and $u(x,t)$ is performed by coarse graining. To this end the pipe is divided into M sections (boxes) of equal lengths. While there have to be enough particles in each box to allow for accurate computation of field quantities, it is advantageous to simulate a large number of boxes to have a high spatial resolution of the inhomogeneities. Therefore, a compromise between both objectives is necessary when choosing the box length.

Below we present an implementation of the simulation algorithm. In the table the variables and constants are introduced:

Constants	
N	Total number of particles in the pipe
length	Length of the pipe measured in meters
itmax	Number of time steps dt to be simulated
tps, tfields	Intervals of snapshots and field data output
B	Number of coarse-graining boxes for the field computation
dt	Time step measured in sec
m	Mass of a grain measured in kg
g	Gravity measured in m/sec²
gamm	Friction coefficient measured in kg/sec
E	In equilibrium systems the thermal energy. Here it is a simulation parameter to be determined from experiments.
Particle variables	
x[], v[]	Positions and velocities of the particles
Fields	
kT[]	Temperature multiplied by Boltzmann's constant k_B
n[]	Density
u[]	Velocity field

The constants in the simulation refer to experiments done by Riethmüller [234] who investigated granular flow in a 6-mm wide glass pipe. Experimentally, he determined $m = 7.4 \times 10^{-3}\,\text{kg}$, $\gamma = 7 \times 10^{-6}\,\text{kg/sec}$, $C = 6.4 \times 10^{-3}$ (see (8.6)), and $E = 2 \times 10^{-8}\,\text{Nm}$. The flow started to develop clusters at density $n_c = 9{,}000/\text{m}$ (particles per length) where the density inside the clusters was $n_{cl} = 56{,}000/\text{m}$.

First, the program defines the constants and variables, the function prototypes, and the function rand() that generates equally distributed random numbers from the interval $[0, 1)$.

```
                        ─────── Langevin.cc ───────
 1   #include <iostream>
 2   #include <fstream>
 3   #include <vector>
 4   #include <math.h>
 5
 6   const int B=1000, N=52000, tps=10, tfields=100, itend=110000;
 7   const double length=3, m=7.4e-7, dt=0.01, gamm=7e-6, g=9.81, C=6.4e-3, E=2e-8;
 8   vector<double> u(B), kT(B), n(B), x(N), v(N);
 9   ofstream psout,fout;
10
11   void fields(double &, double &);
12   void motion();
13   void psplot(int, ofstream &, double, double);
14   double Gaussian();
15   double ranf(){ return double(rand())/(1+double(RAND_MAX));}
                        ─────── (continued) ───────
```

In `main()` the particle and field variables are initialized. The particles are equally distributed along the pipe and move with velocity $v_i = 1$. The output files for the snapshots of the simulation and the field data are opened.

In each iteration step, `fields()` is called to compute the temperature, density, and velocity fields. The function `motion()` then computes the new positions and velocities of the particles according to (8.14). For every `tps` step and every `tfields` step, Postscript snapshots of the fields and the raw field data are written.

```
                        ─────── Langevin.cc ───────
16   int main()
17   {
18     double umax=1, kTmax=1;
19
20     for(int i=0; i<N; i++) {
21       x[i]=i*length/N;
22       v[i]=1;
23     }
24     psout.open("L.ps");
25     fout.open("Lfields");
26     for(int it=0; it<itend; it++){
27       if(it % 20 == 0) cout << "Step " << it << endl;
28       fields(umax,kTmax);
29       motion();
30       if(it%tps==0) psplot(it/tps,psout,umax,kTmax);
31       if(it%tfields==0){
32         fout << "# " << it << endl;
33         for (int j=0; j<B; j++)   fout<<j<<" "<<n[j]<<" "<<kT[j]<<" "<<u[j]<<endl;
34         fout << endl;
35       }
36     }
37   }
                        ─────── (continued) ───────
```

The function `fields()` computes the hydrodynamic fields using the particle coordinates of the previous time step.

```
                        ─────── Langevin.cc ───────
38   void fields(double & umax, double & kTmax)
39   {
40     int ind;
41
42     for(int i=0; i<B; i++)  u[i]=kT[i]=n[i]=0;
43     for(int i=0; i<N; i++){
44       ind=int(x[i]/length*B);
45       n[ind]+=B/length;
```

```
46        u[ind]+=v[i];
47      }
48      umax=0;
49      for(int i=0; i<B; i++) {
50        u[i] /= n[i]*length/B;
51        umax = max(umax,u[i]);
52      }
53      for(int i=0; i<N; i++){
54        ind=int(x[i]/length*B);
55        kT[ind]+=(v[i]-u[ind])*(v[i]-u[ind]);
56      }
57      kTmax=0;
58      for(int i=0; i<B; i++){
59        kT[i]/=n[i]*length/m/B;
60        kTmax=max(kTmax, kT[i]);
61      }
62    }
```
_____ (continued) _____

The function `motion()` computes the new particle positions and velocities using the fields of local velocity u, density n, granular temperature kT, the deterministic force (8.12), and the stochastic force $\sqrt{2E\gamma}\xi(t)$ (see (8.14)). In this function we also implement the periodic boundary conditions, i.e., if a particle leaves the pipe at the lower side it is reinserted at the top of the pipe and vice versa, preserving its velocity. If the particle position is less than 0 (particle left by the upper end), `length` is added; if it is larger than `length`, `length` is subtracted. If the new position is still outside of the legal range [0, `length`] the particle has moved by more than the system's length in one time step dt, which indicates that the time step is too large or that the simulation has become unstable due to unsuitable choice of parameters. The simulation is then aborted with an error message.

_____ Langevin.cc _____
```
63    void motion()
64    {
65      int ind;
66      double fluct,det,dv;
67
68      for(int i=0; i<N; i++){
69        ind=int(x[i]/length*B);
70        x[i] += v[i]*dt;
71        if(x[i] >= length) x[i]-=length;
72        else if(x[i] < 0) x[i]+=length;
73        if((x[i]<0) || (x[i]>=length)){
74          cout << "COORDINATE ERROR " << x[i]<< endl;
75          exit(-2);
76        }
77        det=g-gamm*v[i]/m-C*kT[ind]*n[ind]/m;
78        fluct = 1/m*sqrt(2*E*gamm)*Gaussian();
79        dv=det*dt+fluct*sqrt(dt);
80        if(v[i]+dv>0) v[i]+=dv;
81      }
82    }
```
_____ (continued) _____

For the random force, Gaussian random numbers of unit standard deviation, as generated by `gasdev()`, are needed. The function is based on the algorithm by Box and Muller [33]. A similar version was already used in Chap. 4 (see p. 203).

```
                          —————— Langevin.cc ——————
83  double Gaussian()
84  {
85    static bool first=true;
86    static double sqrt2logx1,twopix2;
87
88    if(first){
89      first=false;
90      sqrt2logx1=sqrt(-2*log(drand48()));
91      twopix2=2*M_PI*drand48();
92      return sqrt2logx1*cos(twopix2);
93    } else {
94      first=true;
95      return sqrt2logx1*sin(twopix2);
96    }
97  }
                          —————— (continued) ——————
```

Finally **psplot()** writes the fields of density, velocity, and temperature as a sequence of Postscript snapshots. The argument **page**=0 indicates that the function is called for the first time. In this case, the Postscript header is written. The field values are encoded by colors, with blue indicating low values and red representing high values. The optimal color resolution has to be determined by some experimentation. Here we scale the color resolution such that the complete range from red to blue is used. The disadvantage of this method is that snapshots taken at different times cannot be directly compared with each other.

```
                          —————— Langevin.cpp ——————
98   void psplot(int page, ofstream & psout, double umax, double kTmax)
99   {
100    if(page==0){
101      psout << "%!PS-Adobe-2.0" << endl;
102      psout << "%%BoundingBox: 60 6 250 760" << endl;
103      psout << "/h{"<<-N/length<< " add "<<-2/float(N)<<" mul 0.5 add setgray ";
104      psout << "0 -1 rlineto currentpoint stroke moveto}def" << endl;
105      psout << "/u{umul mul 0.7 add 1 1 sethsbcolor 0 -1 rlineto currentpoint "
106              << "stroke moveto} def" << endl;
107      psout << "/t{tmul mul 0.7 add 1 1 sethsbcolor 0 -1 rlineto currentpoint "
108              << "stroke moveto} def" << endl;
109      psout << "/tr{100 10 translate 0.7 dup scale 5 setlinewidth stroke}def"
110              << endl;
111      psout << "/font {/Times-Roman findfont 18 scalefont setfont} def" << endl;
112    }
113    if(kTmax==0) kTmax=1;
114    psout << "%%Page: " << page << " " << page << endl;
115    psout << "tr /umul{"<< -0.8/umax <<"}def /tmul{" << -0.8/kTmax <<"}def font"
116            << endl;
117    psout << "-15 10 moveto (kTmax = " << kTmax << " ) show" << endl;
118    psout << "-15 35 moveto (umax = " << umax << " ) show" << endl;
119    psout << "0 60 translate 0 " << B << " moveto" << endl;
120    for(int i=0; i<B; i++) psout << n[i]<< " h ";
121    psout << "50 0 translate 0 " << B << " moveto" << endl;
122    for(int i=0; i<B; i++) psout << u[i] << " u ";
123    psout << "50 0 translate 0 " << B << " moveto" << endl;
124    for(int i=0; i<B; i++) psout << kT[i] << " t ";
125    psout << "showpage" << endl;
126  }
```

Figure 8.1 shows a sequence of snapshots of the simulation. Each of them shows the three fields of density, velocity, and granular temperature. Starting with a homogeneous initial distribution (left), clusters quickly develop. As

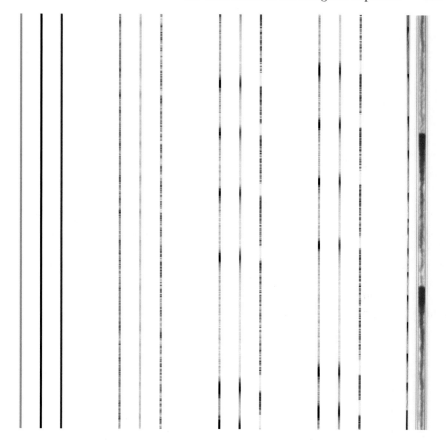

Fig. 8.1. *Left*: numerical solution of the Langevin equation for pipe flow. From *left* to *right* each sequence of three snapshots shows the density, velocity, and granular temperature for different times. The color scheme from `psplot()` has been remapped to grayscale. Dark means high density, low velocity, and high temperature. *Right*: experiment [212] and a closeup thereof.

expected, the velocity field shows minima at positions of high density. At the upper boundary of regions of high density, particles are decelerated rapidly; at the lower boundary they accelerate again until they reach the maximum velocity, which is determined by friction with the wall. The temperature assumes its highest value where particles are decelerated due to the coexistence of fast and slow particles resulting, hence, in a high temperature according to (8.9). For comparison, an experimental result is shown on the right hand side, taken from [212]. Animated sequences of the simulation can be found on the book's the web site.

References

1. B. J. Alder and T. E. Wainwright. Phase transition for a hard sphere system. *J. Chem. Phys.*, 27:1208, 1957.
2. B. J. Alder and T. E. Wainwright. Studies in molecular dynamics. I: General method. *J. Chem. Phys.*, 31:459, 1959.
3. B. J. Alder and T. E. Wainwright. Studies in molecular dynamics. II: Behaviour of a small number of elastic spheres. *J. Chem. Phys.*, 33:1439, 1960.
4. J. R. Allen. Asymmetrical ripple marks and the origin of cross-stratification. *Nature*, 194:167, 1962.
5. M. P. Allen and D. J. Tildesley. *Computer Simulations of Liquids*. Clarendon Press, Oxford, 1987.
6. G. M. Amdahl. Validity of the single processor approach to achieving large-scale computing capabilities. *Proc. Am. Federation of Information Processing Societies*, 30:483, 1967.
7. R. S. Anderson and P. K. Haff. Simulation of eolian saltation. *Science*, 241:820, 1988.
8. B. Andreotti, P. Claudin, and S. Douady. Selection of dune shapes and velocities - Part 1: Dynamics of sand, wind and barchans. *Europ. Phys. J. B*, 28:321, 2002.
9. J. Arvo. *Graphics Gems II*. Academic Press, New York, 1994.
10. T. Aspelmeier, G. Giese, and A. Zippelius. Cooling dynamics of a dilute gas of inelastic rods: A many particle simulation. *Phys. Rev. E*, 57:857, 1997.
11. R Baeza-Yates, M. Marín, and P. Cordero. Analysis of an improved priority queue for discrete event simulation of many moving objects. In C. Isaac and R. Peralta, editors, *Proceedings of the XIV International Conference of the Chilean Computer Science Society*, page 29, Chile, 1994.
12. R. A. Bagnold. *The Physics of Blown Sand and Desert Dunes*. Chapman & Hall, London, 1941.
13. P. Bak. *How Nature Works: The Science of Self-Organized Criticality*. Springer, Berlin Heidelberg New York, 1996.
14. P. Bak, C. Tang, and K. Wiesenfeld. Self-organized criticality: An explanation of $1/f$ noise. *Phys. Rev. Lett.*, 59:381, 1987.
15. P. Bak, C. Tang, and K. Wiesenfeld. Self-organized criticality. *Phys. Rev. A*, 38:364, 1988.

16. D. Baraff. Analytical methods for dynamic simulation of non-penetrating rigid bodies. *Computer Graphics*, 23:223, 1989.

17. D. Baraff. Coping with friction for non-penetrating rigid body simulation. *Computer Graphics*, 25:31, 1991.

18. D. Baraff. Fast contact force computation for nonpenetrating rigid bodies. In *Computer Graphics Proceedings*, Annual Conference Series, page 23, 1994.

19. D. Baraff. Linear-time dynamics using lagrange multipliers. In *Computer Graphics Proceedings*, Annual Conference Series, page 137, 1996.

20. G. C. Barker, A. Mehta, and M. J. Grimson. Comment on "three-dimensional model for particle size segregation by shaking". *Phys. Rev. Lett.*, 70:2194, 1993.

21. G. Baumann, I. Jánosi, and D. E Wolf. Particle trajectories and segregation in a two-dimensional rotating drum. *Europhys. Lett.*, 27:203, 1994.

22. G. Baumann, I. M. Jánosi, and D. E. Wolf. Surface properties and flow of granular material in a 2d rotating drum model. *Phys. Rev. E*, 51:1879, 1995.

23. G. Baumann, E. Jobs, and D. E. Wolf. Granular cocktail rotated and shaken. *Fractals*, 1:767, 1993.

24. G. W. Baxter, R. P. Behringer, T. Fagert, and G. A. Johnson. Pattern formation in flowing sand. *Phys. Rev. Lett.*, 62:2825, 1989.

25. R. Behringer. The dynamics of flowing sand. *Nonlinear Science Today*, 3:1, 1993.

26. R. P. Behringer and J. T. Jenkins, editors. *Powders and Grains'97*, Rotterdam, 1997. Balkema.

27. E. Ben-Naim, J. B. Knight, and E. R. Nowak. Slow relaxation in granular compaction. *J. Chem. Phys.*, 100:6778, 1996.

28. W. Benz and E. Asphaug. Simulations of brittle solids using smooth particle hydrodynamics. *Comp. Phys. Comm.*, 87:253, 1995.

29. S. Bernotat. Size reduction equipment. *Ullmann's Encyclopedia of Industrial Chemistry*, B2:5/14, 1988.

30. G. A. Bird. *Molecular Gas Dynamics*. Clarendon Press, Oxford, 1976.

31. T. Boutreux and P. G. de Gennes. Surface flows if granular mixtures: I. General principles and minimal model. *J. Physique I*, 6:1295, 1996.

32. T. Boutreux and P. G. de Gennes. Surface flow of granular mixtures. In Behringer and Jenkins [26], page 439.

33. G. E. P Box and M. E. Muller. A note on the generation of random normal deviates. *Annals Math. Stat.*, 29:610, 1958.

34. J. J. Brey and D. Cubero. Hydrodynamic transport coefficients of granular gases. In T. Pöschel and S. Luding, editors, *Granular Gases*, volume 564 of *Lecture Notes in Physics*, page 59, Berlin Heidelberg New York, 2001. Springer.

35. F. Bridges, K. Supulver, and D. N. C. Lin. Energy loss and aggregation processes in low speed collisions of ice particles coated with frost or methanol/water mixtures. In T. Pöschel and S. Luding, editors, *Granular Gases*, volume 564 of *Lecture Notes in Physics*, page 153, Berlin Heidelberg New York, 2001. Springer.

36. F. G. Bridges, A. Hatzes, and D. N. C. Lin. Structure, stability and evolution of Saturn's rings. *Nature*, 309:333, 1984.

37. N. V. Brilliantov and T. Pöschel. Rolling friction of a soft sphere on a hard plane. *Europhys. Lett.*, 42:511, 1998.

38. N. V. Brilliantov and T. Pöschel. Rolling as a continuing collision. *Europ. Phys. J. B*, 12:299, 1999.

39. N. V. Brilliantov and T. Pöschel. Deviation from Maxwell distribution in granular gases with constant restitution coefficient. *Phys. Rev. E*, 61:2809, 2000.

40. N. V. Brilliantov and T. Pöschel. Self-diffusion in granular gases. *Phys. Rev. E*, 61:1716, 2000.

41. N. V. Brilliantov and T. Pöschel. Velocity distribution in granular gases of viscoelastic particles. *Phys. Rev. E*, 61:5573, 2000.

42. N. V. Brilliantov and T. Pöschel. *Kinetic Theory of Granular Gases*. Oxford University Press, 2003.

43. N. V. Brilliantov, C. Salueña, T. Schwager, and T. Pöschel. Transient structures in a granular gas. *Phys. Rev. Lett.*, 2004, in press.

44. N. V. Brilliantov, F. Spahn, J.-M. Hertzsch, and T. Pöschel. A model for collisions in granular gases. *Phys. Rev. E*, 53:5382, 1996.

45. R. Brito and M. H. Ernst. Noise reduction and pattern formation in granular flows. *J. Mod. Phys. C*, 8:1339, 1998.

46. R. L. Brown. The fundamental principles of segregation. *J. Inst. Fuel*, 13:15, 1939.

47. W. K. Brown and K. H. Wohletz. Derivation of the Weibull distribution based on physical principles and its connection to the Rosin-Rammler and lognormal distributions. *J. Appl. Phys.*, 78:2758, 1995.

48. U. Brüning. *Realisierungsprinzipien von Mikrorechnern hoher numerischer Leistung*. PhD thesis, Technische Universität Berlin, 1987.

49. V. Buchholtz, J. Freund, and T. Pöschel. Molecular dynamics of comminution in ball mills. *Europ. Phys. J. B*, 16:169, 2000.

50. V. Buchholtz and T. Pöschel. A vectorized algorithm for molecular dynamics of short range interacting particles. *J. Mod. Phys. C*, 4:1049, 1993.

51. V. Buchholtz and T. Pöschel. Numerical investigation of the evolution of sandpiles. *Physica A*, 202:390, 1994.

52. V. Buchholtz and T. Pöschel. Molecular dynamics of arbitrarily shaped granular particles. *J. Phys. I France*, 5:1431, 1995.

53. V. Buchholtz and T. Pöschel. Force distribution and comminution in ball mills. In D. E. Wolf and P. Grassberger, editors, *Friction, Arching and Contact Dynamics*, page 265, Singapore, 1997. World Scientific.

54. V. Buchholtz, T. Pöschel, and H.-J. Tillemans. Simulation of rotating drum experiments using non-spherical particles. *Physica A*, 216:199, 1995.

55. A. J. Burnett. Wear in pneumatic conveying pipelines: A review of past and present work. In *Proc. Symp. on Attrition and Wear in Powder Technology*, page 243, Utrecht, 1992.

56. W. F. Busse and F. C. Starr. Change of a viscoelastic sphere to a torus by random impacts. *Am. J. Phys.*, 28:19, 1960.

57. F. J. Cabrejos and George E. Klinzing. Incipient motion of solid particles in horizontal pneumatic conveying. *Powder Techn.*, 72:51, 1992.

58. E. Caglioti, V. Loreto, H. J. Herrmann, and M. Nicodemi. A "Tetris-like" model for the compaction of dry granular media. *Phys. Rev. Lett.*, 79:1575, 1997.

59. F. Cantelaube, Y. Limon-Duparcmeur, D. Bideau, and G. H. Ristow. Geometrical analysis of avalanches in a 2d drum. *J. Physique I*, 5:581, 1995.

60. H. Caps and N. Vandewalle. Labyrinthic granular landscapes. *Phys. Rev. E*, 64:052301, 2001.

61. N. F. Carnahan and K. E. Starling. Equation of state for nonattractive rigid spheres. *J. Chem. Phys.*, 51:635, 1969.
62. Carlo Cercignani. *Rarefied Gas Dynamics: From Basic Concepts to Actual Calculations.* Cambridge Texts in Applied Mathematics. Cambridge University Press, Cambridge, 2000.
63. S. Chapman and T. G. Cowling. *The Matematical Theory of Non-uniform Gases.* Cambridge University Press, New York, 1970.
64. H. S. Chin. *Stabilization Methods for Simulation of Constrained Multibody Dynamics.* PhD thesis, University of British Columbia, Vancouver, Canada, 1995.
65. M. Y. Choi, D. C. Hong, and Y. W. Kim. Langevin dynamics, scale invariance, and granular flow. *Phys. Rev. E*, 47:137, 1993.
66. R. W. Cottle and C. B. Dantzig. Complementary pivot theory of mathematical programming. *Linear Algebra and its Applications*, 1:103, 1968.
67. R. W. Cottle, J. S. Pang, and R. E. Stone. *The Linear Complementary Problem.* Computer Science and Scientific Computing. Academic Press, Inc., San Diego, 1992.
68. P. A. Cundall and O. D. L. Strack. A discrete numerical model for granular assemblies. *Géotechnique*, 29:47, 1979.
69. P. Deltour and J.-L. Barrat. Quantitative study of a freely cooling granular medium. *J. Physique I*, 7:137, 1997.
70. D. Désérable and J. Martinez. Using a cellular automaton for the simulation of flow of granular material. In C. Thornton, editor, *Powders and Grains'93*, page 345, Rotterdam, 1993. Balkema. Martinez schreibt sich selber falsch.
71. N. E. Dowling. *Mechanical Behavior of Materials: Engineering Methods for Deformation, Fracture, and Fatigue.* Prentice Hall, Englewood Cliffs, 1998.
72. Y. Du, H. Li, and L. P. Kadanoff. Breakdown of hydrodynamics in a one-dimensional system of inelastic particles. *Phys. Rev. Lett.*, 74:1268, 1995.
73. J. Duran, T. Mazozi, S. Luding, É. Clément, and J. Rajchenbach. Discontinuous decompaction of a falling sandpile. *Phys. Rev. E*, 53:1923, 1996.
74. J. Eggers. Sand as Maxwell's demon. *Phys. Rev. Lett.*, 83:5322, 1999.
75. S E. Esipov and T. Pöschel. The granular phase diagram. *J. Stat. Phys.*, 86:1385, 1997.
76. D. J. Evans and S. Murad. Singularity free algorithm for molecular dynamics simulation of rigid polyatomics. *Molecular Physics*, 34:327, 1977.
77. R. Everaers and K. Kremer. A fast grid search algorithms for molecular dynamics simulations with short-range interactions. *Comp. Phys. Comm.*, 81:19, 1994.
78. W. Form, N. Ito, and G. A. Kohring. Vectorized and parallelized algorithms for multi-million particle md simulation. *J. Mod. Phys. C*, 4:1085, 1993.
79. D. Frenkel and B. Smit. *Understanding molecular simulations: From algorithms to applications.* Academic Press, San Diego, 1996.
80. I. Frick, G. Dau, and F. Ebert. Application of numerical simulations for the investigation of the motion of granular media in mixing silos. In *Int. Symp. Reliable Flow of Particulate solids III*, Porsgrunn, Norway, 1999.
81. U. Frisch, B. Hasslacher, and Y. Pomeau. Lattice–gas automata for the navier–stokes equation. *Phys. Rev. Lett.*, 56:1505, 1986.
82. J. A. C. Gallas, H. J. Herrmann, T. Pöschel, and S. Sokołowski. Molecular dynamics simulation of size segregation in three dimensions. *J. Stat. Phys.*, 82:443, 1996.

83. J. A. C. Gallas, H. J. Herrmann, and S. Sokołowski. Convection cells in vibrating granular media. *Phys. Rev. Lett.*, 69:1371, 1992.

84. J. A. C. Gallas, H. J. Herrmann, and S. Sokołowski. Two-dimensional powder transport on a vibrating belt. *J. Physique*, 2:1389, 1992.

85. J. A. C. Gallas and S. Sokołowski. Grain non-sphericity effects on the angle of repose of granular material. *J. Mod. Phys. B*, 7:2037, 1993.

86. A. L. Garcia. *Numerical Methods for Physics*. Prentice Hall, Englewood Cliffs, 2000.

87. C. W. Gear. The numerical integration of ordinary differential equations of various orders. Technical Report ANL 7126, Argonne National Laboratory, 1966.

88. C. W. Gear. *Numerical Initial Value Problems in Ordinary Differential Equations*. Prentice-Hall, Englewood Cliffs, 1971.

89. J. J. Gilvary. Fracture of brittle solids. I. distribution function for fragment size in single fracture (theoretical). *J. Appl. Phys.*, 32:391, 1961.

90. J. J. Gilvary and B. H. Bergstrom. Fracture of brittle solids. II. distribution function for fragment size in single fracture (experimental). *J. Appl. Phys.*, 32:400, 1961.

91. D. Goldberg. What every computer scientist should know about floating-point arithmetic. *ACM Computing Surveys*, 23:5, 1991.

92. I. Goldhirsch. Scales and kinetics of granular flows. *Chaos*, 9:659, 1999.

93. I. Goldhirsch and G. Zanetti. Clustering instability in dissipative gases. *Phys. Rev. Lett.*, 70:1619, 1994.

94. D. Goldman, M. D. Shattuck, C. Bizon, W. D. McCormick, J. B. Swift, and H. L. Swinney. Absence of inelastic collapse in a realistic three ball model. *Phys. Rev. E*, 57:4831, 1998.

95. A. Goldshtein and M. Shapiro. Mechanics of collisional motion of granular materials. Part 1: General hydrodynamic equations. *J. Fluid Mech.*, 282:75, 1995.

96. A. Goldshtein, M. Shapiro, and C. Gutfinger. Mechanics of collisional motion of granular materials. Part 3: Self-similar shock wave propagation. *J. Fluid Mech.*, 316:29, 1996.

97. A. Goldstein, A. Alexeev, and M. Shapiro. Shock waves in granular gases. In Pöschel and Brilliantov [215], page 187.

98. H. Goldstein. *Classical Mechanics*. Addison Wesley, Reading, 1980.

99. Y. Grasselli and H. J. Herrmann. Experimental study of granular stratification. *Granular Matter*, 1:43, 1998.

100. J. M. N. T. Gray and K. Hutter. Pattern formation in granular avalanches. *Cont. Mech. and Thermodyn.*, 9:341, 1997.

101. J. M. N. T. Gray and K. Hutter. Physik granularer Lawinen. *Physikalische Blätter*, 54:37, 1998.

102. J. M. N. T. Gray, Y. C. Tai, and K. Hutter. Shock waves and particle size segregation in shallow granular flows. In A. D. Rosato and D. L. Blackmore, editors, *IUTAM Symposium on Segregation in Granular Materials*, page 269, Dordrecht, 2000. Kluwer.

103. T. Gröger. Dynamic arches in vertical pipe conveyors. In D. E. Wolf and P. Grassberger, editors, *Friction, Arching and Contact Dynamics*, page 225, Singapore, 1997. World Scientific.

104. E. L. Grossman. Effects of container geometry on granular convection. *Phys. Rev. E*, 56:3290, 1997.

105. E. L. Grossman, T. Zhou, and E. Ben-Naim. Towards granular hydrodynamics in two dimensions. *Phys. Rev. E*, 55:4200, 1997.

106. P. K. Haff. Grain flow as a fluid-mechanical phenomenon. *J. Fluid Mech.*, 134:401, 1983.

107. P. K. Haff and B. T. Werner. Computer simulation of the mechanical sorting of grains. *Powder Techn.*, 48:239, 1986.

108. M. F. Handley. *An Investigation onto the Constitutive Behaviour of Brittle Granular Media by Numerical Experiment*. PhD thesis, Univerity of Minnesota, Minneapolis, May 1993.

109. J. P. Hansen and I. R. McDonald. *Theory of Simple Liquids*. Academic Press Limited, London, 1986.

110. J. Hardy, Y. Pomeau, and O. de Pazzis. Time evolution of a two-dimensional model system. I. invariant states and time correlation functions. *J. Math. Phys.*, 14:1746, 1973.

111. D. Helbing. *Verkehrsdynamik*. Springer, Berlin Heidelberg New York, 1997.

112. D. Helbing. Traffic and related self-driven many-particle systems. *Rev. Mod. Phys.*, 73:1067, 2001.

113. R. Henze. Effiziente Kollisionsdetektion zwischen Dreiecks-B-Spline modellierten Objekten für interaktive Computeranimationen. Master's thesis, Friedrich-Alexander-Universität Erlangen-Nürnberg, 1995.

114. H. J. Herrmann. Die wunderbare Welt der Schüttgüter. *Physikalische Blätter*, 51:1083, 1995.

115. H. J. Herrmann, E. Flekkøy, K. Nagel, G. Peng, and G. H. Ristow. Density waves in dry granular flow. In D. E. Wolf, M. Schreckenberg, and A. Bachem, editors, *Traffic and Granular Flow*, page 239, Singapore, 1996. World Scientific.

116. H. J. Herrmann, K. Kroy, and G. Sauermann. Saturation transients in saltation and their implications on dune shapes. *Physica A*, 302:244, 2001.

117. H. J. Herrmann and G. Sauermann. The shape of dunes. *Physica A*, 283:24, 2000.

118. P. Hersen, S. Douady, and B. Andreotti. Relevant length scale of barchan dunes. *Phys. Rev. Lett.*, 89:264301, 2002.

119. H. Hertz. Über die Berührung fester elastischer Körper. *J. f. reine u. angewandte Math.*, 92:156, 1882.

120. C. Hogue. *Computer modelling of the motion of granular particles*. PhD thesis, University of Cambridge, Cambridge, 1993.

121. C. Hogue and D. E. Newland. Efficient computer modelling of the motion of arbitrary grains. In C. Thornton, editor, *Powders and Grains'93*, page 413, Rotterdam, 1993. Balkema.

122. C. Hogue and D. E. Newland. Efficient computer simulation of moving granular particles. *Powder Techn.*, 78:51, 1994.

123. D. C. Hong, S. Yue, J. K. Rudra, M. Y. Choi, and Y. W. Kim. Granular relaxation under tapping and the traffic problem. *Mod. Phys. Lett. B*, 6:761, 1992.

124. W. G. Hoover. *Molecular Dynamics*. Springer-Verlag, Berlin Heidelberg New York, 1986.

125. S. Horikawa, A. Nakahara, T. Nakayama, and M. Matsushita. Self-organized critical density waves of granular material flowing through a pipe. *J. Phys. Soc. Jpn.*, 64:1870, 1995.

126. A. D. Howard and J. L. Walmsley. Simulation model of isolated dune sculpture by wind. In O. E. Barndorff-Nielsen, J.-T. Møller, K. Romer Rasmussen, and B. B. Willetts, editors, *Proceedings of the International Workshop on the Physics of Blown Sand*, volume 8, page 377. University of Aarhus, 1985.

127. K. Hwang. *Advanced Computer Architecture, Parallelism, Scalability, Programmability.* McGraw-Hill, New York, 1994.

128. M. Jean. Frictional contact in collections of rigid or deformable bodies: numerical simulation of geomaterial motions. In A. P. S Salvadurai and M. J. Boulon, editors, *Mechanics of Geomaterial Interfaces*, page 463, Amsterdam, 1995. Elsevier.

129. C. Josserand, A. V. Tkachenko, D. Mueth, and H. M. Jaeger. Memory effects in granular materials. *Phys. Rev. Lett.*, 85:3632, 2000.

130. P. Y. Julien, Y. Q. Lan, and Y. Raslan. Experimental mechanics of sand stratification. In Behringer and Jenkins [26], page 487.

131. R. Jullien and P. Meakin. Simple three-dimensional models for ballistic deposition with restructuring. *Europhys. Lett.*, 4:1385, 1987.

132. R. Jullien and P. Meakin. Ballistic deposition and segregation of polydisperse spheres. *Europhys. Lett.*, 6:629, 1988.

133. R. Jullien and P. Meakin. Three-dimensional model for particle-size segregation by shaking. *Phys. Rev. Lett.*, 69:640, 1992.

134. R. Jullien, P. Meakin, and A. Pavlovitch. Particle size segregation by shaking in two-dimensional disc packings. *Europhys. Lett.*, 22:523, 1993.

135. L. P. Kadanoff, S. R. Nagel, L. Wu, and S. Zhou. Scaling and universality in avalanches. *Phys. Rev. A*, 39:6524, 1989.

136. M. F. Kanninen and C. H. Popelar. *Advanced Fracture Mechanics.* Oxford University Press, 1985.

137. A. Károlyi and J. Kertész. Hydrodynamic cellular automata for granular media. In M. Tomassini, editor, *Computer Physics*, page 675, Lugano, 1994. CSCS.

138. A. Károlyi and J. Kertész. Lattice-gas model of avalanches in a granular pile. *Phys. Rev. E*, 57:852, 1998.

139. J. B. Knight, H. M. Jaeger, and S. R. Nagel. Vibration-induced size separation in granular media: The convection connection. *Phys. Rev. Lett.*, 70:3728, 1993.

140. J. Koeppe, M. Enz, and J. Kakalios. Avalanche segregation of granular media. In Behringer and Jenkins [26], page 443.

141. G. A. Kohring, S. Melin, H. Puhl, H.-J. Tillemans, and W. Vermöhlen. Computer simulations of critical, non-stationary granular flow through a hopper. *Appl. Mechanics and Eng.*, 124:2273, 1995.

142. K. Konrad. Dense-phase pneumatic conveying: A review. *Powder Techn.*, 49:1, 1986.

143. K. Kroy, G. Sauermann, and H. J. Herrmann. Minimal model for aeolian sand dunes. *Phys. Rev. E*, 66:031302, 2002.

144. K. Kroy, G. Sauermann, and H. J. Herrmann. A minimal model for sand dunes. *Phys. Rev. Lett.*, 88:054301, 2002.

145. D. J. Kuck. *The Structure of Computers and Computations*, volume I. John Wiley & Sons, New York, 1978.

146. F. Kun and H. J. Herrmann. Fragmentation of colliding discs. *J. Mod. Phys. C*, 7:837, 1996.

147. F. Kun and H. J. Herrmann. A study of fragmentation process using a discrete element method. *Comp. Meth. in Appl. Mech. and Eng.*, 138:3, 1996.

148. G. Kuwabara and K. Kono. Restitution coefficient in a collision between two spheres. *Jpn. J. Appl. Phys.*, 26:1230, 1987.

149. L. D. Landau and E. M. Lifshitz. *Theory of Elasticity*. Pergamon Press, New York, 1986.

150. P. A. Langston, U. Tüzün, and D. M. Heyes. Continuous potential discrete particle simulations of stress and velocity fields in hoppers: transition from fluid to granular flow. *Chemical Engineering Science*, 49(8):1259, 1994.

151. B. R. Lawn and T. R. Wilshaw. *Fracture of Brittle Solids*. Cambridge Univ. Press, New York, 1975.

152. W. Ledermann. Quaternion. In J. Thewlis, R. C. Glass, D. J. Huges, and A. R. Meetham, editors, *Encyclopedic dictionary of physics*, page 751. Pergamon Press, Oxford, 1962.

153. J. Lee. Density waves in the flows of granular media. *Phys. Rev. E*, 49:281, 1994.

154. A. Levy and H. Kalman, editors. *Handbook of Conveying and Handling of Particulate Solids (Handbook of Powder Technology)*. Elsevier, Amsterdam, 2001.

155. K. Liffman, G. Metcalfe, and P. Cleary. Granular convection and transport due to horizontal shaking. *Phys. Rev. Lett.*, 79:4574, 1997.

156. E. M. Lifshitz and L. P. Pitaevskii. *Physical Kinetics*. Pergamon Press, New York, 1981.

157. A. R. Lima, G. Sauermann, H. J. Herrmann, and K. Kroy. Modelling a dune field. *Physica A*, 310:487, 2002.

158. S. J. Linz. Phenomenological modeling of the compaction dynamics of shaken granular systems. *Phys. Rev. E*, 54:2925, 1996.

159. J. Litwiniszyn. Colmatage considered as a certain stochastic process. *Bull. Acad. Polon. Sci., Ser. Sci. Tech.*, 11:81, 1963. Colmatage?

160. J. Litwiniszyn and L. Ci-Tong. The phenomenon of segregation of grains of a loose medium when shaped in the form of a rotational half cone. *Bull. de L'Académie Polonaise des Sciences, Série des sciences techniques*, 11:169, 1963.

161. C. H. Liu, S. R. Nagel, D. A. Schecter, S. N. Coppersmith, S. Majumdar, O. Narayan, and T. A. Witten. Force fluctuations in bead packs. *Science*, 269:513, 1995.

162. P. Lötstedt. Coulomb friction in two-dimensional rigid body systems. *Z. f. Angewandte Math. und Mech.*, 61:605, 1981.

163. P. Lötstedt. Mechanical systems of rigid bodies subject to unilateral constraints. *SIAM J. Appl. Math.*, 42:281, 1982.

164. M. Y. Louge and M. E. Adams. Anomalous behavior of normal kinematic restitution in the oblique impacts of a hard sphere on an elastoplastic plate. *Phys. Rev. E*, 65:021303, 2002.

165. S. Luding, É. Clément, A. Blumen, J. Rajchenbach, and J. Duran. Anomalous energy dissipation in molecular-dynamics simulations of grains: The "detachment" effect. *Phys. Rev. E*, 50:4113, 1994.

166. S. Luding, É. Clément, J. Rajchenbach, and J. Duran. Simulations of pattern formation in vibrated granular media. *Europhys. Lett.*, 36:247, 1996.

167. S. Luding and S. McNamara. How to handle the inelastic collapse of a dissipative hard-sphere gas with the TC model. *Granular Matter*, 1:113, 1998.

168. H. Makse, P. Cizeau, and H. E. Stanley. Possible stratification mechanism in granular mixtures. *Phys. Rev. Lett.*, 78:3298, 1997.

169. H. Makse, P. Cizeau, and H. E. Stanley. Modeling stratification in two-dimensional sandpiles. *Physica A*, 249:391, 1998.

170. H. A. Makse. Stratification instability in granular flows. *Phys. Rev. E*, 56:7008, 1997.

171. H. A. Makse, S. Havlin, P. Ch. Ivanov, P. R. King, S. Prakash, and H. E. Stanley. Pattern formation in sedimentary rocks: Connectivity, permeability, and spatial correlations. *Physica A*, 233:587, 1996.

172. H. A. Makse, S. Havlin, P. R. King, and H. E. Stanley. Novel pattern formation in granular matter. In L. Schimansky-Geier and T. Pöschel, editors, *Stochastic Dynamics*, Lecture Notes in Physics, page 319, Berlin Heidelberg New York, 1997. Springer.

173. H. A. Makse, S. Havlin, P. R. King, and H. E. Stanley. Spontaneous stratification in granular mixtures. *Nature*, 386:379, 1997.

174. H. A. Makse and H. J. Herrmann. Microscopic model for granular stratification and segregation. *Europhys. Lett.*, 43:1, 1998.

175. R. Mannella. A gentle introduction to the integration of stochastic differential equations. In J. A. Freund and T. Pöschel, editors, *Stochastic Processes in Physics, Chemistry, and Biology*, volume 557 of *Lecture Notes in Physics*, page 353, Berlin Heidelberg New York, 2000. Springer.

176. M. Marín. On the pending event set and binary tournaments. In A. Bargiela and E. Kerckhoffs, editors, *Proceedings of the 10th SCS European Simulation Symposium*, page 110, Nottingham, Oct. 1998. Society for Computer Simulation Europe Publishing House.

177. M. Marín, D. Risso, and P. Cordero. Efficient algorithms for many body hard particle molecular dynamics. *J. Comp. Phys.*, 109:306, 1993.

178. A. Masselot and B. Chopard. A lattice Boltzmann model for particle transport and deposition. *Europhys. Lett.*, 42:259, 1998.

179. S. McNamara. Hydrodynamic modes of a uniform granular medium. *Physics of Fluids A*, 5:3056, 1993.

180. S. McNamara and W. R. Young. Inelastic collapse and clumping in a one-dimensional granular medium. *Phys. Fluids A*, 4:496, 1992.

181. S. McNamara and W. R. Young. Inelastic collapse in two dimensions. *Phys. Rev. E*, 50:R28, 1993.

182. P. Meakin. A simple two-dimensional model for particle segregation. *Physica A*, 163:733, 1990.

183. P. Meakin and R. Jullien. Simple models for two and tree dimensional particle size segregation. *Physica A*, 180:1, 1992.

184. B. Meerson, T. Pöschel, and Y. Bromberg. Close-packed floating clusters: Granular hydrodynamics beyond the freezing point? *Phys. Rev. Lett.*, 91:024301, 2003.

185. B. Meerson, T. Pöschel, P. Sasorov, and T. Schwager. Giant fluctuations at a granular phase separation threshold. *Phys. Rev. E*, 69:021302, 2004.

186. A. Mehta, R. J. Needs, and S. Dattagupta. The Langevin dynamics of vibrated powders. *J. Stat. Phys.*, 68:1131, 1992.

187. L. F. Menabrea. Notions sur la machine analytique de M. Charles Babbage. *Bibliothèque Universelle de Genève*, 41:352, 1842.

188. D. P. Miannay. *Fracture Mechanics*. Springer, Berlin Heidelberg New York, 1997.

189. J. J. Moreau. Some numerical methods in multibody dynamics: application to granular materials. *Eur. J. Mech. A*, 13:93, 1994.

190. W. A. M. Morgado and I. Oppenheim. Energy dissipation for quasielastic granular particle collisions. *Phys. Rev. E*, 55:1940, 1997.

191. D. Müller and T. M. Liebling. Detection of collisions of polygons by using a triangulation. In M. Raous and J. J. Moreau, editors, *Proceedings of the 2nd Contact Mechanics International Symposium*, page 369, Carry-le-Rouet, France, 1993.

192. A. Munjiza, N. Bićanić, and D. R. J. Owen. BSD contact detection algorithm for discrete elements in 2D. In Williams and Mustoe [300], page 39.

193. E. P. Muntz. Rarefied gas dynamics. *Ann. Rev. Fluid Mech.*, 21:387, 1989.

194. G. G. W. Mustoe and G. DePorter. A numerical model for the mechanical behavior of particulate media containing non-circular shaped particles. In C. Thornton, editor, *Powders and Grains'93*, page 421, Rotterdam, 1993. Balkema.

195. M. Nicodemi. Logarithmic compaction in a 3d model for granular media. *J. Physique I*, 7:1535, 1997.

196. M. Nicodemi, A. Coniglio, and H. J. Herrmann. The compaction of granular media and frustrated Ising magnets. *J. Phys. A*, 30:L379, 1997.

197. M. Nicodemi, A. Coniglio, and H. J. Herrmann. Frustration and slow dynamics of granular packings. *Phys. Rev. E*, 55:3962, 1997.

198. H. Nishimori and N. Ouchi. Computational models for sand ripple and sand dune formation. *Int. J. of Mod. Phys. B*, 7:2025, 1993.

199. H. Nishimori and N. Ouchi. Formation of ripple patterns and dunes by wind-blown sand. *Phys. Rev. Lett.*, 71:197, 1993.

200. R. O'Connor, M. J. Gill, and J. R. Williams. A linear complexity contact detection algorithm for multi-body simulation. In Williams and Mustoe [300], page 53.

201. J. A. C. Orza, R. Brito, T. P. C. van Noije, and M. H. Ernst. Patterns and long range correlations in idealized granular flows. *J. Mod. Phys. C*, 8:953, 1997.

202. N. B. Ouchi and H. Nishimori. Modeling of wind-blown sand using cellular automata. *Phys. Rev. E*, 52:5877, 1995.

203. P. R. Owen. Pneumatic transport. *J. Fluid Mech.*, 39:407, 1969.

204. P. Pacheco. *Parallel Programming With MPI*. Morgan Kaufmann, San Francisco, 1996.

205. E. J. R. Parteli and H. J. Herrmann. A simple model for a transverse dune field. *Physica A*, 327:554, 2003.

206. W. Paul and J. Baschnagel. *Stochastic Processes: From Physics to Finance*. Springer, Berlin Heidelberg New York, 1999.

207. G. Peng and H. J. Herrmann. Density waves of granular flow in a pipe using lattice-gas automata. *Phys. Rev. E*, 49:1796, 1994.

208. G. Peng and H. J. Herrmann. Density waves and $1/f$ density fluctuations in granular flow. *Phys. Rev. E*, 51:1745, 1995.

209. G. Peng and T. Ohta. Velocity and density profiles of granular flow in channels using a lattice gas automaton. *Phys. Rev. E*, 55:6811, 1997.

210. G. Peng and T. Ohta. Logarithmic density relaxation in compaction of granular materials. *Phys. Rev. E*, 57:829, 1998.

211. X. Peng, Y. Tomita, and H. Tashiro. Effect of particle-particle collision and particle rotation upon floating mechanism of coarse particles in horizontal pneumatic pipe. *JSME Int. J. B, Fluids Therm. Eng.*, 37:485, 1994.

212. T. Pöschel. Recurrent clogging and density waves of granular material flowing in a narrow pipe. *J. Physique*, 4:499, 1994.

213. T. Pöschel. *Dynamik granularer Systeme – Theorie, Experimente und numerische Experimente*. Logos, Berlin, 1999.
214. T. Pöschel, N. Brilliantov, and T. Schwager. Long-time behavior of granular gases with impact-velocity dependent coefficient of restitution. *Physica A*, 325:274, 2003.
215. T. Pöschel and N. V. Brilliantov, editors. *Granular Gas Dynamics*, volume 624 of *Lecture Notes in Physics*, Berlin Heidelberg New York, 2003. Springer.
216. T. Pöschel, N. V. Brilliantov, and T. Schwager. Violation of molecular chaos in dissipative gases. *J. Mod. Phys. C*, 13:1263, 2002.
217. T. Pöschel and V. Buchholtz. Static friction phenomena in granular materials: Coulomb law versus particle geometry. *Phys. Rev. Lett.*, 71:3963, 1993.
218. T. Pöschel and S. Luding, editors. *Granular Gases*, volume 564 of *Lecture Notes in Physics*, Berlin Heidelberg New York, 2001. Springer.
219. A. V. Potapov and C. S. Campbell. A hybrid finite-element simulation of solid fracture. *J. Mod. Phys. C*, 7:155, 1996.
220. A. V. Potapov and C. S. Campbell. A tree-dimensional simulation of brittle solid fracture. *J. Mod. Phys. C*, 7:717, 1996.
221. A. V. Potapov, C. S. Campbell, and M. Hopkins. A two-dimensional dynamic simulation of solid fracture. Part II: Examples. *J. Mod. Phys. C*, 6:399, 1995.
222. A. V. Potapov, C. S. Campbell, and M. Hopkins. A two-dimensional dynamic simulation of solid fracture. Part II: Examples. *J. Mod. Phys. C*, 6:399, 1995.
223. A. V. Potapov, M. A. Hopkins, and C. S. Campbell. A two-dimensional dynamic simulation of solid fracture. Part I: Description of the model. *J. Mod. Phys. C*, 6:371, 1995.
224. W. H. Press, W. T. Vetterling, S. A. Teukolsky, and B. P. Flannery. *Numerical Recipes*. Cambridge University Press, Cambridge, 1988.
225. I. Prigogine and R. Herman. *Kinetic Theory of Vehicular Traffic*. Elsevier, New York, 1971.
226. H. Puhl. On the modelling of real sand piles. *Physica A*, 182:295, 1992.
227. K. Pye and H. Tsoar. *Aeolian sand and desert dunes*. Unwin Hyman, London, 1990.
228. T. Raafat, J. P. Hulin, and H. J. Herrmann. Density waves in dry granular media falling through a vertical pipe. *Phys. Rev. E*, 53:4345, 1996.
229. R. Ramírez, T. Pöschel, N. V. Brilliantov, and T. Schwager. Coefficient of restitution of colliding viscoelastic spheres. *Phys. Rev. E*, 60:4465, 1999.
230. D. C. Rapaport. The event scheduling problem in molecular dynamic simulations. *J. Comp. Phys.*, 34:184, 1980.
231. D. C. Rapaport. *The Art of Molecular Dynamics Simulation*. Cambridge University Press, Cambridge, 1995.
232. R. Reichardt and W. Wiechert. Event driven simulation of a high energy ball mill. In *Proceedings ASIM 2003*, page 249. ASIM, 2003.
233. W. Reisner. *Bins and bunkers for handling bulk materials*. Trans. Tech. Publ., Cleveland, 1971.
234. T. Riethmüller. Theoretische Modellierung granularer Ströme in dünnen Röhren mit Langevin-Gleichungen. Master's thesis, Humboldt-Universität, Berlin, 1995.
235. T. Riethmüller, D. Rosenkranz, and L. Schimansky-Geier. Granular flow modelled by Brownian particles. In D. E. Wolf, A. Bachem, and M. Schreckenberg, editors, *Traffic and Granular Flow*, page 293, Singapore, 1996. World Scientific.

236. T. Riethmüller, L. Schimansky-Geier, D. E. Rosenkranz, and T. Pöschel. Langevin equation approach to granular flow in a narrow pipe. *J. Stat. Phys.*, 86:421, 1996.

237. F.-X. Riguidel, A. Hansen, and D. Bideau. Gravity-driven motion of a particle on an inclined plane with controlled roughness. *Europhys. Lett.*, 28:13, 1994.

238. G. Ristow and H. J. Herrmann. Forces on the walls and stagnation zones in a hopper filled with granular material. *Physica A*, 213:474, 1995.

239. G. H. Ristow. Particle mass segregation in a two-dimensional rotating drum. *Europhys. Lett.*, 28:97, 1994.

240. G. H. Ristow. Outflow rate and wall stress for two-dimensional hoppers. *Physica A*, 235:319, 1996.

241. G. H. Ristow and H. J. Herrmann. Density patterns in granular media. *Phys. Rev. E*, 50:R5, 1994.

242. S. Rjasanov and W. Wagner. A stochastic weighted particle method for the Boltzmann equation. *J. Comp. Phys.*, 124:243, 1996.

243. S. Rjasanov and W. Wagner. Simulation of rare events by the stochastic weigthed particle method for the Boltzmann equation. *Mathematical and Computer Modelling*, 33:907, 2000.

244. P. Rosin and E. Rammler. The laws governing the fineness of powdered coal. *J. Inst. Fuel*, 7:29, 1933.

245. J. Roth, F. Gähler, and H.-R. Trebin. A molecular dynamics run with 5,180,116,000 particles. *Int. J. Mod. Phys. C*, 11:317, 2000.

246. B. Rothkegel. *Örtliche Verteilung der Stoßenergien ≥ 22 mJ und die zugeordneten Bewegungszustände von Modellmahlkörpern in einer Modellkugelmühle.* PhD thesis, Technische Universität Berlin, 1992.

247. J.-N. Roux. On the geometric origin of mechanical properties of granular materials. *Phys. Rev. E*, 61:6802, 2000.

248. C. Salueña, S. E. Esipov, D. Rosenkranz, and H. V. Panossian. On modelling of arrays of passive granular dampers. In T. Tupper Hyde, editor, *Proceedings of the SPIE's Conference on Smart Structures and Materials, Passive Damping and Isolation*, volume 3672, page 32, Phoenix, 1999.

249. C. Salueña and T. Pöschel. Dissipative properties of vibrated granular materials. *Phys. Rev. E*, 59:4422, 1999.

250. G. Sauermann, P. Rognonc, A. Poliakov, and H. J. Herrmann. The shape of the barchan dunes of Southern Morocco. *Geomorphology*, 36:47, 2000.

251. S. Savage. Disorder, diffusion and structure formation in granular flows. In D. Bideau and A. Hansen, editors, *Disorder and Granular Media*, page 255, Amsterdam, 1993. North-Holland.

252. J. Schäfer. *Rohrfluß granularer Materie: Theorie und Simulationen.* PhD thesis, Research Center Jülich, 1996.

253. J. Schäfer, S. Dippel, and D. E. Wolf. Force schemes in simulations of granular materials. *J. Physique I*, 6:5, 1996.

254. F. Scheck. *Mechanics: From Newton's Laws to Deterministic Chaos.* Springer, Berlin, 1999.

255. K. L. Schick and A. A. Verveen. $1/f$ noise with a low frequency white noise limit. *Nature*, 251:599, 1974.

256. L. Schimansky-Geier, M. Mieth, H. Rosé, and H. Malchow. Structure formation by active Brownian particles. *Phys. Lett. A*, 207:140, 1995.

257. S. Schöllmann. Simulation of a two-dimensional shear cell. *Phys. Rev. E*, 59:889, 1999.

258. N. Schörghofer and T. Zhou. Inelastic collapse of rotating spheres. *Phys. Rev. E*, 54:5511, 1996.

259. R. Schuhmann. Principles of comminution, I-size distribution and standard calculations. Technical Report TO 1189, Mining Technol, Am. Inst. Mining, Metallurgical and Petroleum Engineers, Inc., 1940.

260. T. Schwager. Dissipative Prozesse und Kinetik granularer Stoffe in viskoelastischer Teilchennäherung. Master's thesis, Humboldt-Universität zu Berlin, 1998.

261. T. Schwager and T. Pöschel. Contact of viscous spheres. In D. E. Wolf and P. Grassberger, editors, *Friction, Arching and Contact Dynamics*, page 293, Singapore, 1997. World Scientific.

262. T. Schwager and T. Pöschel. Coefficient of resitution of viscous particles and cooling rate of granular gases. *Phys. Rev. E*, 57:650, 1998.

263. T. Schwager and T. Pöschel. Rigid body dynamics of railway ballast. In K. Popp and W. Schielen, editors, *System Dynamics and Long-Term Behaviour of Railway Vehicles, Track and Subgrade*, volume 6 of *Lecture Notes in Applied Mechanics*, page 451, Berlin Heidelberg New York, 2003. Springer.

264. T. Schwager and T. Pöschel. Duration of contact and coefficient of restitution. *preprint*, 2004.

265. J. Schwedes. *Fließverhalten von Schüttgütern in Bunkern*. Verlag Chemie GmbH, Weinheim, 1968.

266. K. Shida and T. Kawai. Cluster formation by inelastically colliding particles in one-dimensional space. *Physica A*, 162:145, 1989.

267. H. C. Sorby. On the structures produced by the currents present during the deposition of stratified rocks. *The Geologist*, 2:137, 1859.

268. G. Strassburger, A. Betat, M. A. Scherer, and I. Rehberg. Pattern formation by horizontal vibration of granular material. In D. E. Wolf, M. Schreckenberg, and A. Bachem, editors, *Traffic and Granular Flow*, page 329, Singapore, 1996. World Scientific.

269. W. I. Stronge. Rigid body collision with friction. *Proc. Roy. Soc. A*, 431:169, 1990.

270. B. Stroustrup. *The C++ Programming Language*. Addison Wesley, Reading, 2000.

271. Y. Taguchi. Powder turbulence: Direct onset of turbulent flow. *J. Physique*, 2:2103, 1992.

272. Y-h. Taguchi. New origin of a convective motion: Elastically induced convection in granular materials. *Phys. Rev. Lett.*, 69:1367, 1992.

273. Y-h. Taguchi. Numerical modeling of vibrated beds. *J. Mod. Phys. B*, 7:1839–1858, 1993.

274. T. Tanaka, T. Ishida, and Y. Tsuji. Direct numerical simulatin of granular plug flow in a horizontal pipe: The case of cohesionless particles (in japanese). For an english presentation of this work see [271]. *Trans. Jap. Soc. Mech. Eng.*, 57:456, 1991.

275. H.-J. Tillemans. *Molekulardynamik-Simulationen beliebig geformter Teilchen in zwei Dimensionen*. PhD thesis, Universität Köln, 1994.

276. H.-J. Tillemans and H. J. Herrmann. Simulating deformations of granular solids under shear. *Physica A*, 217:261, 1995.

277. J. M. Ting, M. Khwaja, L. Meachum, and J. Rowell. An ellipse based discrete element model for granular materials. *Int. J. for Numerics and Analytical Methods in Geomechanics*, 17:603, 1993.

278. B. C. Trent and L. G. Margolin. A numerical laboratory for granular solids. *Engineering Computations*, 9:191, 1992.

279. A. Trew and G. Wilson, editors. *Past, Present, Parallel: A survey of available parallel computing systems*. Springer, Berlin Heidelberg New York, 1991.

280. D. J. Unger. *Analytical Fracture Mechanics*. Academic Press, New York, 1995.

281. T. Unger, L. Brendel, D. E. Wolf, and J. Kertész. Elastic behavior in contact dynamics of rigid particles. *Phys. Rev. E*, 65:061305, 2002.

282. T. Unger and J. Kertész. The contact dynamics method for granular media. In *AIP Conference Proceedings*, volume 661, page 116, 2003.

283. T. P. C. van Noije and M. H. Ernst. Velocity distributions in homogeneous granular fluids: the free and the heated case. *Granular Matter*, 1:57, 1998.

284. T. P. C. van Noije, M. H. Ernst, and R. Brito. Ring kinetic theory for an idealized granular gas. *Physica A*, 251:266, 1998.

285. T. P. C. van Noije, M. H. Ernst, R. Brito, and J. A. G. Orza. Mesoscopic theory of granular fluids. *Phys. Rev. Lett.*, 79:411, 1997.

286. W. M. Visscher and M. Bolsterli. Random packing of equal and unequal spheres in two and three dimensions. *Nature*, 239:504, 1972.

287. O. R. Walton. Explicit particle dynamics model for granular materials. In Z. Eisenstein, editor, *Numerical Methods in Geomechanics*, page 1261. Balkema, Rotterdam, 1982.

288. O. R. Walton and R. G. Braun. Simulation of rotary-drum and repose tests for frictional spheres and rigid sphere clusters. Joint DOE/NSF Workshop on flow of particulates and fluids, Ithaka, page 1, 1993.

289. O. R. Walton and R. L. Braun. Viscosity, granular temperature, and stress calculations for shearing assemblies of inelastic, frictional disks. *J. Rheol.*, 30:949, 1986.

290. Y. Wang and M. T. Mason. Two-dimensional rigid-body collisions with friction. *J. Appl. Mech.*, 59:635, 1992.

291. W. A. Weibull. *The phenomenon of rupture in solids*, volume 149. In-geniörvetensskapsakademiens, Handlingar, 1938.

292. W. A. Weibull. A statistical theory on the strength of material. *Ingvetensk. Akad. Handl.*, 139:5, 1939.

293. R. Weichert. Correlation between probability of breakage and fragment size distribution of mineral particles. *Int. J. Mineral Processing*, 22:9, 1988.

294. R. Weichert. Anwendung von Fehlstellenstatistik und Bruchmechanik zur Beschreibung von Zerkleinerungsvorgängen. *Zement-Kalk-Gips*, 45:1, 1992.

295. B. T. Werner and P. K. Haff. A simulation study of the low energy ejecta resulting from single impacts in eolian saltation. In R. E. A. Arndt, editor, *Advances in aerodynamics, fluid mechanics and hydraulics*, page 333, New York, 1986. American Society of Civil Engineers.

296. B. T. Werner and P. K. Haff. The impact process in aeolian saltation: two-dimensional simulations. *Sedimentology*, 35:189, 1988.

297. B. T. Werner, P. K. Haff, R. P. Livi, and R. S. Anderson. The measurement of eolian ripple cross-sectional shapes. *Geology*, 14:743, 1986.

298. J. C. Williams. The segregation of powders and granular materials. *Univ. Sheffield Fuel Soc. J.*, 14:29, 1963.

299. J. C. Williams. The segregation of particulate materials. A review. *Powder Techn.*, 15:245, 1976.

300. J. R. Williams and G. G. W. Mustoe, editors. *Proc. 2nd Intern. Conf. on Discrete Element Methods*, Cambridge, 1993. IESL.

301. J. R. Williams and R. O'Connor. A linear complexity intersection algorithm for discrete element simulation of arbitrary geometries. *Engineering Computations*, 12:185, 1995.

302. J. R. Williams and A. Pentland. Superquadratics and modal dynamics for discrete elements in interactive design. *Int. J. Computer Aided Engineering - Engineering Computations*, 9:2, 1992.

303. K. H. Wohletz, M. F. Sheridan, and W. Brown. Particle size distributions and the sequential fragmentation/transport theory applied to volcanic ash. *J. Geophys. Res.*, 15:15,703, 1984.

304. D. E. Wolf. Modelling and computer simulation of granular media. In K. H. Hoffmann and M. Schreiber, editors, *Computational Physics*, page 64. Springer, Berlin Heidelberg, 1996.

305. S. Wolfram. *Theory and Applications of Cellular Automata*. Elsevier, New York, 1971.

306. T. Worsch and P. Sanders. *Parallele Programmierung mit MPI – Ein Praktikum*. Logos, Berlin, 1997.

307. Y. Yan. Mass flow measurement of bulk solids in pneumatic pipelines. *Meas. Sci. Technol.*, 7:1687, 1996.

308. T. Zhou and L. P. Kadanoff. Inelastic collapse of three particles. *Phys. Rev. E*, 54:623, 1996.

Index